STATISTICAL METHODS
FOR THE ENVIRONMENTAL SCIENCES

STATISTICAL METHODS FOR THE ENVIRONMENTAL SCIENCES

A Selection of Papers Presented at the Conference on Environmetrics, held in Cairo, Egypt, April 4–7, 1989

Edited by

A.H. EL-SHAARAWI
National Water Research Institute,
Canada Centre for Inland Waters, Burlington, Ontario, Canada

Reprinted from *Environmental Monitoring and Assessment,*
Volume 17, Nos. 2/3 (1991)

SPRINGER–SCIENCE+BUSINESS MEDIA, B.V.

Library of Congress Cataloging-in-Publication Data

Statistical methods for the environmental sciences : a selection of
papers presented at the Conference on Environmetrics, held in Cairo,
Egypt, April 4-7, 1989 / edited by A.H. El-Shaarawi.
p. cm.
ISBN 978-0-7923-1273-4 ISBN 978-94-011-3186-5 (eBook)
DOI 10.1007/978-94-011-3186-5
1. Pollution--Environmental aspects--Statistical methods--Congresses. I. El-Sharaawi, A. H. II. Conference on Environmetrics (1989 : Cairo, Egypt)
TD193.S73 1991
628--dc20 91-3680

Printed on acid-free paper

TABLE OF CONTENTS

EDITORIAL

This volume contains a selection of papers presented at the Conference on Environmetrics, held April 4-7, 1989, at Ramses Hilton Hotel, Cairo, Egypt. The main objectives of the conference were to promote the development and application of statistical methods in environmental assessment and to provide a published state-of-the-art summary of the application of statistical methods that are commonly used to deal with environmental problems. The material given here will be useful for research workers, students and decision makers who are involved with the collection, analysis and interpretation of environmental data. The conference would not have been possible without the support from Environment Canada, the Egyptian Ministry of Scientific Research and the Egyptian Academy of Scientific Research. I would like to thank Ian B. MacNeill, Co-chairman of the Conference, and the organizing Committee which included David Brillinger, S. Fayed, S. R. Esterby, Eivind Damsleth, J. Gani, F. El-Gohary and R. A. Vollenweider who all contributed to the success of the Conference. The assistance of John Santolucito, Associate Editor of Environmental Monitoring and Assessment, in publishing the proceedings is also gratefully acknowledged. The contributors and refrees are to be thanked for the fine spirit of cooperation and the prompt handling of the correspondence. Also, thanks to Jocelyne Cantin for handling the correspondence and typing some of the manuscripts.

A. H. El-Shaarawi
National Water Research Institute,
Burlington, Ontario L7R 4A6, Canada

PUBLISHER'S ANNOUNCEMENT

Please note that the page numbers in square brackets apply to this edition of *Environmental Monitoring and Assessment*.
The page numbers without square brackets apply to the journal edition (published earlier).

STATISTICAL METHODS TO APPORTION THE SOURCES OF PARTICLES IN THE INDUSTRIAL REGION OF ESTARREJA – PORTUGAL

C. BORREGO and C. A. PIO

Departamento de Ambiente, Universidade de Aveiro 3800 AVEIRO – Portugal

(Received July 1990)

Abstract. Factor analysis models are very attractive for source apportionment and have been widely applied. They do not require *a priori* knowledge of the number and composition of the sources, and they can actually uncover previously unsuspected sources and estimate the composition of the sources using only ambient monitoring data.

Aerosol particles were collected from an industrial atmosphere and analyses for water soluble and carbon components. Principal components analysis permitted the evaluation of the contribution due to industries, soil fraction, secondary pollutants and sea spray particles of the total suspended aerosol mass.

It can be concluded that the atmospheric aerosol in the Industrial Area of Estarreja (Portugal) contains a relative important fraction that is water soluble. Ammonium sulphates and nitrates are the main components of this fraction. Carbon compounds constitute about 30% of the total aerosol mass. These compounds are mainly formed by organic matter emitted by the industries. Due to the mutagenic and carcinogenic characteristics of some organic compounds processed in the Industrial Area (vinyl chloride, benzene, aniline, etc.), the concern exists of negative human health effects as a result of prolonged inhalation. Soil compounds is another important fraction of the aerosol mass, mainly in summer with dry, sunny and windy weather conditions.

A more conclusive idea of the sources and effects of aerosol matter can only be obtained with the specific analysis of organic compounds and the determination of trace elements, characteristic of each particular source.

Introduction

Portugal, a semi-industrialized country in the western corner of Europe, is not usually affected by air pollution imported from more developed European Nations. Therefore, atmospheric pollution exists only in restricted and well localized regions, which include the main urban centres, three industrial complexes and the areas surrounding some large industrial units. In these zones, air pollution episodes are usual, resulting in complaints from the local population. As a result of social pressures, the authorities have established 'Regional Air Management Commissions' in the most polluted zones with the aim of assessing and improving local air quality.

The region of Estarreja, located on the west coast, 40 km south of Oporto (Figure 1) contains a complex of inorganic and organic chemical industries which inevitably give rise to air pollutant emissions. Metal corrosion experiments have indicated that the local atmosphere is one of the most aggressive in the country (Justo, 1984). The first measurements of an air monitoring programme taking place under the supervision of the 'Estarreja Air Management Commission' have shown that total suspended particle

Environmental Monitoring and Assessment **17**: 79–87, 1991.

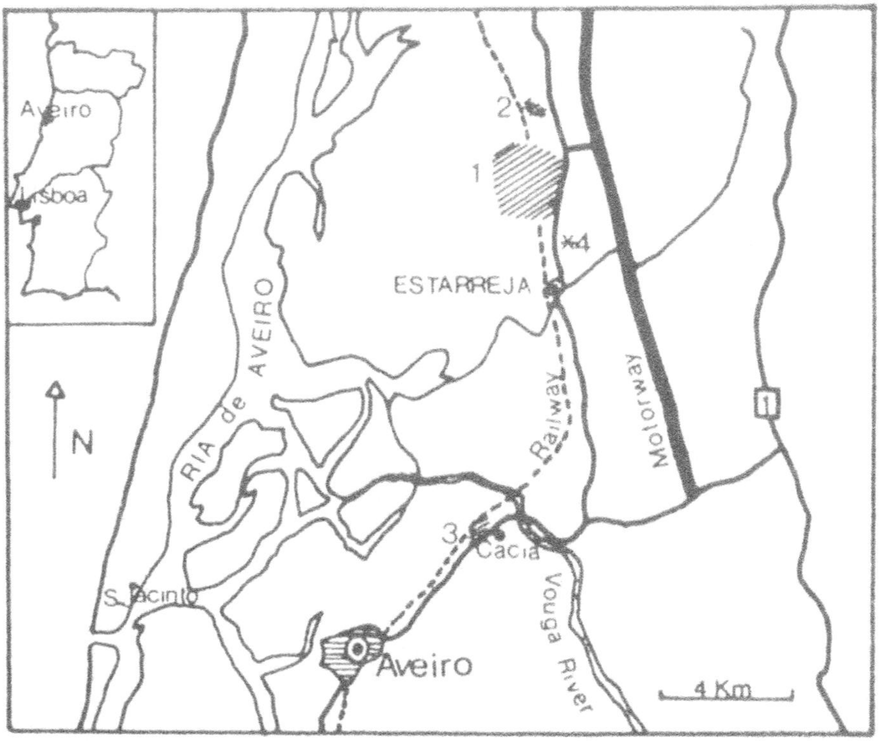

Fig. 1. Map of the Estarreja Area: (1) Main Industrial Complex; (2) 'Nestlé'; (3) Pulp and paper Plant; (4) Monitoring site.

concentrations are higher than the limits recommended by the Economic European Community (EEC).

In this work, an investigation of the major ion composition and carbon content of aerosol particles collected from Estarreja atmosphere is reported.

THE INDUSTRIAL COMPLEX AND ITS INDUSTRIES

The first industries were installed in the 1940's; a more recent expansion occurred at the beginning of this decade. Four independent companies, 'Quimigal', 'Uniteca', 'Cires' and 'Isopor', are located together in an area of 2-3 km^2. 'Quimigal' is the largest installation containing 12 factories which manufacture a variety of compounds, namely sulphuric acid by the contact process from the utilization of pyrites, ammonia through the syntheses of nitrogen and hydrogen, nitric acid from the fertilizers, hydrogen and carbon monoxide from the thermal cracking of naphtha, oxygen and hydrogen through the electrolysis of water, nitrobenzene from the nitrification of benzene and aniline by catalytic reduction of nitrobenzene. Presently the ammonia factory is inactive. 'Uniteca' is a chlor-alkali industry, producing chlorine and caustic soda from the electrolysis of sodium chloride brines by the mercury cathode process. 'Cires' lodges a factory producing vinyl chloride monomer from acetylene and two units producing polyvinyl chloride by the suspension and emulsion processes. 'Isopor' manufactures MDI (methyl diphenyl-isocyanate) using aniline, chlorine and carbon monoxide as reagents. Located 3 km north from the main complex is 'Nestlé' a food processing industry. Ten km to the south is a pulp and paper mill (Figure 1).

Experimental

The survey was carried out over the period November 1983-September 1985. Most of the samples were collected in one of each three day period for a period of 24 hrs, starting at 9 a.m. The sampling was conducted at the old installation of the 'Uprel' factory, located 1 km from the complex. The site is located in open country between the complex and the town of Estarreja, in the direction of prevalent winds. Therefore, the measurements should represent the worst pollution conditions in the zone. Aerosol samples were obtained with a Hi-Volume. Gaseous pollutants were collected in parallel with aerosol samples and analysed by manual wet chemistry methods.

METEOROLOGICAL DATA

Continuous records of wind speed and direction, relative humidity (R.H.), temperature, hours of sunshine and precipitation were obtained from the weather stations at the University of Aveiro Campus and at the S. Jacinto Airfield, situated 15 and 18 km respectively from the industrial complex. From the records, arithmetic mean values of R.H., temperature and wind speed were calculated for each sampling period. Wind direction data was used to determine the fraction of the sampled air which has been blown through the industrial area.

STATISTICAL ANALYSIS OF THE RESULTS

Factor analysis models are based on the principal component analysis of the data which are usually centered and normalized in some manner. The principal components are unique (up to sign reversals); however, there are an infinite number of factor models that can be derived from the principal components. All factor analysis models rely on a linear transformation of the principal components to produce a 'best' factor model. The two most widely used transformations are the orthogonal VARIMAX rotation and the oblique target transformation.

Principal Component Factor analysis is a technique frequently applied to aerosol component concentrations for the identification of class sources contribution to TSP (Total Suspended Particles) levels (Cooper and Watson, 1980). Incorporation of meteorological data on the analysis has been done, permitting the clarification of the influence of weather conditions on the aerosol formation mechanisms (Sexton *et al.*, 1985; and Henry and Hidy, 1979).

Using Principal Components, or concentrations with higher factor scores, as independent variables and TSP as dependent variables, Multilinear Regression methods have been applied to estimate the contribution of each source to the aerosol burden (Kleinman *et al.*, 1980; Wolf *et al.*, 1985).

Results and Discussion

Pollutant mean concentrations and ranges as well as meteorological data are presented in Table I. Table II gives the correlation matrix between the variables.

The aerosol particles have a water soluble component which seems to be formed mainly by the analysed compounds. The sums of total cation and anion concentrations measured reveal a reasonable electroneutrality, although cation concentrations are generally greater than anion concentrations. A closer look to the individual data shows that in many of the samples in which electroneutrality is poorer, calcium concentrations are important, suggesting that this element is associated with non measured anions, possibly carbonates, or phosphates from the fertilizer factory.

TABLE I

Summary of pollutants and meteorological data

Variables	No. of Measures	Units	Mean	Standard deviation	Range	
					Min	Max
NH_3	23	μ m^{-3}	0.47	0.29	0.05	1.09
SO_2	60	,,	27.9	20.5	1	88
NO_2	55	,,	17.4	10.7	2.5	53
NH_4	70	,,	4.42	3.73	0.11	17.7
K	70	,,	0.51	0.25	0.16	1.92
Ca	70	,,	1.53	0.95	0.24	5.59
Mg	70	,,	0.317	0.17	0.09	0.91
Na	70	,,	2.68	1.57	0.76	8.53
Cl	70	,,	4.24	2.70	0.64	15.9
SO_4	70	,,	11.1	7.29	2.59	32.8
NO_3	70	,,	4.10	1.98	0.68	9.55
OC	70	,,	26.0	20.3	3.6	106
EC	70	,,	7.8	4.1	2.2	23.0
TSP	68	,,	132.3	57.0	22	283
TSP_{in}	68	,,	72.7	44.7	0.20	180
ΣCat	70	nEq m^{-3}	482	213	131	182
ΣAni	70	,,	418	170	110	889
pH	64	–	6.4	0.67	4.6	7.4
Rad	67	hr	7.9	3.8	0.0	12.8
Temp	69	°C	19.5	5.0	9.1	26.9
RH	69	%	75.1	10.7	45.5	90.7
Wind	63	knots	6.6	3.3	1.5	17
Prec.	70	mm	15.5	39.9	0.0	205

OC: organic carbon: EC: elemental carbon; TSP; total suspended particulate; TSP_{in}: non analysed mass of TSP; ΣCat: sum of total soluble cations; ΣAni: sum of total soluble anions; Rad: sunshine; Temp: temperature; RH: relative humidity; Wind: wind speed; Prec: precipitation; ions are presented here and throughout the text without the respective sign.

TABLE II

Correlation matrix

	NH_3	SO_2	NO_2	NH_4	K	Ca	Mg	Na	Cl	SO_4	NO_3	OC	EC	TSP	TSP_{in}	Temp	RH	Wind	Prec	
NH_3	1.00																			
SO_2	0.11	1.00																		
NO_2	0.60	−0.01	1.00																	
NH_4	0.15	0.09	0.02	1.00																
K	0.20	0.28	0.28	0.18	1.00															
Ca	0.53	0.05	0.34	0.02	1.00															
Mg	0.01	0.12	−0.07	−0.02	0.22	1.00														
Na	0.07	0.06	−0.05	−0.24	0.13	−0.01	0.68	1.00												
Cl	0.27	0.01	0.23	−0.15	−0.02	0.07	0.64	0.71	1.00											
SO_4	−0.00	0.14	−0.14	0.89	0.23	0.12	0.10	−0.14	−0.21	1.00										
NO_3	0.30	0.18	0.21	0.44	0.63	0.40	0.05	−0.11	−0.15	0.49	1.00									
OC	0.39	0.23	0.57	0.14	0.58	0.15	−0.02	−0.07	0.04	0.08	0.48	1.00								
EC	0.17	0.19	0.56	0.04	0.52	0.29	0.10	0.01	0.11	0.01	0.49	0.48	1.00							
ISP	0.32	0.22	0.19	0.34	0.56	0.57	0.32	−0.08	−0.02	0.46	0.66	0.46	0.53	1.00						
ISP_{in}	0.12	0.14	−0.03	0.24	0.33	0.57	0.31	−0.15	−0.12	0.39	0.46	0.06	0.36	0.90	1.00					
Rad	0.45	0.01	0.06	−0.05	0.08	0.31	0.15	−0.04	−0.10	0.07	0.29	0.03	0.28	0.42	0.51	1.00				
Temp	−0.33	0.06	−0.64	0.14	−0.02	0.23	0.23	−0.11	−0.25	0.33	0.08	−0.50	−0.11	0.27	0.55	0.27	1.00			
RH	−0.46	0.09	−0.52	0.12	−0.16	−0.19	0.03	0.29	0.06	0.25	−0.16	−0.42	−0.51	−0.28	−0.17	−0.17	−0.35	0.28	1.00	
Wind	0.33	0.13	0.08	−0.36	−0.06	0.09	0.29	0.38	0.36	−0.31	−0.38	−0.26	−0.15	−0.16	−0.07	−0.12	−0.10	0.04	1.00	
Prec	−0.17	−0.07	−0.16	−0.26	−0.22	−0.30	0.11	0.30	0.31	−0.33	−0.40	−0.20	−0.27	−0.47	−0.46	−0.36	−0.24	0.14	0.48	1.00

The aerosol is highly neutralized. Attempts were made to measure strong and weak acids using Gran titration (Brosset and Ferm, 1978) but due to the low quantities present, the method couldn't be applied with a minimum precision. Measurement of some rainwater samples taken simultaneously with aerosol collection showed the same pattern.

One important part of the water soluble fraction is formed by ammonium salts of sulphate and nitrate. The concentrations of gaseous ammonia are very high. Ammonia levels show a positive correlation with NO_2, a primary pollutant from the nitric acid factory, in accordance with the existence of industrial emissions (Table 2). The correlation with relative humidity and calcium concentrations shows that natural contribution is also important.

Chlorides exist mainly associated with sodium and magnesium. The analysis of each sample, reveals frequently greater concentrations of chlor in relation with sodium and magnesium concentrations. In several samples, the inverse happens. Higher concentrations of NH_4 not balanced by nitrates plus sulphates and correspond to the presence of ammonium chloride compounds. This can be due to the escape of HCl from the industries and reaction with gaseous NH_2. The presence of higher concentrations of sodium in relation to chlorides seems to be more the consequence of chloride volatilization from sea spray aerosols than to the emission of other sodium compounds (seen by the sodium/magnesium ratio in these samples). Although the atmosphere is neutral, punctual emissions of acidic gases of particles from the industries can react with the marine aerosol, with votalization of HCl and formation of sodium sulphate or nitrate (Clegg and Brimblecome, 1985).

Carbon compounds form on the average, 26% of the total particle mass. Less than 33% of carbon is elemental, the rest being organic carbon. Both the fraction of total carbon and the percentage of organic carbon in aerosols are greater than mean values observed in urban and rural conditions and are the result of local industrial emissions.

Table III presents the Principal Component Analysis of measured data after Varimax rotation. A five Principal Components solution was chosen. A comparison of non-rotated and Varimax rotated Components showed that rotated values give a better understanding of sources and formation processes. Ammonia concentrations data was not used because of the small number of measurements.

The Principal Components Analysis are interpreted in the following way. PC 1 includes NO_2, elemental and organic carbon, and represents primary pollutants emitted directly from the complex. The Component level is inversely related with ambient temperature. There is no correlation with sunshine, indicating that organic carbon is a primary rather than a secondary aerosol. An attempt to relate PC 1 with wind direction failed. The sampling time fraction of wind blowing through the Industrial Area and sequently through the sampling points couldn't be related with any other variable. The same problem has been previously detected (Harrison, 1983) and is attributed to the variation of low altitude wind direction with ground roughness and to the use of non-local meteorological data. Statistical comparison between wind direction at Aveiro and S. Jacinto stations had already shown a poor intersite correlation (Martins, 1983).

PC 2 contains secondary pollutants originating from the neutralization of nitric and

TABLE III

Principal Component pattern for Varimax rotated components

Variable	Component number					Communalities
	PC1	PC2	PC3	PC4	PC5	
SO_2					0.85	0.72
NO_2	0.88					0.80
NH_4		0.91				0.83
K	0.40	0.27		0.26	0.62	0.70
Ca				0.69		0.52
Mg			0.84	0.32		0.83
Na			0.88			0.80
Cl	0.20		0.88			0.84
SO_4		0.89				0.88
NO_3	0.33	0.55		0.43	0.34	0.72
OC	0.78	0.23			0.36	0.79
EC	0.61			0.45	0.24	0.64
TSP	0.22	0.38		0.77	0.32	0.89
TSP_{in}		0.22		0.88		0.88
Rad				0.74		0.57
Temp	−0.75			0.53		0.87
RH	−0.65	0.25	0.21	−0.38		0.69
Wind		−0.54	0.48			0.57
Prec.		−0.35	0.36	−0.52		0.53
Eigen values	5.16	3.14	2.73	1.90	1.12	
Fraction of variance explained by each component	0.17	0.15	0.14	0.19	0.09	

sulphuric acids by gaseous ammonia. The negative wind intensity factor loading is due to the fact that stronger wind intensities are usually related to clean north-western masses transported from the Atlantic.

PC 3 is associated with maritime aerosol incorporation. The direct relation with wind intensity and rainfall is a consequence of windy and rainy weather associated with air masses transported from the sea.

PC 4 represents the soil component, as indicated by the calcium factor loading. The high value of the elemental carbon factor score in this Component is due to the chemical analytical method used and corresponds to the existence of carbonate compounds. The meteorological data show that the importance of the soil component is larger with dry warm weather conditions. The high factor score in TSP_{in} means that an important percentage of the total suspended aerosol mass, not specifically analysed, is of soil origin.

PC 5 is harder to explain. The component contains a high loading in SO_2. Sulphur dioxide concentrations are small, frequently in the lower precision range of the analytical process. Therefore this Component can be an artifice as a result of analytical unprecision. Punctual continuous measurements with a SO_2 analyser showed that concentration levels are generally constant, with higher peaks when the air comes directly from the Industrial Area. The existence of a positive organic carbon loading may also suggest that this

Component represents both industrial and background pollution.

In accounting for the contribution of each of the first four Principal Components to the aerosol burden a linear Multicorrelation Analysis was applied, using as independent variables, the compounds with higher factor scores in each Component. The resultant equation:

$$TSP = (31.0 \pm 36.6) + (1.01 \pm 0.22)\,OC + (2.92 \pm 0.62)\,SO4 - \\ - (0.12 \pm 2.84)\,Na + (28.0 \pm 4.7\,Ca,$$

where the concentrations are expressed in $\mu g\, m^{-3}$, has a multiple correlation coefficient of 0.77. From the equation and Table 1 it is possible to calculate that $23 \pm 28\%$ of the aerosol mass is not explained by the four Principal Components and that the contributions of Components 1, 2, 3 and 4 to the aerosol mass are respectively $20 \pm 4\%$, $25 \pm 5\%$, $0 \pm 6\%$ and $33 \pm 6\%$.

Conclusions

The atmospheric aerosol in the Industrial Area of Estarreja contains a relative important fraction that is water soluble. Ammonium sulphates and nitrates are the main components of this fraction. The particles are highly neutralized and consequently, problems of acidic deposition don't exist in the medium term. However short time existence of acidic gases and aerosols is not excluded.

Carbon compounds constitute about 30% of the total aerosol mass. These compounds are mainly formed by organic matter emitted by the Complex. Due to the mutagenic and carcinogenic characteristics of some organic compounds processes in the Industrial Area (vinyl chrloride, benzene, aniline, etc.) the preoccupation exists of negative human health effects as a result of prolonged inhalation.

Soil component is another important fraction of the aerosol mass, mainly in summer with dry, sunny and windy weather conditions.

A more conclusive idea of the sources and effects of aerosol matter can only be obtained with the specific analysis of organic compounds and the determination of trace elements, characteristic of each particular source. At the present this investigation is in the preliminary stages.

Acknowledgements

The authors gratefully acknowledge the assistance of Miss Albertina Fernandes in preparation of the manuscript.

References

Brosset, C. and Ferm, M.: 1978, 'Man-Made Airborne Acidity and its Determinations', *Atmos. Environ.* **12**, 909-16.

Clegg, S. L. and Brimblecome, P.: 1985, 'Potential Degassing of Hydrogen, Chloride from Acidified Sodium

Chloride Droplets', *Atmos. Environ.* **19**, 465-70.

Cooper, J. A. and Watson, J. G. Jr.: 1980, 'Receptor Oriented Methods of Air Particulate Source Apportionment', *J. Air Pollut. Control Assoc.* **30**, 1116-25.

Harrison, R. M.: 1983, 'Ambient Air Quality in the Vicinity of a Works Manufacturing Sulphuric Acid, Phosphoric Acid and Sodium Tripolyphosphate, *The Sci. Total Environ.* **27**, 121-31.

Henry, R. C. and Hidy, G. M.: 1979, 'Multivariate Analysis of Particulate Sulphate and Other Air Quality Variables by Principal Components - Part. I. Annual Data from Los Angeles and New York', *Atmos. Environ.* **13**, 1581-96.

Justo, M. J.: 1984, 'Corrosão e Protecção de Materiais', *LNETI* **3**, 15-21.

Kleinman, M. T., Pasternack, B. S., Eisendud, M., and Kneip, T. J.: 1980, 'Identifying and Estimating the Relative Importance of Sources of Airborne Particles', *Environ. Sci. Technol.* 62-5.

Martins, J. M.: 1983, *Internal Project Report,* Departamento de Ambiente, Universidade de Aveiro.

Sexton, K., Liu, K., Hayward, S. B., and Spengler, J. D.: Characterization and Source Apportionment of Wintertime Aerosol in a Wood-Burning Community', *Atmos. Environ.* **19**, 1225-36.

Wolf, G. T., Korsog, P. E., Kelly, N. A., and Ferman, M. A.: 1985, 'Relationships Between Fine Particulate Species, Gaseous Pollutants and Meteorological Parameters in Detroit', *Atmos. Environ.* **19**, 1341-9.

THE DISTRIBUTION OF MONTHLY RAINFALL INTENSITY AT SOME SITES IN SAUDI ARABIA

A. M. ABOUAMMOH*

Distribution of Rainfall in Saudi Arabia

(Received March 1990)

Abstract. The analysis of rainfall intensity is useful in various fields of life e.g. agricultural planning, hydrology and transmission of microwaves and high-voltage electricity. The monthly precipitation totals for a 21 year period are used to compare the rainfall regimes at seven sites in Saudi Arabia. These sites differ in their latitude, longitude and elevation above sea level. Some basic monthly statistics of data from these sites are presented to identify the nature of the rainfall at each site. The trend of the number of dry months per year is also used to compare between the sites. The probability of dry months for each month of the 21 yr is used for comparison between the rainfall regimes. Plots for mean and maximum rainfall at these sites are presented. The Fisher-Cornish proposed model for rainfall in arid regions is considered and a simple empirical method for estimating its parameters is used for the twelve month period rainfall data from the seven sites.

1. Introduction

The Kingdom of Saudi Arabia is mainly an arid country with an estimated population of over 12 million in an area of about 2150 000 km^2. The area includes about three fourths of the Arabian peninsula, (Hamza, 1968; Tayeb, 1978). In the past fifteen years the country has seen the establishment of large agricultural projects. Current agricultural production has reached 4.6 million tons of which over 2.5 million tons was wheat in 1986 compared to about 39 000 tons in 1972. Ground water is the main water source for irrigating the majority of these agricultural projects as well as for all water authorities in various cities, towns and villages. The ground water level above the water table at any site depends upon the amount of precipitation through the surface runoff intensity and the infiltration capacity (the passage of water through the soil), (Linsay *et al.,* 1982). In fact, detailed analysis of rainfall intensities is useful in different aspects of earth sciences. Rainfall intensity determines the contribution of rainfall to the water budget, the soil profile, the ecological niche and the watershed. Recently, in 1986, the Saudi National Commission for Wildlife Protection and Development was established. Regions of Saudi Arabia which are put under the management of the Commission for wildlife development are expected to have relatively higher monthly rainfall.

Therefore, it is of great importance to researchers in agricultural planning and forest management to study the longterms runoff of rainfall and its effect on water levels. In fact rainfall which is used is the only input that varies, from year to year, to estimate the water balance or the crop yields (Stern and Coe, 1982), even in countries of arid nature.

Data analysis of rainfall intensity is also useful to researchers in electrical engineering.

* Postal address: Department of Statistics, College of Sciece, King Saud University, P.O. Box 2455, Riyadh 11451, Saudi Arabia.

Environmental Monitoring and Assessment **17**: 89–100, 1991.

Its attenuating effect on microwave transmission in telecommunications has been considered by Bodtmann and Ruthroff (1976) and Drufuca (1977). Its connection to power loss along high-voltage transmission lines, generating audible and radio noise, has been studied by Krikham (1980).

Rainfall intensity (mm hr^{-1}, month or year) and duration or frequency of rain (in hours or days y^{-1}) are the major features used to characterize rainfall. In This study, we are concerned with the accumulated total rainfall in mm $month^{-1}$ rather than the time scale in mm yr^{-1}. This emphasis is based on the available data and the arid nature of the considered regions which have relatively small monthly accumulated rainfall totals.

The main theme of this paper is to present an overall estimate and comparison of the rainfall regimes at seven different sites in Saudi Arabia. These sites, located in different provinces in Saudi Arabia are Dhahran, Jeddah, Khamis Mushait, Madina, Riyadh, Tabouk and Taif. Their latitude, longitude and elevation range from 8′ 18″ to 28′ 22″, 36′ 35″ to 50′ 10″ and 17 to 2057 m, respectively, (Table I).

The paper considers monthly rather than the daily rainfall since most daily rainfall records are zeros. Thus daily do not give reasonable background information for most inferential procedures.

The sequel of the paper is the following. Some basic statistics of the monthly rainfall for the seven sites are discussed in Section 2. In Section 3, the probability of no rain in each month and the probability of dry months in the year are plotted in two figures for each of the seven meteorological sites. In addition figures for the number of dry months in the year, the maximum amount of rainfall and the mean amount of rainfall are used for possible comparison between the rainfall regimes of these sites. In Section 4, the Fisher-Cornish proposed model, (Fisher and Cornish, 1960), used as a probability distribution to rainfall data with zero observations, is considered. Ozturk's method (1984) is used to estimate the model parameter and finally in Section 5 some concluding remarks and comments presented.

2. Data and Basic Statistics

Daily rainfall data are provided for the seven meteorological stations Dhahran, Jeddah, Khamis Mushait, Madina, Riyadh, Tabouk and Taif by the Meteorology and Environmental Protection Administration of the Ministry of Defence and Aviation, Saudi Arabia.

A 21-yr record from 1966 to 1986 is used for this study. Data from other stations are not available at present for a sufficient number of years. Table II presents the number of rainy months at the seven location. Table III includes some basic statistics of the monthly precipitation rainfall, the mean $\bar{X}$ and the standard deviation s^2 are presented. The minimum monthly rainfall are zero except for May in Khamis Mushait (2.9 mm).

The data indicates differences between the seven sites with respect to the annual amount of precipitation during the 21-yr period (1966-1986). Khamis Mushait has the maximum amount of annual mean rainfall (38.23 mm in May). Taif has the next largest amount of annual rainfall in April (37.76 mm). The maximum total amount of rainfall during the 21 year (1966 to 1986) is in Khamis Mushait (2494.68 mm) whereas the total amount of

TABLE I

Latitude, longitude and elevation of seven meteorological sites

Site	Latitude		Longitude		Elevation (m)
	Deg	Min	Deg	Min	
Dhahran	26	16	50	10	17
Jeddah	21	30	39	12	17
Khamis Mushait	18	18	42	48	2057
Madina	24	33	39	43	636
Riyadh	24	42	46	44	611
Tabouk	28	22	36	35	776
Taif	21	29	40	32	1454

TABLE II

Number of rainy months for each month at the seven locations

Month	Locations						
	Dhahran	Jeddah	Khamis Mushait	Madina	Riyadh	Tabouk	Taif
Jan.	19	14	13	10	16	12	12
Feb.	18	3	10	6	16	10	10
Mar.	19	3	18	12	20	6	17
May	8	5	21	11	16	5	19
Jun.	0	0	16	3	1	0	12
Jul.	0	1	17	1	2	0	9
Aug.	0	0	18	4	2	1	11
Sep.	0	1	10	3	1	0	16
Oct.	3	2	9	5	3	5	13
Nov.	10	14	12	12	9	13	16
Dec.	12	13	6	10	17	9	12

rainfalls in Taif and in Riyadh are 2016.48 and 1274.04 mm, respectively.

The arid nature is apparent even in Khamis Mushait where the minimum monthly rainfall was zero for all months in the 21-yr period except for May which had a minimum rainfall of 2.9 mm in 1972.

Dry months with no rainfall from 1966 to 1986 are from June to September in Dhahran, June and August in Jeddah and June, July and September in Tabouk. Months with rainfall less than 5 mm are: May in Dhahran, March, July, September and October in Jeddah, August in Madina, June and September in Riyadh and August in Tabouk.

The probability of rainfall is greater than 0.85, (based on 1966-1986 data) in the period January to March for Dhahran, March to May and August for Khamis Mushait, April for Riyadh and May for Taif.

TABLE III

Summary statistics for each month and location

Month	Statistic	Dhahran	Jeddah	Khamis Mushait	Madina	Riyadh	Tabouk	Taif
Jan.	Max.	59.9	129.1	72.0	21.0	111.1	127.0	77.4
	$\overline{X}$	14.09	17.74	15.37	5.36	11.84	13.60	9.5
	s	17.59	35.35	24.06	7.10	24.29	30.59	20.25
Feb.	Max.	66.7	98.1	100.6	8.2	39.2	36.6	48.2
	$\overline{X}$	14.59	7.99	18.41	0.98	7.01	4.08	4.91
	s	20.81	25.43	29.23	2.03	9.47	36.90	11.55
Mar.	Max.	208.5	3.0	189.7	47.0	99.6	56.2	52.4
	$\overline{X}$	25.25	0.3	36.27	7.79	27.51	7.17	25.25
	s	47.73	0.78	46.26	11.83	30.59	14.86	14.07
Apr.	Max.	57.8	93.0	114.9	79.0	124.3	25.2	289.0
	$\overline{X}$	12.18	4.74	27.23	14.82	32.80	3.81	37.76
	s	20.05	20.25	29.36	18.36	40.21	6.25	66.79
May	Max.	2.3	20.0	84.6	39.6	69.0	31.0	116.0
	$\overline{X}$	0.66	1.13	38.23	6.51	9.81	3.28	34.41
	s	1.15	4.35	24.89	10.15	16.05	8.20	35.89
Jun.	Max.	0.0	0.0	18.3	6.8	2.2	0.0	27.4
	$\overline{X}$	0.0	0.0	5.47	0.62	0.10	0.0	4.27
	s	0.0	0.0	5.22	16.70	0.48	0.0	7.90
Jul.	Max.	0.0	2.0	75.8	6.0	8.8	0.0	33.3
	$\overline{X}$	0.0	0.1	20.62	0.31	0.43	0.0	2.75
	s	0.0	0.45	20.40	1.31	1.92	0.0	7.29
Aug.	Max.	0.0	0.0	72.8	1.1	17.4	0.2	112.0
	$\overline{X}$	0.0	0.0	22.41	0.10	0.87	0.01	9.45
	s	0.0	0.0	20.16	0.25	3.79	0.04	24.83
Sep.	Max.	0.0	1.0	24.5	5.8	4.2	0.0	55.1
	$\overline{X}$	0.0	0.05	3.96	0.37	0.20	0.0	7.52
	s	0.0	0.22	6.88	2.87	0.92	0.0	11.82
Oct.	Max.	18.1	4.0	31.5	12.8	27.1	18.6	72.6
	$\overline{X}$	1.01	0.35	3.27	1.11	1.66	1.72	11.74
	s	3.92	1.20	7.16	2.37	6.50	4.56	19.22
Nov.	Max.	51.5	83.0	73.9	62.0	16.0	89.0	104.1
	$\overline{X}$	5.26	14.13	11.35	10.37	2.43	13.27	14.27
	s	11.65	22.20	18.96	17.40	9.98	25.23	26.87
Dec.	Max.	94.0	55.5	84.8	15.3	45.7	59.3	33.4
	$\overline{X}$	11.33	8.34	5.3	3.13	11.51	4.66	6.21
	s	21.84	14.57	18.36	5.39	13.55	12.94	10.73

3. Rainfall Probabilities

Probability plots of rainfall occurrence and rainfall amounts are very useful in choosing suitable probability models for fitting these variables. In arid regions rainfall occurrence can be studied more appropriately through the occurrence of dry spells or through the probability of no rain during specific period. These probability plots can be used to compare the rainfall regimes of various sites.

Figure 1 illustrates the probability of the occurrence of dry months based on 21 observations for each month of the year for the sites. Although Dhahran has the maximum probability of having dry spells in the period June to September, Tabouk and Jeddah also have a high probability of having dry spells for the same months of the year. Dhahran has the minimum probability of having dry spells in the period Januarv to March. Riyadh is the second site that has a small probability of having dry spells for the same months of the year. There is a zero probability that May will be a dry month in Khamis Mushait. Khamis Mushait has minimum variation of probability values of dry spells and minimum averaged probability over the 12 months.

Figure 1 illustrates the probability of occurrence of a dry month in any year from 1966 to 1986. The maximum number of dry months (11) are in Tabouk (1973, 1978, and 1979) and Madina (1973), whereas the minimum number of dry months (2) are in Khamis

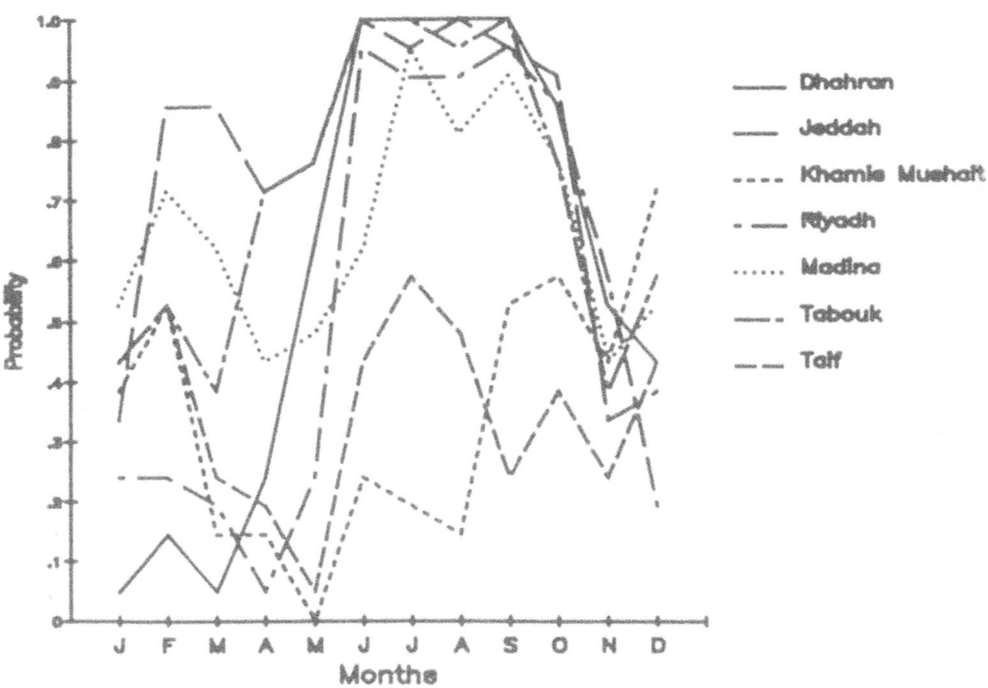

Fig. 1. Probability of dry months for every month at seven sites.

Mushait (1982), Madina (1978) and Taif (1986). Whereas the probability of having a dry month in any of the years 1966 through 1986 is illustrated in Figure 2.

Figures 3 and 4 represent the maximum and the mean amount of rainfall, respectively at the seven sites. In these figures, the axis (month) begins in July and ends in June for an easier comparison between the sites. These figures show that Khamis Mushait has a higher mean rainfall than any other site in five months of the year. Madina has the lowest mean rainfall in four months of the year. Taif received maximum rainfall in seven months of the year, whereas Madina received the lowest maximum rainfall for four months of the year. Other comparisons with different sites can be based on these figures.

4. Fitting a Probability Distribution

Let the rainfall occurrence follow a Poisson distribution with mean $1/\mu$ i.e. the time interval between any two rainfall events has an exponential distribution with mean μ Thus, the number of rainfall events in a time interval t is Poisson distributed with mean $\theta = t/\mu$. Rainfall amounts are assumed to be independent of their occurrences, mutually independent and exponentially distributed with mean $1/\rho$.

Such a model was used by Buishand (1977) for monthly totals. De Boer (1958) used a similar procedure to describe the distribution of rainfall months for monthly periods. The

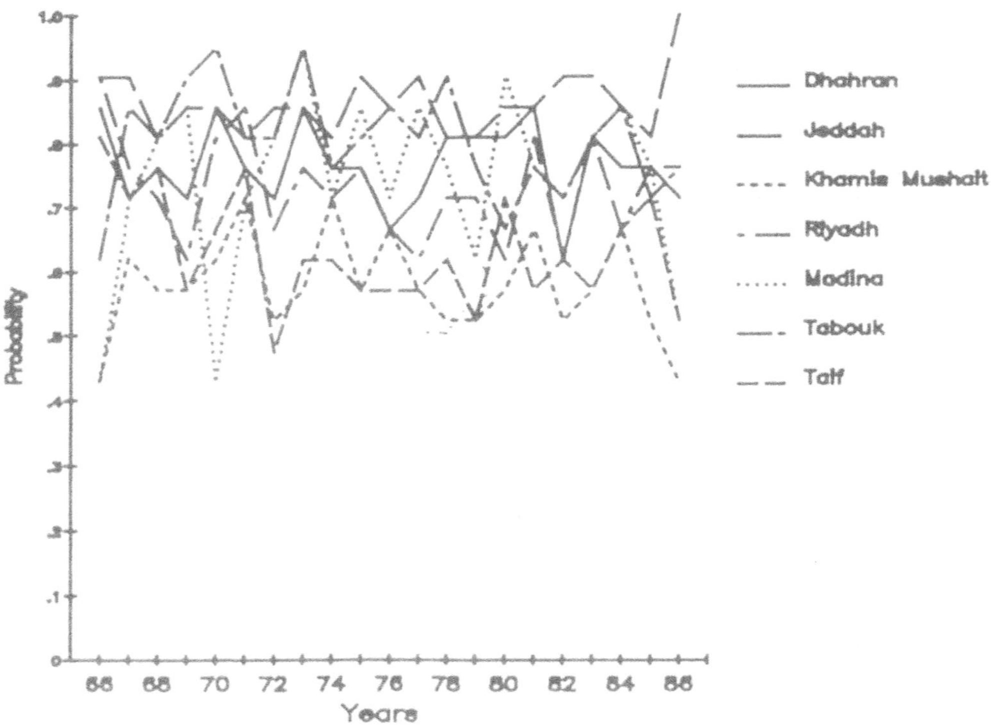

Fig. 2. Probability of dry months for each year (1966-86).

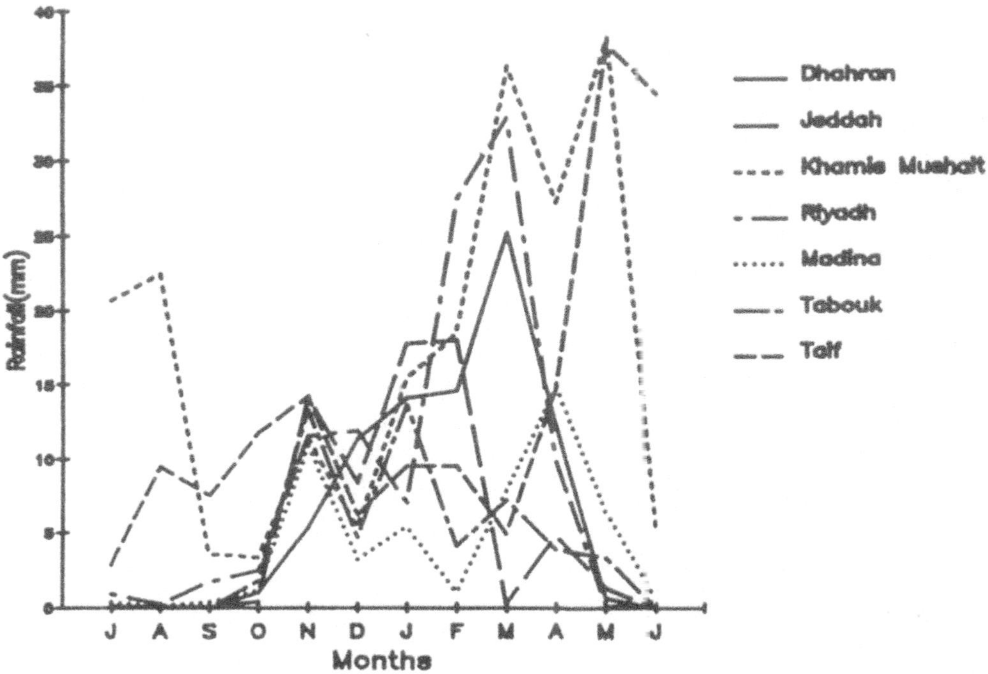

Fig. 3. Mean amount of monthly rainfall at seven sites.

probability of receiving a rainfall of an amount less than or equal x during the time interval $(0, t)$, is

$$P(X \leq x) = P(N=0) + \sum_{k=1}^{\infty} P(X \leq x \mid N=k)\, P(N=k), \tag{1}$$

where the number of rainfalls events N follows a Poisson distribution during the time interval with parameter $\theta t / \mu$. Thus the probability distribution function is given by

$$F(x) = e^{-\theta} + e^{-\theta} \sum_{k=1}^{\infty} \frac{\theta^k}{k!} \int_0^x \frac{(\lambda/\theta)^k}{(k-1)!} y^k e^{-(\lambda/\theta)y}\, dy, \tag{2}$$

where $\lambda = \rho\theta$

Integrating (2) over the total rainfall x gives

$$P(N \geq 1) = \int_0^\infty f(x)\, dx$$

$$= e^{-\theta} \sum_{k=1}^{\infty} \frac{\lambda^k}{k!\,(k-1)!} \int_0^\infty x^{k-1} e^{-(\lambda/\theta)x}\, dx$$

$$= e^{-\theta} \sum_{k=1}^{\infty} \theta^k / k!$$

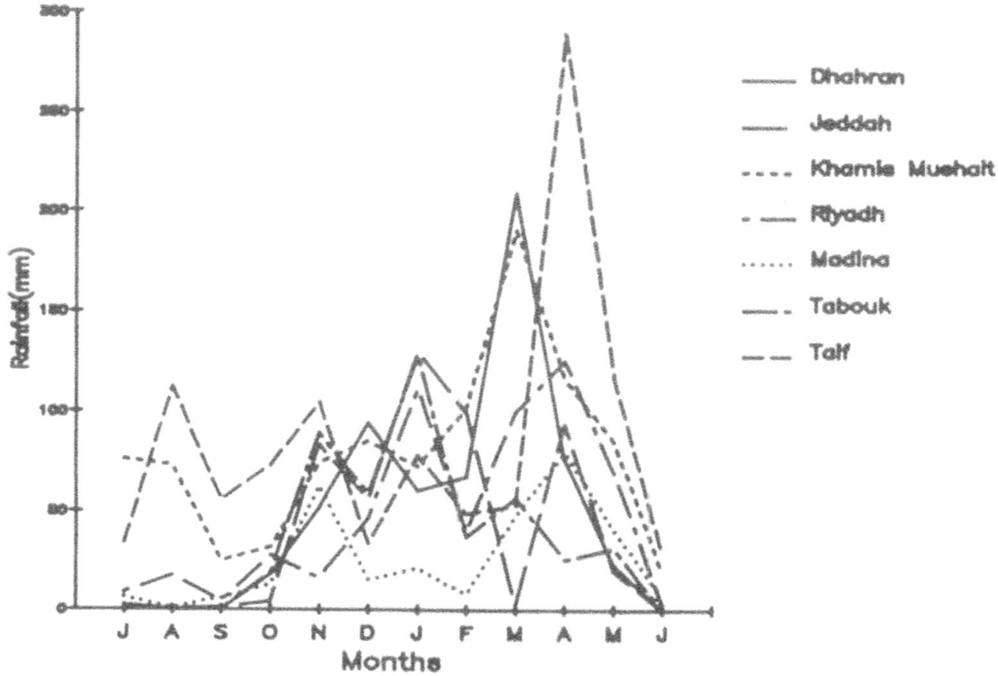

Fig. 4. The maximum amount of rainfall at seven sites.

This implies that the number of rainfall during a specific time interval is Poissin distributed with a probability mass function

$$P(N=k) = e^{-\theta}\, \theta^k/k!, \quad k=0, 1, \ldots .$$

The probability density function of precipitation totals for a given period is given by

$$f(x) = e^{-\theta-(\lambda/\theta)x} \sum_{k=1}^{\infty} \lambda^k\, x^{k-1}/k!\,(k-1)!, \quad x>0. \tag{3}$$

The random variable X has a positive probability at x=0 with

$$P(X=0) = e^{-\theta}. \tag{4}$$

Fisher and Cornish (1960) proposed this mixture of a gamma distribution for precipitation over arid regions. In fact Ozturk (1981) has used this model for fitting the monthly precipitation data of Whitestown meteorological station in Indiana. Abouammoh (1986) had found that the model used with probability density (3), is reasonable for fitting monthly data at Dhahran, Riyadh and Jeddah. Discussion of different estimation methods on fitting the monthly data in particular moment method, maximum likelihood method, empirical method and approximate maximum likelihood method have also been considered by the author.

The simple empirical method is adopted for estimating the parameters of model (3). The goodness of fit of the model is evaluated by the maximum absolute distance between the empirical and the estimated distribution. Some properties of parameter estimates such as their means, variances, efficiencies relative to moment estimates and bias are discussed briefly in Ozturk (1984). Abouammoh (1986) found that estimating model (3) by using the empirical method leads to results similar to those results obtained by other methods of estimation.

The probability density (1) can be expressed as

$$f(x) = \exp\left(-\theta - \frac{\lambda x}{\theta}\right)\left(\frac{\lambda}{x}\right)^{1/2} I_1\,[2(\lambda\, x)^{1/2}], \tag{5}$$

where I_1 stands for a modified Bessel function of order one.

Fisher and Cornish (1960) have shown that the r-th cumulant of the distribution (2) is

$$K_r = r!\, \theta^{3+1}/\lambda^r.$$

Thus the mean, the variance and the coefficient of skewness are

$$\mu = \theta^2/\lambda,\ \sigma^2 = 2\,\theta^3/\lambda^2, \quad \text{and} \quad \propto = 3/(2\,\theta)^{1/2},$$

respectively.

The probabilities of having no precipitation or some precipitation during a given time interval are $q = P(X=0) = \exp(-\theta)$ and $p = 1-q = P(X>0) = 1-\exp(-\theta)$, respectively. Also, the number of non-zero observations m, in a random sample of size n is a binomial random variable with parameters (n, p). The mean and the variance of m are

$$np = n(1-e^{-\theta}) \quad \text{and} \quad npq = ne^{-\theta}\,(1-e^{-\theta}),$$

respectively.

The probability $q = \exp(-\theta)$ can be estimated for a random sample $x_1, \ldots, x_n$, of which m are non-zero observations by $\hat{q} = 1-(m/n)$. Hence θ can be empirically estimated by

$$\hat{\theta} = -\log(1 - (m/n)). \tag{6}$$

Therefore, using the mean of the distribution $F(x)$ implies

$$\hat{\lambda} = (\hat{\theta})^2/\bar{x}. \tag{7}$$

Estimation of θ and λ given by relations (6) and (7) are easily calculated. The main drawback of these estimates are $\hat{\theta} = \hat{\lambda} = 0$, for $m=0$ and $\hat{\theta} = \hat{\lambda} = \infty$ for $m=n$.

Table IV gives the estimated values for θ and λ in the seven sites.

5. Concluding Remarks

To test the goodness of fit one may evaluate the statistic D^*_n which is defined by

$$D^*_n = \max_n |F_n(x) - \hat{F}(x)|,$$

where $n=21$, $\hat{F}(x)$ is given by (5) whose parameters are estimated by the empirical method

TABLE IV
Empirical estimation of parameters of $f(x)$ for seven meteorological stations

Site	Month	Jan	Feb	Mar	Apr	May	Jun	Jul	Aug	Sep	Oct	Nov	Dec
Dhahran	$\hat{\theta}$	2.351	1.946	2.351	1.435	0.480	–	–	–	–	0.154	0.647	0.847
	$\hat{\lambda}$	0.392	0.260	0.219	0.169	0.349	–	–	–	–	0.023	0.080	0.063
Jeddah	$\hat{\theta}$	1.099	0.154	0.154	0.337	0.272	–	0.049	–	0.049	0.100	1.099	0.965
	$\hat{\lambda}$	0.068	0.003	0.079	0.024	0.065	–	0.024	–	1.050	0.026	0.085	0.112
Khamis	$\hat{\theta}$	0.965	0.647	1.846	1.946	–	1.435	1.658	1.946	0.647	0.560	0.847	0.337
Mushait	$\hat{\lambda}$	0.061	0.023	0.104	0.139	–	0.377	0.136	0.169	0.118	0.096	0.063	0.021
Madina	$\hat{\theta}$	0.647	0.337	0.480	0.847	0.742	0.154	0.049	0.211	0.154	0.272	0.847	0.647
	$\hat{\lambda}$	0.078	0.116	0.030	0.048	0.085	0.038	0.008	0.447	0.642	0.067	0.069	0.134
Riyadh	$\hat{\theta}$	1.435	1.435	1.658	3.045	1.435	0.049	0.100	0.100	0.049	0.154	0.560	1.658
	$\hat{\lambda}$	0.174	0.294	0.100	0.2826	0.210	0.024	0.023	0.012	0.012	0.014	0.129	0.239
Tabuk	$\hat{\theta}$	0.847	0.647	0.965	0.337	0.272	–	–	0.049	–	0.272	0.965	0.560
	$\hat{\lambda}$	0.053	0.103	0.130	0.029	0.023	–	–	0.238	–	0.043	0.070	0.067
Taif	$\hat{\theta}$	0.847	0.647	1.435	1.658	2.351	0.847	0.560	0.742	1.435	0.965	1.435	0.847
	$\hat{\lambda}$	0.076	0.085	0.082	0.073	0.162	0.168	0.114	0.058	0.274	0.079	0.144	0.116

in Table IV and $F_n(x)$ is the empirical distribution function. Specifically

$$F_n(x) = \frac{j}{n}, \quad x_{(j)} < x \leq x_{(j+1)}, \quad j=0, 1, \ldots, n$$

and $x_{(j)}$ is the jth order statistic, $j=1, \ldots, n$. In fact there are not tables for evaluating the goodness of fit for this model but smaller values for D* give some positive indication for better fitting.

Table V contains the values of D^* for every month of the year at the seven meteorological sites. The distribution of D^* is not available. Therefore, observed values of D^* can not be compared with its standard. Nonetheless, we can note that most of the observed values of D^* are less than 0.2. It is also noted that relatively high values D^* correspond to high maximum, or range, and high coefficient of variation. Specifically, values of D* for March at Dhahran (0.380) and January at Tabouk (0.302) are the highest values in Table 10. For other months at all sites, except where the model is not applicable, the proposed model fits the given data satisfactorily.

The empirical method for estimating the parameters of the proposed model is simple to carry out by using any calculator. The main drawback of this method is that it is not applicable for May at Khamis Mushait since it was a wet month during 1966 through 1986. For this month, other known models for fitting precipitation data such as gamma, log normal and kappa types can be used, (Suzuki, 1980).

Figures illustrating the probability of dry months and maximum and mean amount of rainfall show a large variation in the distribution of monthly rainfalls at these seven

TABLE V

The value of D^* statistics

Site	Dhahran	Jeddah	Khamis Mushait	Madina	Riyadh	Tabouk	Taif
Month							
Jan	0.189	0.237	0.135	0.074	0.176	0.302	0.179
Feb	0.255	0.043	0.115	0.071	0.098	0.126	0.143
Mar	0.380	0.034	0.114	0.068	0.109	0.291	0.235
Apr	0.237	0.127	0.114	0.145	0.238	0.235	0.218
May	0.070	0.109	–	0.122	0.165	0.087	0.142
Jun	–	–	0.061	0.030	0.017	–	0.199
Jul	–	0.18	0.106	0.018	0.042	–	0.113
Aug	–	–	0.085	0.082	0.037	0.019	0.173
Sep	–	0.000	0.080	0.034	0.018	–	0.175
Oct	0.062	0.035	0.101	0.055	0.043	0.081	0.124
Nov	0.093	0.147	0.140	0.179	0.096	0.156	0.185
Dec	0.143	0.139	0.146	0.116	0.154	0.152	0.172

sites. These plots are useful for quick reference to the rainfall regimes at the sites. Thus when planning for agricultural projects, variation in the rainfall behaviour has to be considered.

This is not an exhaustive study but it forms an additional data base for studying the distribution of monthly precipitation totals at the main meteorological sites in Saudi Arabia.

Acknowledgement

The author is grateful to the referees for their useful comments.

References

Abouammoh, A. M.: 1986, 'On Probability Distribution of Monthly Precipitation Totals in Arid Regions', *J. Appl. Statist.* **2**, 51-63.

Bodtmann, W. R. and Ruthroff, C. L.: 1976, 'The Measurement of 1-min Rain Rates from Weighing Rainage Recordings', *J. Appl. Meteor.* **15**, 1160-1166.

Buishand, T. A.: 1977, *Stochastic Modelling of Daily Rainfall Sequences*, Meded. Landbouwhogesch. Wageningen.

DeBoer, H. J.: 1958, 'On the Cumulative Frequency Distribution of k-Day Period Amounts of Precipitation for any Station in the Netherlands while $k = 30$', *Arch. Met. Geophys Bioklim* **B9**, 244-253.

Drufuca, G.: 1977, 'Radar-Derived Statistics on the Structure of Precipitation Patterns', *J. Appl. Meteor.* **16**, 1029-1035.

Fisher, R. A. and Cornish, E. A.: 1960, 'The Percentile Points of Distributions Having Known Cumulants', *Technometrics* **2**, 209-225.

Hamza, F.: 1968, *The Heart of Arabian Peninsula.* (In Arabic), Nasr Riyadh.

Kirkham, H.: 1980, 'Instantaneous Rainfall Rate, its Measurement and its Influence on High-Voltage Transmission Lines', *J. Appl. Meteor.* **19**, 35-40.

Linslay, Jr. R. K., Kohler, M. A. and Paulhus, J. L. H.: 1982, Hydrology for Engineers, McGraw, Auckland.

Stern, R. D. and Coe, R.: 1982, 'The Use of Rainfall Models in Agricultural Planning', *Agric. Meteorol.* **26**,35-40.

Suzuki, E.: 1980, 'A Summarized Review of Theoretical Distributions Fitted to Climatic Factors and Markov chain models of Weather Sequences, with Some Examples', in S. Ikeda *et al.*, (eds.), *Statistical Climatology, Developments in Atmospheric Sciences,* 13, pp. 1-20.

Tayeb, F. A.: 1978, 'The Role of the Ground Water in Irrigation and Drainage of the Al-Hasa Region of Eastern Saudi Arabia', Unpublished M. A. Thesis, Eastern Michigan University, Ypsilanti Michigan.

Ozturk, A.: 1981, 'On the Study of a Probability Distribution of Precipitation Totals', *J. Appl. Meteor.* **20**, 1499-1505.

Ozturk, A.: 1984, 'A Simple Method for Estimating the Parameters of Probability Distribution for Precipitation Totals', Proc. Contributed papers 44th Sers., ISI, pp. 46-49.

THE ECONOMICS OF TRANSBOUNDARY AIR POLLUTION IN EUROPE

EKKO C. VAN IERLAND

Department of General Economics, Wageningen Agricultural University, P.O. Box 8130, 6700 EW Wageningen, The Netherlands

(Received March 1990)

Abstract. Acid rain is causing substantial damage in all Eastern and Western European countries. This article presents a stepwise linear optimisation model, that places transboundary air pollution by SO_2 and NO_X in a game theoretical framework. The national authorities of 28 countries are perceived as players in a game in which they can choose optimal strategies. It is illustrated that optimal national abatement programmes may be far from optimal if considered from an international point of view. Several scenarios are discussed, including a reference case, full cooperation, Pareto optimality and a critical loads approach. The need for international cooperation and regional differentiation of abatement programmes is emphasised.

1. Introduction

Transboundary air pollution of sulphur dioxide, nitrogen oxides and ammonia (acid rain) is causing substantial damage to materials, buildings and ecosystems, particularly forests and aquatic systems in European countries. The acidifying emissions are caused by industrial activity, power plants, refineries and traffic, as well as by the combustion of fossil fuel by households and government. Ammonia is mainly emitted by intensive agriculture. A large number of technologies exist to reduce the emissions, such as flue gas desulphurisation and the application of catalysts. In addition, changes in fuel mix and industrial structure are possible.

At present political decision-making for optimal abatement strategies is going on at the European level, including Eastern Europe. This paper describes an optimisation model that could be used to assist in finding optimal abatement strategies.* The model offers the possibility to: (1) minimize the total abatement and damage costs, given a fixed production structure: (a) at a national level for each country individually, (b) at a European level for all countries involved in transboundary air pollution; (2) minimize the total abatement and damage costs assuming Pareto optimality, and (3) minimize abatement costs, given a fixed production structure and regional target levels for the deposition of acidifying compounds (critical loads). Furthermore, a large number of specific constraints may be set, for example regionally differentiated critical loads or maximum abatement costs per country.

In accordance with Mäler (1989), the paper shows that international cooperation may reduce the total abatement and damage costs of acidification. It is also shown that a system of regionally differentiated reduction plans (as opposed to fixed percentage

* I am greatly indebted to Hans Coenen who assisted in performing the model calculations.

reductions for all countries) will contribute to optimal abatement strategies.

The model presented in this paper is meant to illustrate the most important international policy aspects of acidification in Europe (see also Van Ierland and Oskam, 1988). As the empirical information on emissions, abatement costs and damage in the countries in question is rather weak, the model cannot pretend to be completely empirically validated. In choosing the parameters of the model, some heroic assumptions have been made and further empirical research is needed to strengthen the empirical foundation of the model. Nevertheless it is felt that the quantitative relationships in the model reflect some of the basic relationships of acidification in Europe.

2. Optimal Abatement Strategies

Transboundary air pollution is characterised by the fact that most European countries are emitting large quantities of SO_2 and NO_x that are transported via the atmosphere to other countries. In this way acidifying substances are causing damage in countries other than where they originate.

For national authorities, this poses a severe problem in formulating optimal abatement strategies. If, for example, a country would like to reduce the damage caused by acidification in its own country it can only reduce the domestic emission. As a large part of the deposition is the result of *foreign* emissions, the impact of domestic emission reduction on the damage within the country will be modest. As a result, the present policy of most countries tends to modest abatement plans as the national authorities are inclined to look only at the domestic benefits of their abatement plans. This implies that the abatement strategies of individual nations (although being optimal from a national point of view) are not at all optimal from an international point of view.

In Section 3 of this paper, the structure of an optimisation model is described that analyses the most important aspects of acidification in Europe. In Section 4 the model is applied to calculate the average annual deposition of acidifying compounds per country for the year 1995 assuming the offically planned reduction targets for all countries as published in ECE (1987). This projection is hereafter referred to as 'the reference projection'.

As pointed out by Mäler (1989), international cooperation on emissions reduction could be beneficial to most countries involved and if side payments (i.e. transfers of money from one country to another to abate pollution) are allowed, all countries (even countries that are suffering little environmental damage due to acidification) could be better off through more stringent abatement policy.

From an international point of view the optimal policy in the tradition of welfare economics would be the policy that equates marginal abatement costs to marginal damage costs or one that minimizes the total cost of pollution abatement and the total damage costs for all countries involved. If sufficient empirical data on abatement cost functions per country and damage cost functions per country are available, a stepwise linear programming model can calculate the optimal level of abatement for each country, taking into account the international dispersion of emissions via the atmosphere. This analysis is followed in Section 5.

Next, in Section 6, the model is used to calculate the required abatement percentages per country to limit the deposition of acidifying compounds to a given critical load of 1400 acid equivalent[†] per hectare per year. This approach, that is compatible with a strategy for sustainable economic development, differs considerably from the welfare economic approach. Instead of minimising the sum of damage and abatement costs, it takes an ecological deposition constraint as a guideline for abatement strategy. This ecological approach leads to more stringent abatement than the welfare economic approach.

In Section 7, regionally differentiated critical loads are specified. As an example, more stringent deposition constraints are set for Scandinavian countries. The maximium deposition for these countries is set at 400 acid equivalents per year.

Section 8 deals with a coalition in which the Western European countries participate, while the Eastern European countries are assumed to emit SO_2 and NO_x at the levels of the reference projection.

Finally, Section 9 contains the conclusions and some suggestions for further research.

3. The Structure of the Optimisation Model

To describe the policy dilemmas for acidification in Europe a stepwise linear optimisation model was developed. For 28 countries in Eastern and Western Europe the model specifies the expected level of emissions of SO_2 and NO_x in the year 1995, assuming that no abatement takes place. These levels have been calculated by means of energy projections (Guilmot *et al.*, (1986)), the expected rates of growth of gross domestic product (GDP) and changes in fuel mix. Next, the emissions after abatement are calculated by taking the difference between the emissions before abatement and the level of abatement:

$$SO_2^* = SO_2 - SA \tag{1}$$

where

SO_2^* vector of emission of SO_2 after abatement for all countries i; $i = 1 \ldots. 28$;
SO_2 vector of emission of SO_2 before abatement for all countries i; $i = 1 \ldots. 28$;
SA vector of abatement of SO_2 for all countries i; $i = 1 \ldots. 28$.

The abatement cost curve for each country is specified as a stepwise linear cost function that shows the total abatement costs of SO_2 per country as a function of the quantity of SO_2-abatement (Figure 1). The cost curves are chosen such that the first trajectory shows the cost efficient abatement measures that can be realised at relatively low costs, for example 1000 Hfl. per ton (thousand Dutch Guilders; 1 Hfl. (0.5 US \$ or 1 Hfl. (0.9 DM). The second trajectory shows the more expensive abatement options, for example at 1500 Hfl. per ton and the third trajectory shows the most expensive abatement techniques, for example, 2500 Hfl. per ton. For abatement cost curves see for example OECD (1988).

[†] 1 acid equivalent is defined as 1 mol H^+ potential acid; 1 ton $SO_2 = 31\,500$ acid equivalents; 1 ton $NO_x = 21\,500$ acid equivalents; 1 ton $NH_3 = 59\,000$ acid equivalents.

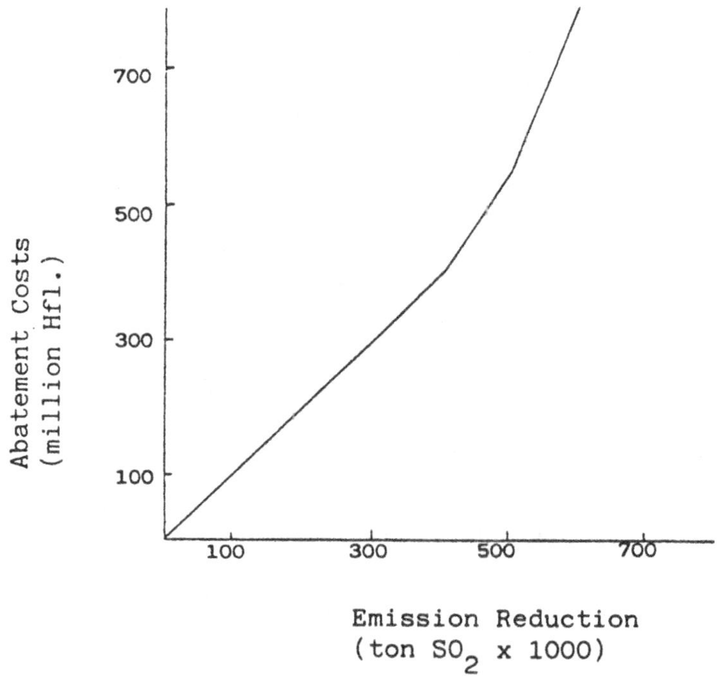

Fig. 1. An example of the stepwise linear abatement cost function for SO_2 for Belgium in 1995.

In the model, the dispersion of acidifying substances via the atmosphere is calculated by means of the emitter-receptor matrices of EMEP (1981 and 1988). These matrices have been established for 28 European countries on the basis of past meteorological patterns. The emitter-receptor matrix for SO_2 is generally accepted. The NO_x matrix however is a preliminary result of EMEP calculations and will probably be modified in the future, as declared by EMEP (1988). Each row of an emitter-receptor matrix for SO_2 indicates for each country the quantity of SO_2 that the country in question receives from the other countries. Each column shows what quantities of SO_2 the relevant country is emitting to other countries (see Table I for an example). The first row of Table I shows that country 1 receives 10 tons of SO_2 from its own territory, 20 tons from country 2 and 40 tons from country 3. See ANNEX I for the SO_2 matrix of EMEP.

TABLE I

A hypothetical and simplified emitter-receptor matrix for SO_2 for three countries

Receptor	Emitter	Country 1	Country 2	Country 3	Total
Country 1		10	20	40	70
Country 2		25	40	25	90
Country 3		8	10	20	38
	Total	43	70	85	

By means of the emitter-receptor matrices for SO_2 and NO_x the deposition of SO_2 can be calculated for each country according to:

$$DS = AS * SO_2^* \tag{2}$$

where
DS = vector of deposition of SO_2 (28 × 1);
AS = emitter-receptor matrix of SO_2 (28 × 28);
SO_2^* = vector of sulphur dioxide emissions after abatement; (28 × 1).

Similarly the deposition of NO_x is calculated:

$$DN = AN * NO_x^* \tag{3}$$

where
DN = vector of deposition of NO_x; (28 × 1);
AN = emitter-receptor matrix of NO_x (28 × 28);
NO_x^*= vector of nitrogen oxides emissions after abatement; (28 × 1).

The total annual deposition of SO_2, NO_x and ammonia (NH_3) can be expressed in a single unit, i.e. 1 unit of acid equivalent:

$$DA = \delta_1 \, DS + \delta_2 \, DN + \delta_3 \, DNH_3 \tag{4}$$

where
DA = vector of acid deposition measured in units of acid equivalent;
DS = vector of deposition of sulphur oxide;
DN = vector of deposition of nitrogen oxides;
DNH_3 = vector of deposition of NH_3;
$\delta_1 \ldots \delta_3$= parameters expressing units of acid equivalent per ton of each compound.

Furthermore, the average annual deposition of acid equivalents per ha is calculated for each country. So far ammonia emissions have not been taken into account in the model calculations. In a more elaborate study they should also be included.

Finally, we could formulate for each country, a damage function that expresses the monetary value of the damage costs that are caused by the deposition of acidifying substances. Although in reality it is extremely difficult to establish the monetary value of the damage costs of acidification, we have assumed a linear damage cost function that 'guestimates' the annual monetary value of the damage as a function of the average annual deposition measured in units of acid equivalents. No damage is assumed up to 300 acid equivalents per hectare. Above 300 acid equivalents the damage is assumed to be equal to 2500 Hfl. per ton SO_2 and 1700 Hfl. per ton NO_x.

4. The Reference Projection

To calculate the effects of the proposed policy measures by the different countries we could fix the emissions of SO_2 and NO_x at the target levels for 1995 as expressed in the publication 'National strategies and policies for air pollution abatement' of the Economic Commission for Europe (ECE, 1987). In doing so, the model calculates the quantities of abatement of SO_2 and NO_x and the abatement costs for SO_2 and NO_x for all countries. It also calculates the deposition of acid equivalents per hectare per year for the year 1995 for each country and also the damage cost for each country on the basis of the assumed damage cost functions.

In Table II the calculated average annual deposition of acid equivalents per hectare is shown for each country. We can compare these projected figures with the standard of 1400 acid equivalents per hectare per year that is often assumed to reflect a deposition level that avoids major damage to pine forests (see also Section 6). It is then obvious that this standard of 1400 acid equivalents per hectare is still largely exceeded in the following countries: Austria, Belgium, Bulgaria, Czechoslovakia, German Democratic Republic, Federal Republic of Germany, Hungary, Italy, The Netherlands, Poland, Switzerland, The United Kingdom and Yugoslavia. This shows that the projected development of SO_2 and NO_x emissions in the reference projection, which reflects official policy, is by no means capable of limiting the deposition per hectare to a target level of 1400 acid

TABLE II

Average annual deposition (excluding NH_3), percentage SO_2 reduction and percentage NO_x reduction for the reference projection in 1995

Country	Deposition (acid equivalents ha^{-1})	% Reduction of SO_2	% Reduction of NO_x
ALB	867	42	0
AUS	2213	73	42
BEL	2561	21	1
BUL	1823	19	0
CZE	6054	42	34
DEN	1218	47	1
FIN	428	61	9
FRA	1093	39	19
GDR	5055	17	0
FRG	2170	62	40
GRE	1097	43	0
HUN	2851	42	0
ICE	86	14	25
IRE	587	6	0
ITA	1857	34	0
LUX	1126	52	29
NET	1855	55	6
NOR	404	44	0
POL	3841	0	0
POR	730	37	0
ROM	1410	17	0
SPA	1105	23	0
SWE	519	73	39
SWI	1609	62	37
TUR	398	0	0
USSR	1002	42	6
UK	1694	36	11
YUG	2036	0	0

TABLE III

Abatement costs for SO_2 and N_x and damage costs per country in 1995 for the reference projection[a]

Country	Total abatement costs (million Hfl)	Total abatement costs as % GDP 1985	Damage costs (million Hfl.)	Total abatement costs + damage costs
ALB	50	*	129	179
AUS	692	0.3	1274	1966
BEL	146	0.1	547	693
BUL	241	*	1340	1581
CZE	2866	*	5842	8708
DEN	207	0.1	314	521
FIN	465	0.3	344	809
FRA	1540	0.1	3425	4965
GDR	800	*	4084	4884
FRG	5396	0.3	3705	9101
GRE	494	0.5	835	1329
HUN	820	1.2	1883	2703
ICE	10	0.1	0	10
IRE	16	0.1	157	173
ITA	1262	0.1	3721	4983
LUX	23	0.2	17	40
NET	305	0.1	506	811
NOR	74	0.0	268	342
POL	20	0.0	8788	8807
POR	119	0.2	262	381
ROM	40	0.0	2093	2133
SPA	900	0.2	3194	4094
SWE	925	0.3	781	1706
SWI	345	0.1	429	774
TUR	1	0.0	360	361
USSR	6589	*	18737	25326
UK	2010	0.1	2700	4710
YUG	0	0.0	3524	3524
Total	26356		69258	95614

[a] average exchange rate 1987: 1 US$ = 2.03 Hfl.
*no GDP estimate available.

equivalents in all countries. Particularly in Central Europe, this deposition target is largely exceeded, even while ammonia emissions are not taken into account.

Table III shows the abatement costs per country for the reference projection expressed in Hfl. and as a percentage of the 1985 GDP for each country. The abatement effort ranges from 0% of GDP to 1.2% of GDP for Hungary, according to the model calculations. The total abatement costs amount to 26 billion Hfl., while the total damage costs are estimated at 69 billion Hfl. It is evident that the abatement costs as a percentage of GDP are in general, very modest. Figure 2 shows the average annual deposition as acid equivalents per hectare for the reference projection and within brackets for the full cooperative solution, which is discussed below.

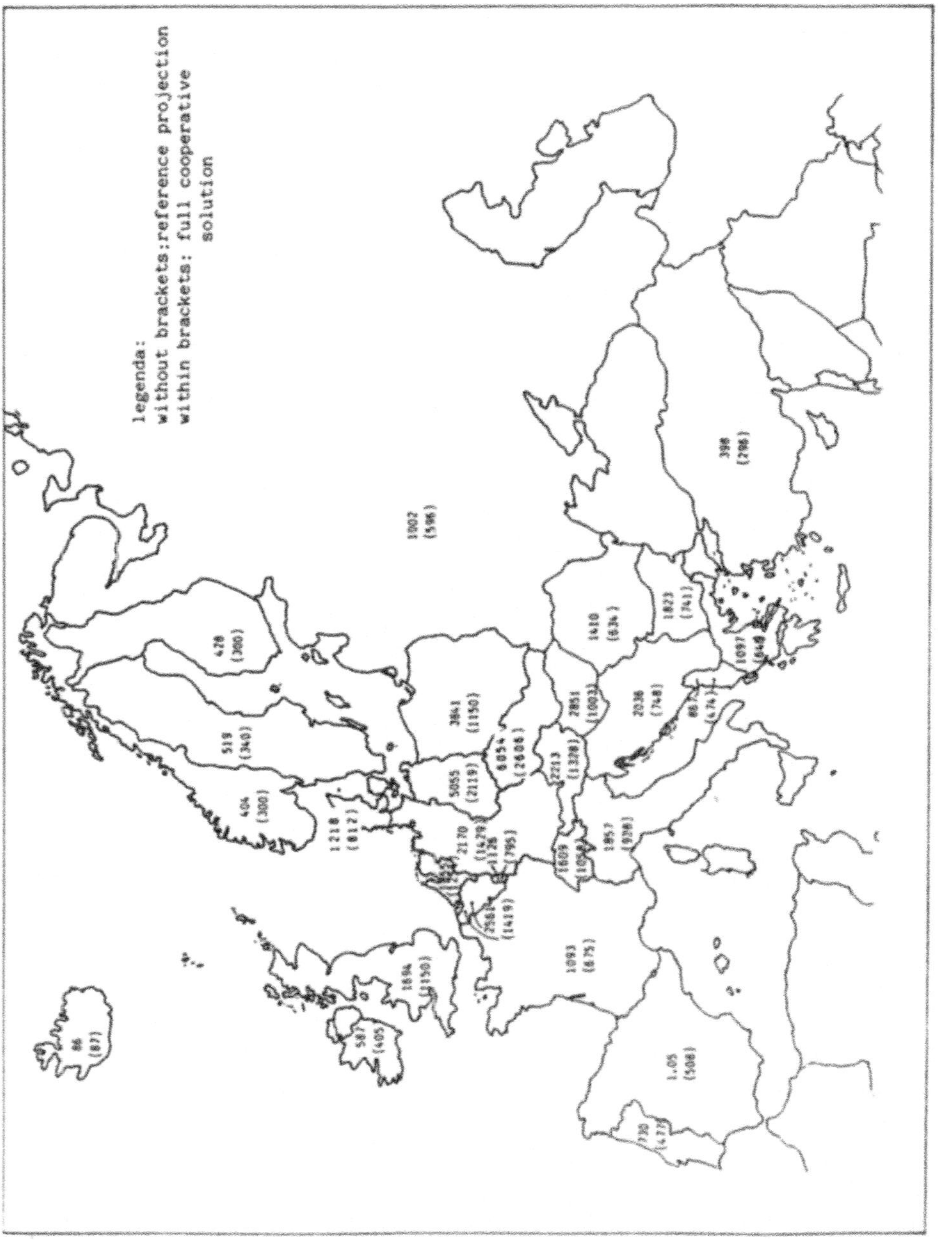

Fig. 2 Average annual deposition of acid equivalents per hectare (excluding ammonia) for the reference projection and for the full cooperative solution (within brackets) in 1995.

5. Full Cooperation: Abatement and Damage Cost Minimisation

Instead of setting emission targets for SO_2 and NO_x, as we did in the reference projection, we could formulate an optimisation problem in which we assume that all countries are willing to cooperate in order to reduce the negative effects of acidification. We therefore could minimize for all countries, the sum of abatement costs and damage costs due to acidification. In doing so we find the 'optimal abatement strategy for Europe', because the minimum of abatement costs and damage costs for all countries is established. It should, however, be noted that the distribution of net benefits from the optimal abatement strategy may differ considerably per country. Some countries will have to intensify their abatement strategy and will be faced with additional abatement costs. Other countries may benefit from the full cooperation solution because there is less need to reduce emissions or because the damage is reduced due to abatement measures in other countries. Therefore, the distributional aspect of costs and benefits is not automatically solved in the full cooperative solution. Below we pay more attention to these aspects, when discussing the Pareto dominant case. Table IV shows, for the full cooperative solution, the average annual acid deposition per hectare and the abatement of SO_2 and NO_x as a percentage of 'emissions before abatement' in 1995. In the full cooperative solution the countries in general pay more attention to the abatement of SO_2 as compared to the central projection. This phenomenon is explained by the fact that the contribution of NO_x to acidification per ton is smaller than that of SO_2 and by the fact that abatement cost of NO_x per ton are generally higher than for SO_2.

In Table V the economic aspects of the full cooperative solution are summarized. Compared to the central projection, all countries are better off as far as the total abatement costs and damage costs are concerned, with the exception of Belgium, Denmark, the German Democratic Republic, Hungary and Ireland. For these countries, the damage is reduced but the abatement costs have increased by a larger amount in the full cooperative solution as compared to the reference projection. For all other countries either the reduction of the damage costs exceeds the increase in abatement costs or the abatement costs have also decreased. The total abatement costs and damage costs is smaller in the full cooperative solution than in the reference projection, i.e. 72.1 billion Hfl. versus 95.6 billion Hfl. (Tables III and V). It is clear that these results depend heavily on the shape of the damage costs functions. In practice it is extremely difficult to estimate the damage costs of acidification. The damages relate, inter alia, to ecosystems, forests, lake acidification, buildings, health effects and loss of agricultural production. Although damage cost estimates have been made for a number of countries, no solid empirical foundation for these estimates is available. This is usually the case, because the damage is often related to public goods for which no market price exists. In Section 6 we discuss the critical loads approach that avoids the complications of damage costs estimation. For the time being we have assumed that the damage costs are a linear function of acid deposition. When calculating the results of Table V the damage cost function is assumed to be a linear function of the acid deposition and the damage cost is guestimated at 2500 Hfl. per ton SO_2 and 1700 Hfl. per ton NO_x, where the ratio between these amounts corresponds to the different contribution of a ton of SO_2 and NO_x to acidification. Furthermore, it is assumed

TABLE IV

Average annual deposition (excluding ammonia), percentage SO_2 reduction and percentage NO_x reduction for the full cooperative solution in 1995

Country	Deposition (acid equivalent ha^{-1})	% Reduction of SO_2	% Reduction of NO_x
ALB	474	60	0
AUS	1328	75	0
BEL	1419	75	8
BUL	741	75	0
CZE	2606	75	8
DEN	812	60	8
FIN	300	66	8
FRA	675	75	8
GDR	2119	75	8
FRG	1429	75	8
GRE	646	60	0
HUN	1003	90	8
ICE	87	0	0
IRE	405	60	0
ITA	928	75	0
LUX	795	76	10
NET	1254	75	8
NOR	300	21	0
POL	1150	90	8
POR	477	60	0
ROM	634	75	8
SPA	508	75	0
SWE	340	75	8
SWI	1058	76	8
TUR	296	0	0
USSR	596	60	0
UK	1150	60	0
YUG	748	90	7

that no damage costs occur up to 300 acid equivalents per hectare.

When comparing the total of abatement costs and damage costs in the full cooperative solution and the reference projection, we find that Belgium, Denmark, the German Democratic Republic, Hungary and Ireland are in a worse situation in the full cooperative solution. As is shown by Mäler (1989) we could set constraints on the net benefits of each country in such a way that no country is worse off by participating in the cooperative solution. In this so-called Pareto dominant outcome all countries gain by participating in the cooperative solution or at least they do not deteriorate as compared to the reference projection (Tables VI and VII). Therefore we have added in the Pareto dominant case for each country, the restriction that the sum of damage and abatement costs is smaller or equal to this sum in the reference projection. This leads to a more effecient allocation of abatement measures among countries as compared to the reference projection. Furthermore, abatement will take place at least costs and at those places where it contributes most to the avoidance of damage costs. At the same time, it is guaranteed that all countries

TABLE V

Abatement costs for SO_2 and NO_x and damage costs per country in 1995 for the full cooperative solution[a]

Country	Total abatement costs (million Hfl.)	Total abatement costs as % GDP 1985	Damage costs (million Hfl.)	Total abatement costs + damage costs
ALB	72	*	40	112
AUS	351	0.2	685	1036
BEL	584	0.2	271	855
BUL	1023	*	388	1411
CZE	3133	*	2341	5474
DEN	273	0.1	175	448
FIN	498	0.3	0	498
FRA	2240	0.1	1620	3860
GDR	4010	*	1562	5572
FRG	2530	0.1	2236	4766
GRE	686	0.6	362	1048
HUN	2368	3.5	519	2887
ICE	0	0.0	0	0
IRE	160	0.3	58	218
ITA	3101	0.3	1501	4602
LUX	19	0.2	10	29
NET	459	0.1	310	769
NOR	36	0.0	0	36
POL	5991	2.8	2110	8101
POR	191	0.3	108	299
ROM	219	0.1	630	849
SPA	3216	0.6	825	4041
SWE	495	0.2	144	639
SWI	136	0.0	249	385
TUR	0	0.0	0	0
USSR	9207	*	7906	17113
UK	2827	0.2	1646	4473
YUG	1704	1.2	910	2614
Total	45529		26606	72135

[a] average exchange rate 1987: 1 US$ = 2.03 Hfl.
*no GDP estimate available.

gain or have at least no losses by participating in the cooperation.

6. Critical Loads Approach

In the previous sections the optimisation was based on the minimisation of the sum of abatement costs and damage costs for all countries. This welfare economic approach differs considerably from the ecological orientated critical loads approach. We could define the critical load as the level of deposition of acid equivalents that avoids major damage to ecosystems, health and economic assets in the long term. The critical loads approach therefore is in accordance with the concept of sustainable economic develop-

TABLE VI

Average annual deposition (excluding ammonia), percentage SO_2 reduction and percentage NO_x reduction for the Pareto dominant case in 1995

Country	Deposition (acid equivalent ha^{-1})	% Reduction of SO_2	% Reduction of NO_x
ALB	479	60	0
AUS	1307	75	0
BEL	1682	52	0
BUL	745	75	0
CZE	2428	85	8
DEN	810	60	8
FIN	300	67	8
FRA	668	75	8
GDR	2637	60	0
FRG	1287	90	8
GRE	646	60	0
HUN	1072	85	8
ICE	87	0	0
IRE	519	20	0
ITA	926	75	0
LUX	733	76	10
NET	1207	75	8
NOR	300	19	0
POL	1192	90	8
POR	476	60	0
ROM	643	75	8
SPA	505	75	0
SWE	342	75	8
SWI	1037	76	8
TUR	297	0	0
USSR	600	60	0
UK	1079	64	8
YUG	756	90	7

ment that saves the environment and that avoids overexploitation of the natural environment and natural resources.

In the Netherlands a deposition of 1400 acid equivalents per hectare is considered to be a critical load that avoids damage to forests of pine trees and other vulnerable ecosystems (RIVM, 1988). For many types of ecosystems, critical loads could be specified depending on vegetation, soil characteristics, weather conditions etc. In the near future, the average annual deposition in the Netherlands should be limited to a level below 1400 acid equivalents per hectare per year to avoid major damage to pine forests. For more vulnerable ecosystems an average annual deposition of 700 acid equivalents per hectare per year is considered to be an upper limit. In the critical loads approach we therefore set

TABLE VII

Abatement costs for SO_2 and NO_x and damage costs per country in 1995 for the Pareto dominant case[a]

Country	Total abatement costs (million (Hfl.)	Total abatement costs as % GDP 1985	Damage costs (million Hfl)	Total abatement costs + damage costs
ALB	72	*	41	113
AUS	351	0.2	671	1022
BEL	359	0.1	335	694
BUL	1023	*	392	1415
CZE	4037	*	2160	6197
DEN	272	0.1	174	446
FIN	509	0.3	0	509
FRA	2240	0.1	1589	3829
GDR	2877	*	2007	4884
FRG	3628	0.2	1955	5583
GRE	686	0.6	363	1049
HUN	2133	3.2	569	2702
ICE	0	0.0	0	0
IRE	53	0.1	120	173
ITA	3101	0.3	1496	4597
LUX	19	0.2	9	28
NET	458	0.1	295	753
NOR	31	0.0	0	31
POL	5991	2.8	2213	8204
POR	191	0.3	107	298
ROM	219	0.1	647	866
SPA	3216	0.6	813	4029
SWE	495	0.2	150	645
SWI	136	0.0	241	377
TUR	0	0.0	0	0
USSR	3201	0.2	1508	4709
YUG	1704	1.2	925	2629
Total	46209		26792	73001

[a] average exchange rate 1987: 1 US$ = 2.03 Hfl.
*no GDP estimate available.

an upper limit on the average deposition of acid equivalents per hectare per year due to the deposition of SO_2 and NO_x*.

In the critical loads projection we have set an upper limit on the average annual deposition per hectare per year of 1400 acid equivalents per hectare. As can be seen from Table II, this level is largely exceeded in a large number of countries. Table IV also shows that this level is still exceeded in the case of the full cooperative solution for Czechoslovakia and the German Democratic Republic. This is also true in the Pareto dominant case for a number of countries as shown in Table VI, even if we do not take into

*As stated above, in a more elaborated study also the deposition of ammonia should be taken into account. For the purpose of simplicity it has not been considered in the present paper.

account the acidifying effects of ammonia. The results for the critical loads approach are shown in Tables VII and IX. From the tables it becomes clear that deposition in a large number of countries remains below the critical load of 1400 acid equivalents per hectare per year. Czechoslovakia, however, reaches the level of 1400. As Czechoslovakia uses its maximum capacity to reduce its emissions, many other countries have to contribute to the reduction of the deposition in Czechoslovakia. Therefore, the deposition in other countries, particularly those that are a great distance from Central Europe fall far below the level of 1400.

As can be seen from Table VIII, many countries in Central Europe use the maximum emission reduction capacity for SO_2 (90%) and for NO_x (51%). Countries at the periphery, such as Ireland, Norway, Portugal and Turkey have relatively modest abatement percentages, particularly for NO_x. This is an indication that the emission reduction percentages should be regionally differentiated and that a fixed percentage reduction as compared to a particular base year will not be cost effective. In practice, this

TABLE VIII

Average annual deposition (excluding ammonia), percentage SO_2 reduction and percentage NO_x reduction for the critical loads approach in 1995

Country	Deposition (acid equivalent ha^{-1})	% Reduction of SO_2	% Reduction of NO_x
ALB	402	60	0
AUS	855	90	30
BEL	800	90	29
BUL	466	90	0
CZE	1400	90	51
DEN	437	90	8
FIN	300	46	8
FRA	461	90	8
GDR	1085	90	51
FRG	852	90	30
GRE	566	60	0
HUN	720	90	51
ICE	86	0	0
IRE	352	60	8
ITA	572	90	8
LUX	546	76	33
NET	701	90	30
NOR	246	0	8
POL	780	90	33
POR	476	60	0
ROM	447	90	8
SPA	482	75	0
SWE	300	53	8
SWI	701	90	30
TUR	264	0	0
USSR	419	75	8
UK	505	90	8
YUG	541	90	30

TABLE IX

Abatement costs for SO_2 and NO_x and damage costs per country for the critical loads approach in 1995[a]

Country	Total abatement costs (million Hfl.)	Total abatement costs as % GDP 1985	Damage costs (million Hfl.)	Total abatement costs + damage costs
ALB	72	*	23	95
AUS	671	0.3	370	1041
BEL	1063	0.4	121	1184
BUL	1491	*	146	1637
CZE	7142	*	1117	8259
DEN	533	0.3	47	580
FIN	337	0.2	0	337
FRA	3232	0.2	696	3928
GDR	7885	*	675	8560
FRG	5275	0.3	1093	6368
GRE	686	0.6	278	964
HUN	2952	4.4	310	3262
ICE	0	0.0	0	0
IRE	164	0.3	28	192
ITA	4592	0.4	649	5241
LUX	32	0.3	5	37
NET	972	0.2	131	1103
NOR	12	0.0	0	12
POL	7363	3.4	1190	8553
POR	191	0.3	107	298
ROM	309	0.2	278	587
SPA	3216	0.6	721	3937
SWE	323	0.1	0	323
SWI	328	0.1	131	459
TUR	0	0.0	0	0
USSR	12822	*	3183	16005
UK	5757	0.4	397	6154
YUG	1836	1.3	488	2324
Total	69256		12184	81440

[a] Average exchange rate 1987: 1 US$ = 2.03 Hfl.
*no GDP estimate available.

differentiation in abatement plans is taking place, in so far as countries in Central Europe, for example the Federal Republic of Germany, intend to implement more abatement programs than other countries. The regional differentiation of abatement policy should therefore play an important role in the formulation of the environmental policy of the European Community. As far as the cost of the critical loads approach are concerned it is clear that it is rather expensive to reach the 1400 target in Czechoslovakia. The total abatement and damage costs are for example, in the full cooperative case 72.1 billion Hfl.; in the critical load approach they rise to 81.4 billion Hfl.

7. A Differentiated Critical Loads Approach

It is often stated that the ecosystems in Scandinavia are more vulnerable to acidification than ecosystems in the southern parts of Europe. This suggests that it could be worthwhile to specify regionally differentiated deposition targets or to specify damage costs functions that are clearly differentiated per region.

In this section we show the results of setting more stringent deposition targets in Scandinavian countries, in particular in Denmark, Finland, Norway and Sweden. In these countries, deposition should not exceed the maximum of 400 acid equivalents per hectare; in other countries the maximum is kept at 1400. Clearly, this requires a more intensive abatement strategy, not only for the Scandinavian countries but also for other countries that contribute to the deposition in Scandinavia. The results are shown in Tables X and XI. The total abatement and damage costs increase to 82.3 billion Hfl. if the target level is 400 in Scandinavia as compared to 81.4 billion Hfl. if the limit is set at 1400 acid

TABLE X

Average annual deposition (excluding ammonia), percentage SO_2 reduction and percentage NO_x reduction for the Scandinavian case in 1995

Country	Deposition (acid equivalent ha^{-1})	% Reduction of SO_2	% Reduction of NO_x
ALB	420	60	0
AUS	849	90	30
BEL	772	90	29
BUL	666	75	0
CZE	1400	90	51
DEN	400	90	51
FIN	300	46	8
FRA	453	90	8
GDR	1068	90	51
FRG	831	90	30
GRE	604	60	0
HUN	732	90	30
ICE	84	0	0
IRE	338	60	30
ITA	571	90	8
LUX	543	76	33
NET	667	90	37
NOR	237	0	8
POL	780	90	30
POR	474	60	0
ROM	473	90	8
SPA	480	75	0
SWE	300	45	29
SWI	698	90	20
TUR	274	0	0
USSR	426	75	0
UK	474	90	30
YUG	565	90	7

TABLE XI

Abatement costs for SO_2 and NO_x and damage costs per country in 1995 for the Scandinavian case in 1995[a]

Country	Total abatement costs (million Hfl.)	Total abatement costs as % GDP 1985	Damage costs (million Hfl.)	Total abatement costs + damage costs
ALB	72	*	27	99
AUS	671	0.3	365	1036
BEL	1063	0.4	114	1177
BUL	1023	*	323	1346
CZE	7142	*	1117	8259
DEN	1029	0.5	34	1063
FIN	335	0.2	0	335
FRA	3232	0.2	660	3892
GDR	7885	*	659	8544
FRG	5275	0.3	1052	6327
GRE	686	0.6	319	1005
HUN	2556	3.8	319	2875
ICE	0	0.0	0	0
IRE	214	0.4	21	235
ITA	4592	0.4	648	5240
LUX	32	0.3	5	37
NET	1212	0.3	119	1331
NOR	12	0.0	0	12
POL	7032	3.3	1189	8221
POR	191	0.3	106	297
ROM	309	0.2	326	635
SPA	3216	0.6	714	3930
SWE	505	0.2	0	505
SWI	268	0.1	130	398
TUR	0	0.0	0	0
USSR	12662	*	3365	16027
UK	6930	0.5	337	7267
YUG	1704	1.2	538	2242
Total	69848		12487	82335

[a] average exchange rate 1987: 1 US $ = 2.03 Hfl.
*no GDP estimate available.

equivalents per hectare per year for all countries. However, even in this more stringent case, the abatement costs as a percentage of 1985 GDP remain for most countries below 1% of GDP, with the exception of some Eastern European countries.

If we compare the regionally differentiated case with the previous case we can see that the deposition reduction can be reached by additional reduction of NO_x in Denmark, Ireland, The Netherlands, Sweden and The United Kingdom. At the same time some other countries could reduce the abatement of SO_2 and NO_x; for example Bulgaria (SO_2), Hungaria (NO_x), Sweden (SO_2) and Switzerland (NO_x). This shows that the European countries are heavily interdependent because of the transboundary character of SO_2 and NO_x pollution. The additional abatement costs to reach the more stringent Scandinavian target is about 0.6 billion Hfl.

8. An Example of a Coalition

The assumption of full cooperation is certainly a strong one. In real political life it seems unlikely that all countries will participate in a full cooperative solution. As an example of a coalition in which all countries do not participate, we assume that all countries in Eastern Europe will reduce emissions as is given in the reference projection, while all countries in Western Europe cooperate in a coalition. In this case, Eastern Europe will abate far less than in the full cooperative solution. As a result, the damage in Western Europe will also be higher. At the same time the incentives for Western European countries to abate pollution, decreases as compared to the full cooperative solution, because these countries no longer take account of the damage that they are causing in Eastern Europe. The marginal damage of 1 unit of SO_2 is therefore lower and the marginal abatement costs will also be lower.

Tables XII and XIII show that acid deposition is higher in all countries than in the

TABLE XII

Average annual deposition (excluding ammonia), percentage SO_2 reduction and percentage NO_x reduction for coalition formation in 1995

Country	Deposition (acid equivalent ha^{-1})	% Reduction of SO_2	% Reduction of NO_x
ALB	987	42	0
AUS	2436	0	0
BEL	1711	60	8
BUL	1765	19	0
CZE	5962	42	34
DEN	1499	0	0
FIN	569	0	0
FRA	763	75	8
GDR	5019	17	0
FRG	1935	75	0
GRE	1441	0	0
HUN	2863	42	0
ICE	88	0	0
IRE	589	0	0
ITA	1355	60	0
LUX	816	76	10
NET	1522	60	0
NOR	388	60	0
POL	3835	0	0
POR	526	60	0
ROM	1406	17	0
SPA	678	60	0
SWE	552	60	0
SWI	1274	76	0
TUR	357	60	0
USSR	1004	42	6
UK	1221	60	0
YUG	1969	0	0

TABLE XIII

Abatement costs for SO_2 and NO_x and damage costs per country for coalition in 1995[a]

Country	Total abatement costs (million Hfl.)	Total abatement costs as % GDP 1985	Damage costs (million Hfl.)	Total abatement costs + damage costs
ALB	50	*	129	179
AUS	0	0.0	1422	1422
BEL	429	0.2	342	771
BUL	241	*	1289	1530
CZE	4129	*	5748	9877
DEN	0	0.0	410	410
FIN	0	0.0	720	720
FRA	2240	0.1	1997	4237
GDR	800	*	4053	4853
FRG	2401	0.1	3239	5640
GRE	0	0.0	1195	1195
HUN	820	1.2	1892	2712
ICE	0	0.0	0	0
IRE	0	0.0	158	158
ITA	2255	0.2	2521	4776
LUX	19	0.2	11	30
NET	315	0.1	398	713
NOR	101	0.1	226	327
POL	20	0.0	8773	8793
POR	191	0.3	138	329
ROM	40	0.0	2650	2690
SPA	2338	0.4	1498	3836
SWE	347	0.1	901	1248
SWI	126	0.0	319	445
TUR	199	0.1	208	407
USSR	6589	*	18787	25376
UK	2827	0.2	1785	4612
YUG	0	0.0	3389	3389
Total	26477		64198	90675

[a] average exchange rate 1987: 1 US $ = 2.03 Hfl.
*no GDP estimate available.

full cooperative solution and abatement costs are lower in almost all countries. The total abatement costs and damage costs is generally much higher than in the full cooperative solution. This clearly illustrates the negative aspects of non-cooperation. Not only in Eastern Europe but also in Western Europe do these negative effects become evident. Only a limited number of countries benefit from non-cooperation: Belgium, Denmark, the German Democratic Republic, Hungary, Ireland, The Netherlands and Spain. The net benefits of these countries are, however, largely exceeded by the net losses of the other countries.

9. Conclusions

In this paper, an optimisation model is presented for optimal abatement of SO_2 and NO_x emissions in Europe. Although the empirical basis of the model should be improved, particularly with regard to abatement cost curves and damage cost curves, the model is capable of analyzing the problems of acidification from several points of views.

Firstly, we calculated the average annual deposition of acid equivalents per hectare per year for each country, assuming that the officially projected emission reduction, as formulated in the ECE-document, will be realised in 1995. The calculations make it clear that the critical load of 1400 acid equivalents is still broadly exceeded in a large number of countries. If ammonia (that contributes about 30% to acidification in some countries) had been taken into account, the results would have been worse.

Secondly, the calculations suggest that a full cooperative solution could improve the results by stimulating a more efficient abatement strategy among countries. The total abatement and damage costs could be lowered for almost all countries by choosing more stringent abatement targets for each country, as was illustrated in the full cooperative solution.

Even more interesting is the Pareto dominant case which shows that with only a few exceptions, all countries could benefit from cooperation. In the Pareto dominant case, no country is worse off and most improve their position. This result strongly pleads for more international cooperation and optimisation of abatement policy.

A different analysis is chosen in the case of the critical loads approach. Instead of minimizing abatement costs and damage costs for all countries, upper limits have been set for the average annual deposition of acid equivalents per hectare per year for all countries. These calculations are compatible with a strategy of sustainable development that aims at avoiding major ecological damage in the long term. This study shows that in many countries, the critical loads are exceeded in the reference projection and the full cooperative solution. Therefore, a larger abatement effort is required to reach a critical load, set at 1400 acid equivalents per hectare per year. If we realise that in the long term a critical load of about 700 acid equivalents or even less is required, it becomes clear that more stringent abatement is necessary. Regionally differentiated deposition targets are worthwhile when it is shown that some ecosystems, for example the ecosystems in Scandinavia, are more vulnerable than others.

The example of a coalition shows how important cooperation between Eastern and Western Europe is in order to abate transboundary air pollution. If some countries do not participate in a coalition the results are heavily adversely influenced and the solution is far from optimal.

Not only is further improvement of abatement techniques necessary, increased application of existing abatement techniques is required. In addition, improvement of energy efficiency is needed or even a limitation of the use of polluting energy resources. Therefore, a major effort should be made to stimulate pollution abatement and to develop renewable, non-polluting energy resources that could be introduced in the coming decades.

References

ECE: 1987, *National Strategies and Policies for Air Pollution Abatement*, United Nations, New York.

EMEP: 1981, 'Cooperative Programme for Monitoring and Evaluation of the Long-Range Transmission of Air Pollutants in Europe', Fourth Routine technical report from the Western Metereological Synthesizing Centre, EMEP/MSC-W report 1/81.

EMEP: 1986, *Emission of Sulphur Dioxide in Europe in 1980 and 1983*, Dovland, H. and J. Saltbones. EMEP/ECE-report 1/86, Norway.

EMEP: 1988, *Estimates of Airborne Transboundary Transport of Sulphur and Nitrogen over Europe*, A. Eliassen, Ø. Hov, T. Iversen, J. Saltbones and D. Simpson. The Norwegian Metereological Institute, Metereological Synthesizing Centre, West (MSC-W) of EMEP.

Guilmot, J. F. *et al.*: 1986, *Energy 2000, Commission of the European Community*, Cambridge University Press, London.

Ierland, E. C. van, and Oskam, E. A.: 1988, *A Regional Sectoral Scenario Model for SO_2 and NO_x Emissions in European Countries*, in Dietz, F. J. and W. J. M. Heijman (eds.), Environmental Policy in a Market Economy. Pudoc, Wageningen.

IIASA: 1987, Rains, Enem version 3.b, Laxenburg.

Mäler, K. G.: 1989, 'The Acid Rain Game', in Folmer, H. and E. C. van Ierland (eds.), *Valuation Methods and Policy-Making in Environmental Economics*, North Holland, Amsterdam.

OECD: 1988, *The Compendium of Emission Control Techniques and Their Costs*, OECD Environment Directorate, Paris.

RIVM: 1988, (Public Institute for Health and Environment in the Netherlands). *Zorgen voor morgen* (in Dutch), Samsom H.D. Tjeenk Willink, Alphen aan de Rijn.

ANNEX I: Emitter-receptormatrix for SO_2

MONTHLY MEAN VALUES FOR THE DEPOSITION OF SULPHUR – FOR THE PERIOD: 78 10 01 – 80 09 01
HORIZONTAL – EMITTERS → VERTICAL – RECEIVERS UNIT – 100. TONS SULPHUR

	1	2	3	4	5	6	7	8	9	10	11	12	13	14	15	16	17	18	19	20	21	22	23	24	25	26	27	28	29	UND	SUM	
1	10	0	0	3	0	0	0	0	0	0	2	2	0	0	8	0	0	0	0	0	2	0	0	0	0	0	0	13	0	12	67	1
2	0	52	3	0	35	0	0	20	24	32	0	14	0	0	56	0	2	0	13	0	0	2	0	3	0	0	8	35	0	30	341	2
3	0	0	67	0	0	0	0	28	2	24	0	0	0	0	0	0	4	0	0	0	0	0	0	0	0	0	18	0	0	9	161	3
4	0	0	0	153	8	0	0	2	6	4	7	12	0	0	8	0	0	0	7	0	34	0	0	0	3	12	2	45	0	29	346	4
5	0	22	10	2	483	2	0	45	195	108	0	68	0	0	39	0	7	0	95	0	14	4	0	2	0	14	32	53	0	91	1361	5
6	0	0	0	0	3	39	0	3	12	11	0	0	0	0	0	0	0	0	4	0	0	0	0	0	0	0	11	0	0	11	109	6
7	0	0	2	0	8	3	77	5	19	15	0	3	0	0	2	0	2	2	15	0	2	0	14	0	0	43	14	3	0	55	293	7
8	0	0	33	0	7	0	629	0	20	98	0	0	0	2	41	4	12	0	4	0	0	66	0	7	0	0	99	6	0	166	1212	8
9	0	0	8	0	56	3	0	20	497	85	0	4	0	0	3	0	6	0	23	0	0	0	0	0	0	3	22	6	0	26	778	9
10	0	8	37	0	48	4	0	108	118	561	0	6	0	0	24	5	23	0	18	0	0	6	0	6	0	3	72	15	0	80	1158	10
11	3	0	0	34	4	0	0	3	3	3	93	5	0	0	14	0	0	0	2	0	8	2	0	0	4	4	0	27	0	31	253	11
12	0	9	0	0	46	0	0	7	18	13	0	194	0	0	27	0	0	0	21	0	16	0	0	0	0	4	3	75	0	20	467	12
13	0	0	0	0	0	0	0	0	0	0	0	0	0	0	0	0	0	0	0	0	0	0	0	0	0	0	0	0	0	18	24	13
14	0	0	0	0	0	0	0	2	0	0	0	0	0	18	0	0	0	0	0	0	0	0	0	0	0	0	12	0	0	26	65	14
15	0	6	2	0	12	0	0	53	11	22	0	11	0	0	793	0	0	0	6	0	3	17	0	7	0	2	9	65	5	93	1132	15
16	0	0	0	0	0	0	0	3	0	2	0	0	0	0	0	3	0	0	0	0	0	0	0	0	0	0	0	0	0	0	11	16
17	0	0	16	0	2	0	0	15	6	45	0	0	0	0	0	0	40	0	0	0	0	0	0	0	0	0	0	27	0	11	173	17
18	0	0	4	0	8	6	2	11	22	20	0	2	0	0	0	0	3	20	10	0	0	0	10	0	0	8	40	2	0	74	255	18
19	0	7	7	3	136	8	0	26	213	80	0	43	0	0	18	0	7	0	565	0	13	2	4	0	0	34	32	40	0	77	1330	19
20	0	0	0	0	0	0	0	0	0	0	0	0	0	0	0	0	0	0	0	20	0	17	0	0	0	0	0	0	0	29	73	20
21	0	3	0	30	37	0	0	7	26	16	4	71	0	0	27	0	0	0	38	0	287	0	0	0	4	48	6	115	0	63	797	21
22	0	0	2	0	0	0	0	38	4	20	0	0	0	0	2	0	3	0	0	9	0	367	0	0	0	0	15	0	2	111	583	22
23	0	0	5	0	18	16	10	15	42	35	0	5	0	0	4	0	5	8	31	0	3	0	83	0	0	24	35	7	0	113	472	23
24	0	0	0	0	2	0	0	23	3	13	0	0	0	0	47	0	0	0	0	0	0	2	0	14	0	0	5	3	0	17	141	24
25	0	0	0	28	5	0	0	3	6	5	20	7	0	0	13	0	0	0	5	0	12	3	0	175	17	0	22	0	0	79	416	25
26	3	14	18	69	189	25	50	67	283	190	22	159	0	2	94	3	23	5	386	0	205	10	38	3	45	3610	103	208	3	1061	6901	26
27	0	0	7	0	4	0	0	27	11	21	0	0	0	7	2	0	5	0	2	0	0	4	0	0	0	0	675	0	0	72	847	27
28	3	14	2	32	32	0	0	20	25	22	6	72	0	0	131	0	2	0	21	0	26	7	0	0	0	8	7	557	2	83	1093	28
29	10	13	51	85	106	52	30	295	214	252	79	68	0	20	448	6	53	11	130	9	76	175	45	10	85	226	611	216	21	1074	4486	29
	37	165	292	448	1272	171	175	1490	1793	1712	241	760	0	58	1812	38	213	50	1414	44	713	703	206	63	323	4072	1877	1526	44	3571	2530	0
	1	2	3	4	5	6	7	8	910	11	12	13	14	15		16	17	18	19	20	21	22	23	24	25	26	27	28	29	UND	SUM	

TRUE AND FALSE POSITIVE RATES IN MAXIMUM CONTAMINANT LEVEL TESTS

JACQUELINE OLER

Drexel University, Philadelphia, U.S.A.

(Received April 1990)

Abstract. The U.S. Environment Protection Agency (EPA) is promulgating a revised national primary drinking water regulation (NPDWR) which includes a monthly sample size and maximum contaminant level (MCL) for total coliform bacteria in public water systems. No previous quantification has been made of the coliform content that must be present in the sampled water in order for an MCL to be exceeded. This paper presents a method for evaluating the coliform level an MCL will detect with likelihood P.

Our approach is to treat an MCL as a decision rule, with Type I (false positive) and Type II (false negative) error rates. The stringency of an MCL is quantified as the mean coliform level in the sampled system that it will detect with likelihood P. MCLs are contrasted on stringency by comparing the mean coliform level each targets for detection, with fixed error rates.

Interim rules (NIPDWR), in effect since 1975, are shown to vary widely on the coliform content each targets for detection, that is, on stringency. Yes/no decisions on contamination have not been decided based on mean coliform content. Coliform levels permitted in monitored public water systems have been determined by the particular MCL used for testing. The same coliform level will test positively with one MCL 90 times in 100 yet be guaranteed 95% nondetection by a second MCL.

EPA's reasonably safe standard for drinking water is reformulated on our stringency criteria. Its proposed monthly MCL is evaluated on its capability for maintaining this standard. Smaller systems will not provide its users this level of protection under the new rule.

In addition, our evaluation of the safe water standard on stringency and the rationale for a monthly MCL require that coliform levels be identically distributed (i.d.) across month and sampled system. Empirical data strongly refute this model and question the utility of a monthly MCL.

This work suggests an alternative, single sample MCL, with repeat sampling for verification, which can be configured to provide monitoring to discover mean coliform values at any level, in any size of system, at minimal extra cost.

1. Introduction

The U.S. Environmental Protection Agency (EPA) published an amended national primary drinking water regulation (NPDWR) for total coliform bacteria in 1989 (USEPA, 1989). Applicable to all public water systems, the revisions include a monthly maximum contaminant level (MCL) as well as monitoring and analytical requirements, with a stated maximum contaminant level goal (MCLG) of zero. Although the MCLG is a non-enforceable health goal, the EPA is charged with promulgating a NPDWR which includes either an MCL set as close to the MCLG as feasible, or, a required treatment technique (USEPA, 1987).

No previous quantification has been made of the coliform content that must be present in the sampled water in order for each MCL to be exceeded. Here, a method for evaluating the coliform level an MCL will detect with likelihood P is presented (Oler, 1987). The method is used to show that the current MCLs, promulgated in 1975 as national interim

Environmental Monitoring and Assessment **17**: 123–136, 1991.

primary drinking water regulations (NIPDWR), vary widely on the coliform level each targets for detection, that is, on stringency (USEPA, 1975). We demonstrate that the coliform levels allowed to exist in our monitored public water systems are actually determined by the particular MCL used for testing, and vary over a broad range.

The NIPDWRs were not derived from the decision theory perspective. Our approach here is to treat them as classical decision rules (Mood *et al.*, 1974).

An MCL is considered to be a decision rule by which a system is judged to be contaminated (tests positively) if its MCL is exceeded. This introduces the potential for measuring the likelihood of the following two error types:

(1) Type I or false positive, incorrectly judging clean water to be contaminated, and
(2) Type II or false negative, incorrectlty judging dirty water to be clean.

The true positive rate, 1 - Type II likelihood, is called the power of the decision rule, since it measures the likelihood that the testing will achieve what it has been designed for; in this case, that contaminated water will be detected.

These two error rate concepts are used to contrast the MCLs on stringency. In an adaptation of the statistical decision theory approach, we fix the likelihood that an MCL will be exceeded at P. Numerical values are then determined for the population parameters which would provide sampled portions exceeding the MCL with P probability.

Choosing $P = 0.90$, for example, we determine the lower limit on the mean coliform level required to ensure that an MCL will be exceeded with at least 0.90 likelihood. When two MCLs are contrasted on their 0.90 mean values, $\mu^{0.90}$, one is said to be more stringent than the second if its limiting 0.90 mean value is lower. It detects lower contamination levels than the second with equal likelihood.

We also determine the upper limit on the probability that a single sample portion will test positively which ensures its MCL will be exceeded with no more than 0.05 probability. The mean coliform level ensuring the ≤ 0.05 portion probability is calculated and interpreted as the upper limit in the statement of an implied null hypothesis. These $\mu^{0.05}$ levels are interpreted as being targeted for detection by the MCL at an alpha $= 0.05$ significance level. That is, one is 95% certain that the mean coliform level in the tested water is at least $\mu^{0.05}$ if that MCL is exceeded. A comparison of two MCLs on their 0.05 mean coliform level indicates that one targets lower levels for detection than the second at that level if its $\mu^{0.05}$ value is smaller than that of the second.

Introducing true and false positive rate calculations for the current MCL decision rules allows their comparison with respect to stringencies and targeted contamination levels, when it is assumed that the number of positive sampled portions is binomially distributed. This implies that the tested portions are independently and identically distributed (i.i.d.) Bernouilli trials. The probability that a sample is coliform-positive is assumed to be constant across the system and over the time period during which sample outcomes are accumulated for comparison with the MCL. Coliform counts in the 10 mL and 100 mL tested portions are assumed to be Poisson distributed with a constant mean across sampled sites and time frame.

Empirical data from field studies indicate this assumption is incorrect (El-Shaarawi *et al.*, 1981; Pipes and Christian, 1982; Christian and Pipes, 1983; USEPA, 1983). Muenz (1978) stressed the critical aspect of the homogeneity assumption in his analyses of statistical issues in water quality control. Nevertheless, the frequently referenced statistical methodology which EPA uses to describe 'reasonably safe water' or an acceptable 'protection reliability standard' makes this assumption (Pipes, 1983).

The first application of our decision theory approach for evaluating the stringency of an MCL is to reformulate EPA's clean water standard by calculating the mean coliform level it targets for ≥ 0.95 and ≤ 0.05 likelihood of detection.

Our second application of this approach is a first evaluation of the newly proposed NPDWR. Again, assuming the i.i.d. Bernouilli model, we calculate mean coliform levels which ensure that the monthly MCL will be exceeded with ≥ 0.95 and $\leq$ likelihoods. These μ^P values are then compared with the analogous μ^P implied in EPA's safe water standard.

In a final application, we evaluate EPA's proposed MCL on the coliform content a single test with 'repeats' targets for 0.95 detection. If a sample is coliform-positive, the revised NPDWR requires three repeat samples to be taken, approximating the original positive sample closely in time and space. MCL violation is determined on the evidence from both original and repeat samples.

Only the initial and repeat samples, which can more reasonably be assumed to be Bernouilli with constant probability are combined in this second evaluation of the proposed MCL on stringency. That is, the new MCL is evaluated without assuming that all monthly samples are i.i.d. over both time and space. Of course, the potential increased sensitivity that accrues from a larger (homogeneous) sample size, if a reality, is lost. The long-range stringency may be underestimated although the collected field data evidence does not suggest this.

A review of the basic components in decision rule formulation is available in the Appendix.

2. Current National Interim Primary Drinking Water Regulations

Current NIPDWR analytical methodology for coliform analyses falls into the nine categories described in Table I. The categories distinguish:

(1) Two types of laboratory procedures used to detect coliforms in the tested portion,

(2) Size (volume) and number of tested portions per sample,

(3) Number of monthly samples tested, and,

(4) Test statistic with monthly critical limit.

The two laboratory procedures determine either a membrane filter colony (forming) count, cfu, (MF1-3) or the number of fermentation tubes testing positively for coliform contamination (FT1-6). One hundred millilitre volumes are used in MF protocols. In FT protocols, the number of tubes tested monthly is $N = 5n$. The volume of water in the fermentation tube tests is either 10 mL (FT1-3) or 100 mL (FT4-6). The number of samples required in monthly testing is determined by the size of the population which the public water system serves. Monthly sample sizes arbitrarily selected here are $n = 3, 5, 10$,

20, 100, and 200.

Excepting MF1, the test statistic is the number of positive monthly samples. MF1 uses the monthly sample average number of coliform counts. It is discussed briefly here as the new MCL is based only on presence/absence of coliforms in 100 mL volumes, not on mean coliform counts/100 mL.

A water system is in violation if the MCL it elects for testing is exceeded. The NIPDWR MCLs are stated in the middle column of Table I.

3. A Method for Evaluating Current NIPDWR MCL Stringencies

We choose a numerical value P for the likelihood that an MCL will be exceeded. Then we calculate the mean coliform level which ensures that the MCL will be exceeded with likelihood P.

In MF testing, the single sample (Bernouilli trial) probability which ensures that an MCL is exceeded with at least 0.90 (at most 0.05) likelihood is indicated as $\pi_M^{0.90}$ ($\pi_M^{0.05}$) for each n value.

Single sample portion probabilities in FT1-3 and FT4-6 tests which ensure the MCL is exceeded at least 90 times in 100 are expressed respectively as $\pi_{F+}^{0.90}$ and $\pi_{F++}^{0.90}$. The analogous probabilities ensuring the MCL is exceeded at most 5 times in 100 are indicated as $\pi_{F+}^{0.05}$ and $\pi_{F++}^{0.05}$. These Bernouilli trial probabilities are reported in Tables II, III, and IV.

Continuing the assumption that monthly tested portions are identically distributed across sites, the 10 mL and 100 mL test volumes are treated as Poisson trials with a common mean parameter indicated as μ^P. Population mean levels, $\mu^{0.90}$ and $\mu^{0.05}$, which

TABLE I

National Interim Primary Drinking Water Regulations (NIPWDR)

Test protocol	Maximum Contamination Level (MCL)	Reference
I. Membrane filter		
(a) Monthly average MCL	$\bar{x} > 1/100$ mL	MF1
(b) Single sample MCL		
$n < 20$ samples month^{-1}	$> 4/100$ mL in more than one sample	MF2
$n \geq 20$ samples month^{-1}	$> 4/100$ mL in more than 5% of samples	MF3
II. Fermentation tube (10 mL)		
(a) Monthly average MCL	$> 10\%$ of $5n$ tubes are positive (+)	FT1
(b) Single sample MCL		
$n < 20$ samples month^{-1}	≥ 3 + tubes/5 in more than one sample	FT2
$n \geq 20$ samples month^{-1}	≥ 3 + tubes/5 in more than 5% of samples	FT3
III. Fermentation tube (100 mL)		
(a) Monthly average MCL	$> 60\%$ of $5n$ tubes are positive (+)	FT4
(b) Single sample MCL		
$n < 5$ samples month^{-1}	5 + tubes in more than one sample	FT5
$n \geq 5$ samples month^{-1}	5 + tubes in more than 20% of samples	FT6

Adapted from: USEPA, 1987.

ensure that single sample portions test positively with the Bernouilli probabilities discussed above, are readily calculated.

For example, the probability that a 10 mL tube tests positively must be $\geq \pi_{F+}^{0.90}$ in order that the FT1 is exceeded with ≥ 0.90 likelihood. A tube presumably tests positively if there is at least one coliform. The calculated mean level, $\mu^{0.90}$, is that which ensures $x \geq 1$ in a Poisson count with probability $\geq \pi_{F+}^{0.90}$.

The calculations for FT2-3 and FT5-6, in which subsamples of five 10 mL and 100 mL tubes are evaluated, require an additional step in which 10 mL and 100 mL single tube positive test probabilities are obtained as Bernouilli probabilities in five trials.

In MF protocols, the probability of a ≥ 0.05 Poisson count is equated to π_M^P and the expression solved for μ^P.

A numerical lower (upper) bound on the mean coliform count 100 mL^{-1}, $\mu^{0.90}$ ($\mu^{0.05}$), which ensures a positive test, that is, an MCL exceeded with ≥ 0.90 (≤ 0.05) likelihood, is reported for FT1-6 and MF2-3, for each of the selected sample sizes in Tables II, III, and IV.

For the MF1 protocol, sampled coliform counts 100 mL^{-1} are assumed to be i.i.d. in order to calculate an upper (lower) limit on $\mu^{0.05}$ ($\mu^{0.90}$). The central limit theorem provides the equations reported and estimated upper (lower) bounds are reported in Table IV for sample sizes ≥ 10. The limiting values are functions of σ, the standard deviation of the coliform counts 100 mL^{-1} in the sampled water system.

TABLE II

Lower and Upper Limits on Targeted Total Coliforms 100 mL^{-1}: FT1-FT3

	n	3	5	10	20	100	200
	N	15	25	50	100	500	1000
MCL Protocol							
FT1	$\mu^{0.05}$	0.24	0.34	0.58	0.67	0.84	0.90
	$\pi_{F+}^{0.05}$	0.02	0.03	0.06	0.06	0.08	0.09
	$\mu^{0.90}$	2.69	2.22	1.98	1.64	1.27	1.20
	$\pi_{F+}^{0.90}$	0.24	0.20	0.18	0.15	0.12	0.11
FT2	$\mu^{0.05}$	3.24	2.51	1.84			
	$\pi_{F+}^{0.05}$	0.28	0.22	0.17			
	$\mu^{0.90}$	11.30	7.87	5.30			
	$\pi_{F+}^{0.90}$	0.68	0.54	0.41			
FT3	$\mu^{0.05}$				1.35	1.65	1.75
	$\pi_{F+}^{0.05}$				0.13	0.15	0.16
	$\mu^{0.90}$				3.76	2.73	2.52
	$\pi_{F+}^{0.90}$				0.31	0.24	0.22

TABLE III

Lower and Upper Limits on Targeted Total Coliforms 100 mL^{-1}: FT4-FT6

MCL Protocol	n N	3 15	5 25	10 50	20 100	100 500	200 1000
FT4	$\mu^{0.05}$	0.55	0.61	0.68	0.74	0.83	0.86
	$\pi_{F++}^{0.05}$	0.42	0.46	0.49	0.52	0.56	0.58
	$\mu^{0.90}$	1.49	1.33	1.17	1.10	0.99	0.97
	$\pi_{F++}^{0.90}$	0.78	0.74	0.69	0.66	0.63	0.62
FT5	$\mu^{0.05}$	1.10					
	$\pi_{F++}^{0.05}$	0.67					
	$\mu^{0.90}$	3.16					
	$\pi_{F++}^{0.90}$	0.96					
FT6	$\mu^{0.05}$		0.92	0.95	1.01	1.18	1.23
	$\pi_{F++}^{0.05}$		0.60	0.61	0.64	0.69	0.71
	$\mu^{0.90}$		2.28	1.91	1.69	1.45	1.40
	$\pi_{F++}^{0.90}$		0.90	0.85	0.82	0.76	0.75

TABLE IV

Lower and Upper Limits on Targeted Total Coliforms 100 mL: MF1-MF3

MCL Protocol	n	3	5	10	20	100	200
MF1	$\mu^{0.05}$	NA[a]	NA	1−52 σ	1−0.37 σ	1−0.16 σ	1−0.12 σ
	$\pi_{M}^{0.05}$	NA	NA	NA	NA	NA	NA
	$\mu^{0.90}$	NA	NA	1+0.40 σ	1+0.29 σ	1+0.13 σ	1+0.09 σ
	$\pi_{M}^{0.90}$	NA	NA	NA	NA	NA	NA
MF2	$\mu^{0.05}$	2.67	2.24	1.81			
	$\pi_{M}^{0.05}$	0.13	0.08	0.04			
	$\mu^{0.90}$	6.77	5.15	3.83			
	$\pi_{M}^{0.90}$	0.80	0.58	0.34			
MF3	$\mu^{0.05}$				1.48	1.66	1.73
	$\pi_{M}^{0.05}$				0.02	0.03	0.03
	$\mu^{0.90}$				2.98	2.37	2.24
	$\pi_{M}^{0.90}$				0.18	0.09	0.08

[a] Not available.

CURRENT NIPDWR MCLs CONTRASTED ON STRINGENCY

The mean coliform levels, $\mu^{0.05}$, reported in Tables II, III, and IV provide comparisons across the current NIPDWR MCLs on the mean coliform level each requires to ensure detection with ≥ 0.90 likelihood. The calculations were made under the assumption that coliform counts are i.i.d. across sampled sites across time. The coliform levels required to exceed an MCL with ≥ 0.90 probability are monotone decreasing (with n within test) for the sample sizes investigated here. An anticipated improvement in sensitivity is shown quantitatively; every protocol detects lowering coliform levels with at least 0.90 likelihood as n increases.

Of greater interest, the MCLs can be contrasted within sample sizes across protocols. If we order the $\mu^{0.90}$ within sample size, it is seen that, for all sample sizes studied here, FT4 detects the lowest mean coliform levels with ≥ 0.90 probability. FT1 is the next, followed by FT5-6, MF2-3 and FT2-3.

FT4 is the most stringent in the sense that it will be exceeded (with ≥ 0.90 likelihood) for mean coliform levels uniformly lower than all others, in the range 1 to 1.5 100 mL^{-1}. FT1 is second, detecting mean levels in the range 1.2 to 2.7 100 mL^{-1}. FT1 and FT6 are quite similar with FT6 testing positively for mean levels in the range 1.4 to 2.3 100 mL^{-1}. Levels for ≥ 0.90 likelihood detection by FT2-3 and MF2-3 are seen to be much higher, in the range 2.5 to 11.3 100 mL^{-1} and 2.2 to 6.8 100 mL^{-1}, respectively.

We see that, with a fixed sample size, there is a wide variation in the mean coliform levels which would be detected with ≥ 0.90 probability, depending upon which test protocol is used. When contamination can be described by a common population parameter, μ, these eight test protocols vary broadly in the mean coliform levels they detect with ≥ 0.90 likelihood.

The mean level that an MCL detects with ≤ 0.05 likelihood is interpreted here as (the upper limit on) the coliform level in the implied null hypothesis,

$$H_0\colon \mu \leq \mu^{0.05},$$

which that MCL is testing at an alpha = 0.05 level of significance. This is the contamination level the MCL targets for detection at the 0.05 level. If the MCL is exceeded, one concludes that the mean coliform level is greater than $\mu^{0.05}$. On average, the decision will be correct 95 times in 100.

As illustrated in the tables, the targeted coliform levels, $\mu^{0.05}$, increase with n, with a few exceptions at the smaller sample sizes for FT2, MF2 and FT5. This results from the MCL being fixed at 1 for $1 \leq n \leq 20$. In general, as n increases, MCLs target higher contaminant levels for 0.05 detection.

When comparisons are made within sample size, MCLs are found again to target highly variable coliform levels. FT1 and FT4 target the lowest for 0.05 detection, in the range 0.25 to 0.90 100 mL^{-1}. FT6 targets mean levels in the midrange 0.90 to 1.25 100 mL^{-1}, for the sample sizes studied here. FT2-3 and MF2-3 are targeting mean coliform levels in the range 1.35 to 3.25 100 mL^{-1} and 1.5 to 2.7 100 mL^{-1}, respectively. For $20 \leq n \leq 200$, FT1 and FT4 are targeting very similar, low levels in the range 0.7 to 0.9 100 mL^{-1}. FT3 and MF3 are similar also but target levels twice as high at a 0.05 level.

Although FT4 is the most stringent of all MCLs in the full sample size range studied here, in the decision theoretic sense it is not uniformly powerful. Tests can be compared on power if:

(1) Each tests a common null hypothesis at

(2) The same alpha level of significance.

With alpha fixed at 0.05, the $\mu^{0.05}$ can be seen in the tables to be highly variable across

MCL protocols with a fixed n.

We have shown that even with n held constant, mean levels differing widely across the MCL protocols are implied when a consistent 0.05 level of significance criterion is introduced. In practice, MCLs determine only yes/no with regard to coliform levels being excessive. However, widely differing mean levels have been confirmed when an MCL is exceeded is translated into 'reject the null hypothesis, H_0: $\mu \leq \mu^{0.05}$.'

For $n = 20$, the mean levels in null hypotheses implied by FT3 or MF3 (the levels targeted for rejection with ≤ 0.05 likelihood) are larger than mean levels with ≥ 0.90 likelihood to be positive if FT4 is the test protocol.

When $n = 100$ and $n = 200$, FT4's $\mu^{0.90}$ is smaller than all other $\mu^{0.90}$s, and all but FT1's $\mu^{0.05}$. As discussed above, FT4 will be positive for (will detect) considerably lower levels of contamination with ≥ 0.90 probability than the other MCLs for any size studied here. Now it is seen that FT4 targets mean coliform levels for detection at ≥ 0.90 likelihood that are lower than mean coliform levels guaranteed 0.95 likelihood of no detection by any other protocol except FT1, for $n = 100$ and $n = 200$.

In summary, it has been shown that the current NIPDWR MCL protocols do not target the same mean coliform levels for detection, at an alpha $= 0.05$. These interim regulations have been testing widely varying null hypotheses, that is, targeting quite different levels of contamination for a yes/no decision on excessive coliform content. Positive MCL tests can result from widely disparate mean coliform levels, depending upon which protocol is used. Clearly, no mean coliform level had been targeted for detection a priori.

Also, the mean coliform levels that ensure an MCL will be exceeded with ≥ 0.90 likelihood vary widely across test protocols. Thus, not only do they determine contamination on very diverse criteria, they also range widely on the coliform levels they are able to detect with ≥ 0.90 likelihood.

EVALUATING μ^P FOR EPA's REASONABLY SAFE WATER STANDARD

Statistical methodology in support of the monthly MCL to be promulgated as the final rule by the U.S.E.P.A. Office of Drinking Water is available in the proposed revised NPDWR published on November 3, 1987 (USEPA, 1987, 42228). There, the statistical concept of a 95% confidence interval, assuming monthly samples have a constant probability of being coliform-positive over both time and area, is used to argue that 'if 60 samples are collected and 95% are negative for coliforms, then there is a 95% confidence that the fraction of water with coliforms presents is less than 10% (Pipes, 1983). EPA believes that this level of quality, at this level of confidence, represents reasonably safe water'.

That is, EPA's 'reasonably safe water' standard is described in the context of $n = 60$ samples with a constant probability of testing positively, in which 95% are coliform-negative.

We reformulate this standard on the stringency criteria presented here.

The probability a sample is coliform-positive ensuring that the MCL will be exceeded with ≥ 0.95 likelihood ($n = 60$) is calculated as

$$P(x \geq 4 \mid n = 60, \mu^{0.95}) \geq 0.95 \rightarrow \pi^{0.95} \geq 0.130.$$

Then, the mean coliform level in a 100 mL sample ensuring a Poisson coliform count ≥ 1 with ≥ 0.13 likelihood is calculated as $\mu^{0.95} \geq 0.14$.

With the same sampling model EPA uses to define its 'protection reliability standard' or 'reasonably safe water', we evaluate the implied stringency as follows: the mean coliform targeted for detection with 0.95 likelihood is 0.14 100 mL^{-1}.

In an analogous calculation, the implied null hypothesis is

$$H_0: \mu^{0.05} \leq 0.02.$$

If the MCL is exceeded, it is 95% certain that the mean coliform level is at least 0.02 100 mL^{-1}.

4. Proposed National Primary Drinking Water Regulations

A brief summary of the proposed NPDWR follows, for the reader's convenience.

The total coliform monitoring requirements to be published as EPA's final rule are easily summarized. MCLs are based simply on the presence/absence of total coliforms in a sample rather than on a mean count (Pipes and Christian, 1984). A standard 100 mL volume must be used for all analytical methods, for example, ten, 10 mL, five, 20 mL or one, 100 mL volume. A sample is positive if at least one of the tested volumes is positive. Analytical procedures include membrane filter and multiple tube fermentation techniques, as long as a 100 mL water volume is used in the analysis.

Minimum monthly monitoring requirements are population based, essentially 1/1000 month^{-1} for systems servicing $\leq$ 100 000 people. Sampling frequencies decline to 480 month^{-1} for systems serving populations $>$ 3 960 000.

The monthly MCL is as follows:

(1) For systems analysing between 1 and 39 samples month^{-1}, no more than 1 sample/month may be coliform-positive.

(2) For systems analysing at least 40 samples month^{-1}, no more than 5% of the monthly samples may be coliform-positive.

The few exceptions, for systems with 25 to 1000 people and requiring $<$ 1 month^{-1} are not discussed here.

Three repeat samples are required whenever a positive sample is found in order to temporarily focus increased monitoring on systems where the water quality is suspect and to confirm whether a coliform-positive sample indicates if the contamination is in the distribution system or is localized. The repeats are to include a sample from the original positive source while the others must be near-neighbours in time and location.

If total coliforms are detected in any of the repeat samples, the system must again repeat sampling as above unless the MCL has already been violated and the State notified. However, if total coliform-positive samples occur only at the same tap, the State may invalidate the original positive sample as nondistributional.

If any routine or repeat sample is total coliform-positive, that culture must be analysed to determine if fecal coliforms (or *E. coli*) are present. If present, the State must be notified, although, if the contamination is shown to be limited to the service connection, the State may invalidate the sample. If one or more routine or repeat samples is fecal coliform- or *E. coli*-positive and the original total coliform-positive sample is not invalidated, the system is in violation of the MCL for total coliforms. Thus, for all population sizes the effective MCL is two positive samples if one or more is fecal coliform- or *E. coli*-positive.

The November 1987 revised NPDWR proposed a longterm MCL (in addition to a monthly MCL) specifying that no more than 5% of the most recent 60 samples could be coliform-positive. EPA also proposed a minimum of 5 samples month^{-1}, reasoning that, with $n = 1$ month^{-1}, one is only 95% certain that less than 34% of the water is contaminated if one of the 12 samples accumulated over a year (the long-term MCL) is coliform-positive (Pipes 1983). With this example, EPA shows that its safe water standard cannot be maintained in systems required to take less than five monthly samples, even with the long-term MCL. However, neither the minimum $n = 5$ month^{-1}, nor the long-term MCL are in the final rule which incorporates the public comments and EPA's response to the issues raised therein.

STRINGENCY OF THE PROPOSED NPDWR MONTHLY MCL

Field studies (Pipes and Christian, 1982; Christian and Pipes, 1983) indicate total coliforms are more widely dispersed than EPA's assumption of i.i.d. samples across month and water system. However, we first calculate the stringency of the proposed monthly MCL for small n on this assumption.

When $n = 3$, for example, the proposed MCL will declare a system in violation if more than one 100 mL volume has a positive test, that is, contains one or more coliform. This MCL can be exceeded either by two original coliform-positive samples, or if one original and one or more of three repeat samples test positively. The mean coliform level which ensures this MCL will be exceeded with ≥ 0.95 likelihood is $\mu^{0.95} \geq 1.05$ 100 mL^{-1} (without repeat sampling, $\mu^{0.95} \geq 2.0$ 100 mL^{-1}). The single volume probability of testing positively is $\pi^{0.95} \geq 0.65$.

The capability of the new monthly MCL to ensure EPA's clean water standard in small systems using our stringency 'yardstick' is assessed. The mean coliform level which ensures ≥ 0.95 likelihood that the MCL will be exceeded with the EPA model of $n = 60$ i.i.d. samples, was calculated above as 0.14 100 mL^{-1}. In comparison, the mean coliform level required when $n = 3$, with repeat sampling, is 1.05 100 mL^{-1}. The proposed MCL for systems serving 2501-3300 people, in which $n = 3$, is not as stringent as to provide its users with water meeting EPA's protection reliability standard. The mean level which is targeted for discovery 95 times in 100 in these small systems is 7.5 times that of EPA's reasonably safe water standard.

For systems required to sample $n = 10$ month^{-1}, the proposed MCL is also two or more coliform-positive samples. With repeat sampling, $\mu^{0.95} \geq 0.4$ 100 mL^{-1}, which is three times that which is described as reasonably safe water. When the minimum monthly sample size is 25, $\mu^{0.95} \geq 0.17$ 100 mL^{-1}, with repeat sampling. This is a close approximation to the clean water standard of 0.14 100 mL^{-1}.

These results show that even with EPA's assumption of constant coliform dispersion across sampling sites, the new NPDWR will fail to provide the desired level of safety in smaller systems. Even when sample sizes are incorrectly magnified by combining results across month and system, the reasonably safe water standard is not met.

The stringency of a single sample test, with three repeats if coliform-positive, is now calculated. Its $\mu^{0.95}$ is determined assuming only that the original and repeat samples (which are to be taken as near in time and space as is feasible) are i.i.d.

STRINGENCY OF PROPOSED NPDWR SINGLE-SAMPLE MCL

The proposed NPDWR MCL sampling requirements are to be based on the population size the system serves; approximately 1/1000 month^{-1}. A single sample volume is 100 mL. The stringency of this single sample MCL can be assessed by determining the mean coliform level in the system which produces a positive test with ≥ 0.95 likelihood. The mean coliform level ensuring that a single 100 mL volume will test positively with ≥ 0.95 likelihood is

$$\mu^{0.95} \geq -\ln(1-0.95) = 3.00.$$

The presence of total coliforms is established if the original positive sample is followed, in repeat sampling, with at least one positive test in three. With $\pi^{0.95} \geq 0.95$, this will occur with 0.999 likelihood. The stringency of the proposed single sample MCL, with repeats, can be described as targeting a mean coliform level of 3.0 for 95% detection.

This is less stringent than the single sample MCL of the best (current) NIPDWR, FT4, in which a single sample is five, 100 mL volumes in which four or more must be positive. The calculated $\mu^{0.95}$ is ≥ 2.58. It is surprising that EPA is proposing a rule with a single sample MCL less stringent than one of the NIPDWR, especially in combination with a new MCLG set at zero.

Yet it is easy to formulate an MCL whose $\mu^{0.95}$ is closer to zero, at what would appear to be minimal extra cost. Specifying a test volume of 200 mL would reduce $\mu^{0.95}$, the 95% detectable mean coliform level to 1.5 100 mL^{-1}. Setting a test volume of 300 mL would target 1/100 mL contamination at all sampled sites for ≥ 0.95 likelihood of detection. Our $\mu^{0.95}$ is not synonymous with an MCLG; however, any reduction in $\mu^{0.95}$ will bring the achieved MCL closer to zero.

Summary

EPA is promulgating a monthly MCL. Assessing contamination on a monthly MCL requires the assumption that sample outcomes can be aggregated across sampled sites over the month. Until the distribution of coliform-positive outcomes over time and space has been investigated further, the utility of a monthly MCL is questioned.

Our yardstick for assessing an MCL protocol on its stringency, or μ^{P} value could be used to determine test volumes (or subsamples as used in the NIPDWR) which would make it possible for systems to ensure its customers water with any selected total coliform protection reliability standard. It could be used to determine test volumes for a range of population sizes.

The method is applicable to any presence/absence test in fixed units of time, space and volume, when the distribution of random contaminant-positive outcomes across the time and space of the sampling frame is established. This approach to assessing the stringency of a monitoring program could be used to determine the sampling frames, test portion sizes and maximum contaminant limits with which to ensure protection reliability standards for other environmental contaminants.

We have evaluated the mean total coliform level which can be detected with ≥ 0.95 likelihood in 1 1000 month^{-1} MCL monitoring, for the smaller systems whose monthly MCL (2) coincides with the critical value for a single sample MCL with repeats. For small systems, the new monthly MCL regulation is not sufficiently stringent to provide its users with water which meets EPA's safe water standard even when monthly samples are combined.

This work suggests that the single sample MCL with repeats be used across systems and time. The single sample volume should be determined in order to set the single sample MCL as close to the MCLG of zero as is economically feasible and enforceable.

Appendix

Two major components in decision rule formulation are the quantification of (i) the false positive or Type I error rate, and, (ii) the true positive rate, called the power of the test procedure.

Typically, there is a numerical mean level of substantive concern, $\mu = \theta_0$. Here, this would be a mean coliform level which the monitoring procedure targets for detection on health safety considerations. The burden of proof is to detect coliform contamination above this mean level, which is expressed in a null hypothesis as H_0: $\mu \leq \theta_0$.

Decision rules are constructed as follows. A sample statistic is sought which will provide, with high probability, a counterexample to the stated null hypothesis if it is false. The real number line is partitioned into two zones: one zone is determined as being a likely range for the statistic while the second comprises values that the statistic is unlikely to equal if the null hypothesis is correct. If the statistic, calculated on sample data, falls into the likely zone, no decision can be made with regard to the truth of the null hypothesis. The conclusion can be only: do not reject H_0. However, if the computed statistic falls into the unlikely zone, the decision is made that the alternative statement has been supported. The alternative hypothesis is H_1: $\mu > \theta_0$. Thus, a decision rule is a partitioning of the real line into likely and unlikely areas for the statistic, under the assumption that the null hypothesis is true.

Since the alternative is established when rejecting the null hypothesis, the unlikely zone for the statistic under the null hypothesis should comprise values most likely under, or most supportive of, the alternative, H_1, as well as least likely under H_0.

The percentage of times the null hypothesis will be incorrectly rejected, that is, the false positive or Type I error rate, is indicated as alpha (α). Generally the value for alpha is set in consideration of the risks involved with false positives. Typically, $\alpha = .05$. When the null hypothesis is rejected, the alternative statement is said to have been established at the alpha level of significance. This conclusion will be incorrect at most alpha times in 100.

In contrast, the power or sensitivity of a decision rule is the porportion of times its null hypothesis will be rejected correctly. This is the proportion of samples whose calculated statistics fall into the unlikely range when the samples have been taken from a population whose mean value is one of those hypothesized in the alternative.

The power varies with the true value for the sampled population. Decision rules are selected from those which have increasing sensitivity as mean levels lie further to the right of θ_0.

Decision rules are sought which have a high likelihood (power) that the null hypothesis will be rejected when the effect size is substantial. Effect size is defined as the distance between the true population mean and the mean specified in the null hypothesis (sometimes divided by the population sigma). In tests of the same null hypothesis at the same alpha level of significance, one decision rule is said to be uniformly better than a second if its power is larger than that of the second for all choices for alternative mean values.

Briefly, a good experimental design provides a sampling statistic whose observed value lies in the 'reject H_0' zone at most alpha times in 100 if the hypothesized null value, θ_0, is correct. Alpha is preset by the experimenter and both alpha and θ_0 are determined on substantive reasoning. Specification of the 'reject H_0' interval is made with concern to optimize the second major factor, the power of the decision rule. The test is designed to reject H_0 with sensitivities in the range 0.80 to 0.95, when the true population mean, θ, exceeds θ_0 by an amount which warrants detection. Two test protocols can be compared on sensitivity, or power, only if both are testing the same null hypothesis with the same alpha, or false positive rate.

Acknowledgements

The work described in this paper was made possible through continued faculty research leave funding from the Paul and Gabriella Rosenbaum Foundation. The author is grateful for the financial support and encouragement afforded her by the Foundation and by the Office of Science and Research of the New Jersey Department of Environmental Protection. Her Drexel colleague, Professor Wesley Pipes, posed the question of MCL stringencies. These results would not be in print except for Ms. Madge Goldman.

References

Christian, R. R. and Pipes, W. O.: 1983, 'Frequency Distributions of Coliforms in Water Distribution Systems', *Applied Environmental Microbiology* **45**, 603-606.

El-Shaarawi, A. H., Esterby, S. R. and Dutka, B. J.: 1981, 'Bacterial Density in Water Determined by Poisson or Negative Binomial Distributions', *Applied Environmental Microbiology* **41**, 107-116.

Mood, A. M., Graybill, F. A. and Boes, D. C.: 1974, *Introduction to the Theory of Statistics*, New York, McGraw-Hill.

Muenz, L.: 1978, 'Some Statistical Considerations in Water Quality Control', in C. Hendricks (ed.), *Evaluation of the Microbiology Standards for Drinking Water,* EPA 570/9-78-00C, Washington, D.C., U.S.E.P.A.

Oler, J.: 1987, 'National Interim Primary Drinking Water Regulations as Decision Rules', Technical Report #4, submitted to New Jersey Department of Environmental Protection, Trenton, N. J.

Pipes, W. O.: 1983, 'Monitoring of Microbial Quality', in P. Berger and Y. Argaman (eds.), *Assessment of Microbiology and Turbidity Standards for Drinking Water,* EPA 570/9-83-001, Washington, D.C., U.S.E.P.A.

Pipes, W. O. and Christian, R. R.: 1982, 'Sampling Frequency Microbiological Drinking Water Regulation', EPA 570/9-82-001, Washington, D.C., U.S.E.P.A.

U.S. Environmental Protection Agency: 1975, 'National Interim Primary Drinking Water Regulations', *Federal Register* **40**, 59566-59588.

U.S. Environmental Protection Agency: 1983, 'Assessment of Microbiology and Turbidity Standards for Drinking Water', EPA 570/9/83-001, Washington, D.C., U.S.E.P.A.

U.S. Environmental Protection Agency: 1987, 'Drinking Water; National Primary Drinking Water Regulations; Total Coliforms', *Federal Register* **52**, 42224-42245.

STATISTICAL ANALYSIS OF EFFECTS OF MEASURES AGAINST AGRICULTURAL POLLUTION

HANS VIGGO SÆBØ

Norwegian Computing Center, P.O. Box 114 Blindern, 0314 Oslo 3, Norway

(Received February 1990)

Abstract. The Norwegian Government has initiated a plan to reduce agricultural pollution. One of the projects in this plan is aimed at investigating the effects of different measures in order to evaluate their effects and costs.

A set of experiments has been designed to estimate the effects of measures to reduce or control the use of fertilizers and erosion. The project started in 1985. It comprises continuous measurements in two water courses in each of four counties: one test drainage area where the relevant measures were implemented at the end of 1986, and one reference area where no specific measures are carried out. A series of chemical parameters are measured together with runoff and other hydrological and meteorogical data.

The paper provides a preliminary analysis of the data collected in one of the counties during the period June 1985 to April 1988. It contains examples of analysis of covariance to show possible effects of the measures carried out in the test area.

Natural variations in precipitation and pollution are large, making it difficult to see the effects of the measures without using statistical techniques to take the multivariability of the problem into account. Some effects can be shown with analysis of covariance. However, the relatively short measurement period makes it neccessary to be careful when interpreting the results.

1. Introduction

Agriculture is a major source of pollution to Norwegian rivers and water courses. The Norwegian Government has initiated a programme against agricultural pollution. The programme includes educational work and training of farmers, but is also aimed at the implementation of concrete measures, for example to control and reduce the use of fertilizers. A separate project has been carried out in order to evaluate these measures.

A set of measurements has been designed to estimate the effects of measures to reduce or control the use of fertilizers and erosion. The project started in 1985 and will continue until 1989. It is being carried out in four Norwegian counties and comprises continuous measurements in two water courses in each county: one test drainage area where the relevant measures were implemented at the end of 1986, and one reference area where they were not implemented. A series of chemical parameters are measured together with runoff and other hydrological and meteorological data.

The paper provides a preliminary analysis of the data collected in the county of Roagaland during the period June 1985 to April 1988. It contains examples of analysis of covariance to show possible effects of the measures implemented in the test area. Principal

Environmental Monitoring and Assessment **17**: 137–146, 1991.

component analysis has been used to consider the interference between the measured parameters in order to simplify the presentation of results.

The analysis presented in this paper has been performed by using the SPSS statistical software for microcomputers (Norusis, 1986).

2. Design and Statistical Methods

The measurements in Rogaland have been carried out in the test river Time, where some measures were implemented in August 1986, and the reference river Herikstad where no measures have been implemented. The two neighbouring water courses and the measurement stations are shown on the map in Figure 1. This area is one of Norway's richest agricultural areas. Agricultural pollution is mainly due to the use of fertilizers, resulting in an increased amount of phosphorus and other organic pollutants. The measures implemented in Time comprise control of fertilizing, both in order to avoid excess consumption and to avoid fertilizing during winter (outside the growing season).

Water has been collected continuously in-stream by sampling at a rate proportional to the rate of flow in the rivers. On average the collection bottles have been emptied and their contents analysed once a week. With some exceptions due to technical problems measurements have thus been carried out continuously. The pollutants registered are:

Total phosphorus,
Dissolved, total phosphorus,
Dissolved phosphate,
Total nitrogen,
Nitrite/nitrate,
Potassium,
Suspended matter,
Suspended glow rest,
Total organic carbon.

The data consists basically of measured concentrations and average rate of flow over time periods varying between 1 and 15 days, with about 7 days on average. Analysis has been based on *rate of flow per week,* and we have:

$$\text{Loading/week} = \text{Concentrations} \cdot \text{rate of flow/week}.$$

To account for the varying periods of time of each measurement, each observation has been weighted by the time period when analysing loadings.

The hypothesis to be tested is whether the implemented measures affect the loadings of the various pollutants, and if so to what extent. Possible effects should be detectable when comparing the situation in the two rivers before and after measures are implemented in one of them. This problem calls for an analysis of covariance approach. Another method which can be applied is to compare the differences in concentrations and loadings between the two rivers by forming pairwise differences between measurements taken at the same time. However, many measurements in the two rivers do not coincide in time, and use of this method must be based on fewer observations than the analysis of covariance.

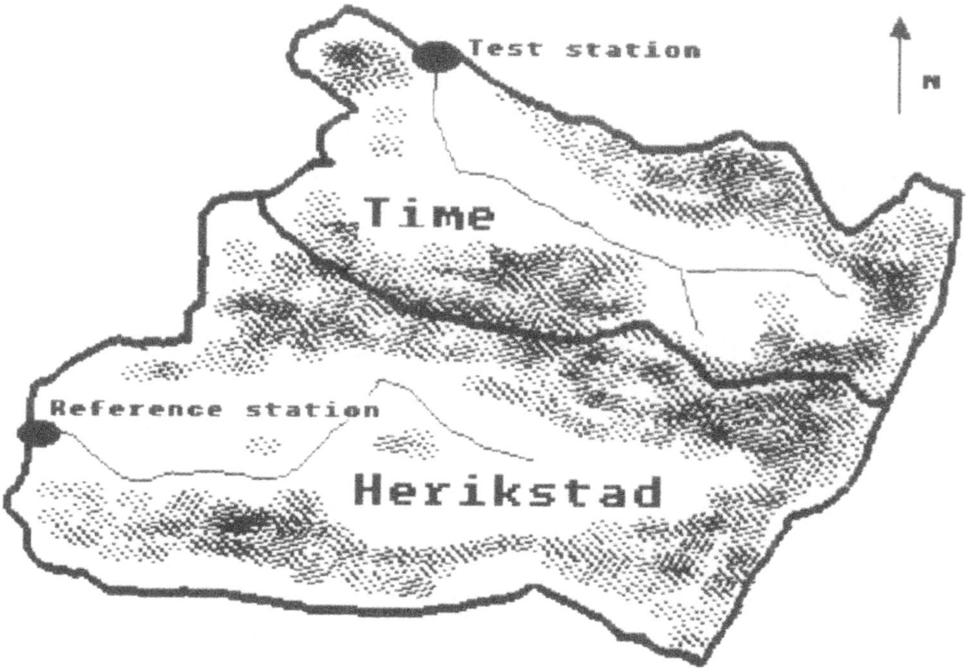

Fig. 1. Test and reference water courses.

The analysis used in this paper is analysis of covariance. The conclusions are supported by using the method of pairwise differences.

There are relatively few examples which use similar statistical methods for solving corresponding problems within the field of water quality studies. Silver (1986) demonstrated the use of covariance analysis for groundwater monitoring. Clifford *et al.* (1986) used analysis of covariance for pairwise differences to try to detect the effects of a ban of detergent phosphorus in Wisconsin Lakes.

Most of the measured parameters are correlated and principal component analysis has been performed to study this correlation and to group the variables. This approach has enabled us to concentrate on analysing a small number of variables (components) which represent the others.

3. Data

3.1. Concentrations, Runoff and Loadings

Since June 1985, when the measurements started, about 80% of the time has been covered. There is no tendency of a systematic pattern in the periods with missing observations, though if an observation is missing this often occurs during several coherent weeks. But there is no reason to believe that missing data will affect the possibilities of drawing conclusions from the project.

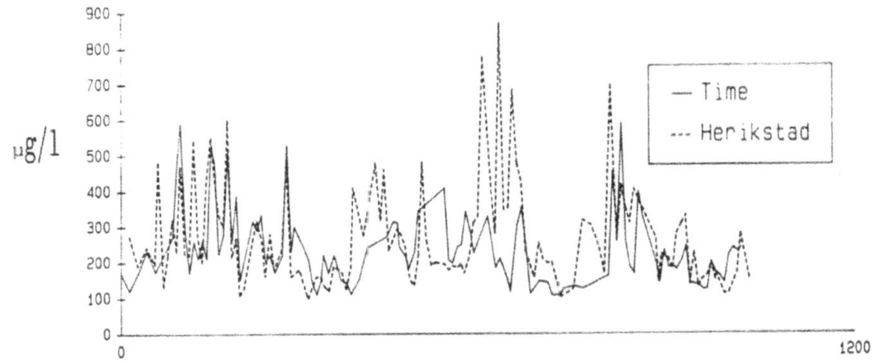

Fig. 2. Concentration of total phosphorus.

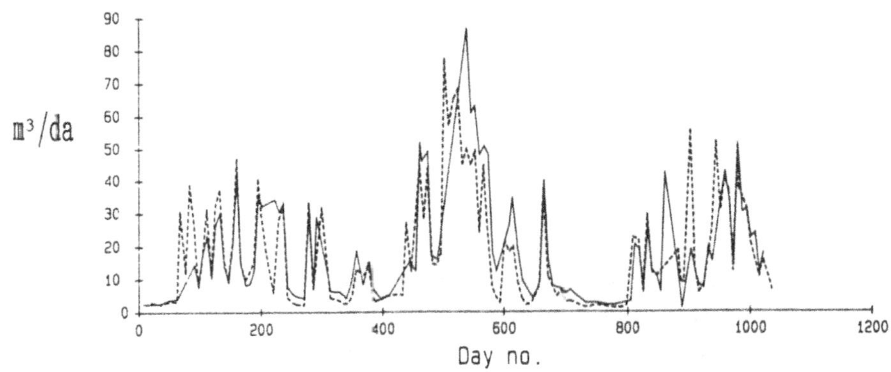

Fig. 3. Runoff per week and area unit.

An example of the measurement results is shown in Figure 2, displaying all the registered concentrations of total phosphorus. These concentrations vary greatly, which is common for such measurement series. The variations make direct interpretations difficult. Figures based on averages (over some relevant periods) are easier to interpret and are shown later.

Figure 3 shows the corresponding runoff (given as runoff per week and unit of area of water course). One should notice that variations are large even between the two matching rivers. We shall see that this makes it difficult to interpret differences in the estimated loadings between the test and reference water courses.

3.2. Quantification of Measures

The measures have consisted of planning and control of fertilizing and safeguarding against erosion. The quantification of such measures for input to statistical models represents a problem. One possible solution is to construct a coefficient based on agricultural expertise, for example with the value 1 in periods when effects are assumed to have an effect and zero elsewhere. This coefficient does not grade possible effects, and a coefficient ranging from 1 to 5 has also been proposed. The problem with this coefficient is of course that even if it is ordered, one does not know whether the step from for example 4 to 5 represents the same as the step from 1 to 2 (interval scale). Both types of coefficients are used as inputs to a simple covariance model.

4. Model

Two versions of a covariance model have been used according to the two alternative choices of measure coefficients:

$$\ln Z_{ijkt} = A_j + B_{jk} + a \cdot C_{kt} + b \cdot \ln Q_{ijkt} + E_{ijkt}$$

where Z denotes loading/week and Q runoff/week.

C is the coefficient of measure. It takes values 0 or 1 in the first version of the model and values 0-5 in the second.

The indices denote:

i = observation number
j = parameter (total phosphorus, nitrogen etc.)
k = river (Time/Herikstad)
t = season (winter/spring, summer, autumn).

A is an average effect, B a river specific effect, S the season effect and E an error term assumed to be normally distributed. Different error terms are assumed to be independent.

The models are based on logarithmic transformations of both loadings and runoff to aim at normally distributed residuals, since measurements of both concentrations, runoff and estimated loadings usually are skewed upwards (some extreme values).

5. Results and Discussion

5.1. Analysis of Covariance

With a coefficient of measure equal to 0 or 1, the model is equivalent to an analysis of covariance model with factors for river, season and measure with runoff as a covariance. No interaction terms are assumed (tests show that such terms do not contribute significantly to either of the models used in this work). The result of this analysis shows that the effects of measures are significant (level $p<0.10$) for all parameters except for nitrite/nitrate ($p<0.15$) and suspended glow rest. The most significant effect is obtained for total nitrogen and potassium.

Table I sums up the results of the analysis when the coefficient of measure varies from 0 to 5. This coefficient is applied as a continuous variable in the model (explanatory variable in regression). The analysis is based on 229 observations or slightly less for all parameters except for suspended glow rest, where we have only 80 observations.

Residual analysis show that the assumed model is adequate.

TABLE I

Level of significance and regression coefficient for effects of measures on loadings

Parameter	Level of significance			Regression coefficients	
	River	Season	Measures	Measures a (ln Z/C)	Runoff b (ln Z/in Q)
Total phosphorus	(0.66)	0.00	0.01	–0.05	(1.01)
Dissolved, tot. P	0.00	0.00	0.01	–0.06	1.25
Dissolved phosphate	0.02	0.00	0.01	–0.07	1.28
Total nitrogen	0.00	(0.01)	0.00	–0.05	0.95
Nitrite/nitrate	0.00	(0.88)	0.02	–0.03	0.92
Potassium	0.01	0.00	0.00	–0.05	(0.98)
Suspended matter	(0.72)	0.00	0.03	–0.05	0.94
Suspended glow rest	(0.11	(0.07)	(0.26)	(0.04)	0.86
Total organic carbon	(0.80)	0.00	(0.35)	(–0.01)	1.21

( ) denotes non-significant at an estimated level of 5%. For b ( ) denotes non-significant difference from 1.00.

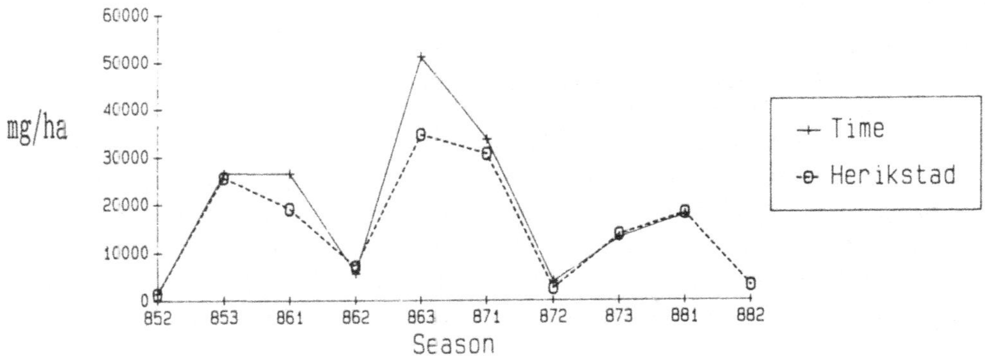

Fig. 4. Transport of dissolved, total phosphorus by season.

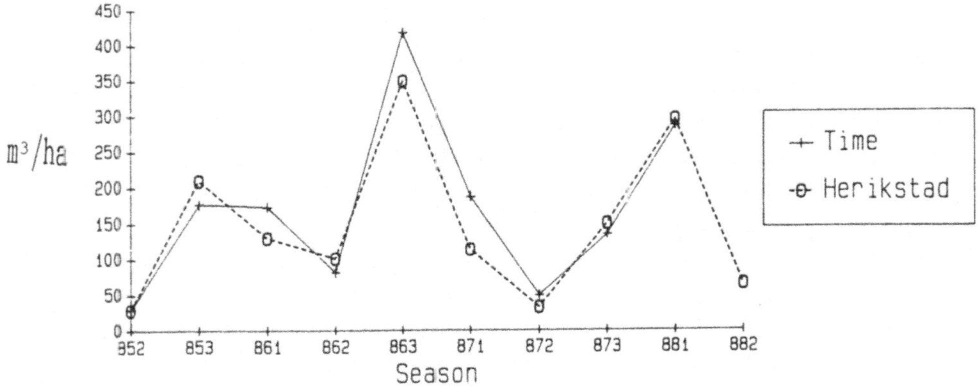

Fig. 5. Runoff per week and area unit by season.

The analysis shows significant effects ($p<0.05$) for all parameters except for suspended glow rest and total organic carbon. As with the simpler model with coefficient of measure equal to 0 or 1, the level of significance is smallest for nitrogen and potassium.

The results indicate a general decrease in concentrations over the test period in both rivers. This decrease may be due to the implemented measures (some effects are also assumed in the reference river Herikstad).

The model parameter a can roughly be intepreted as the ratio between a relative increase in loading of one unit and an absolute increase in the coefficient of measure. $a=-0.05$ (for example for total phosphorus) hence indicates a 5% reduction in loading with an increase in coefficient of measure of 1, or 25% with a change from no measures to full measures (maximum value for C is equal to 5).

Figures 4 and 5 show the development for transport of dissolved, total phosphorus per week and area unit in the two water courses compared with the corresponding runoff. It is not possible to see any particular effect of measures in Figure 4, since the runoff variations between the rivers tend to give major contributions to the variations in loadings. It is easier to consider the effects of measures in figures showing development in concentrations.

We shall see that dissolved, total phosphorus, nitrite/nitrate and suspended matter represent each of three groups of the total of nine parameters. Figures showing the development in concentrations of these three parameters are therefore presented (Figures 6 to 8). The figures indicate that concentrations are clearly lower in the Time River from the beginning of 1987 for dissolved, total phosphorus and suspended matter. For nitrate/nitrate the development in the two rivers has been parallel, following a downward trend. This trend leads to the statistically significant coefficient of measure. This coefficient is as mentioned not significant when representing measures/no measures with 1 or 0 (this model is not affected by trends affecting both rivers).

5.2. Pairwise differences

By forming pairwise differences between measurements of concentrations or loadings carried out at the same day in the two neighbouring water courses, the variation in the data due to season or a possible common trend and runoff variation is eliminated. The disadvantage is that we loose observations where measurements have been carried out only in one of the rivers. Out of 299 observations in the Time or Herikstad Rivers, it is only possible to have 77 pairwise differences. The results of performing this analysis do not indicate any differences compared to the results of analysis of covariance.

5.3. Principal component analysis

Most of the chemical components's concentrations are strongly positively correlated, but some are also negatively correlated, for example phosphorus and nitrate. A principal component analysis has been performed to show the most important relationships. It turns out that three factors explain a total of 81% of the variations in the nine measured variables (log-transformed). Each of these first three principal components roughly represents a group of the originally measured variables.

The principal components respectively represent 41, 26, and 14% of the total variation

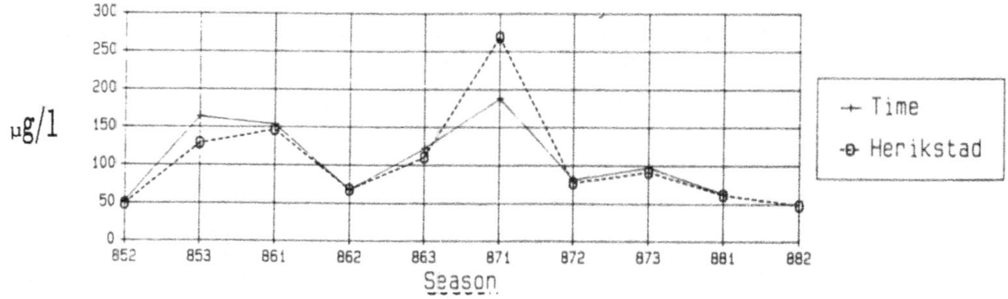

Fig. 6. Concentration of dissolved, total phosphorus by season.

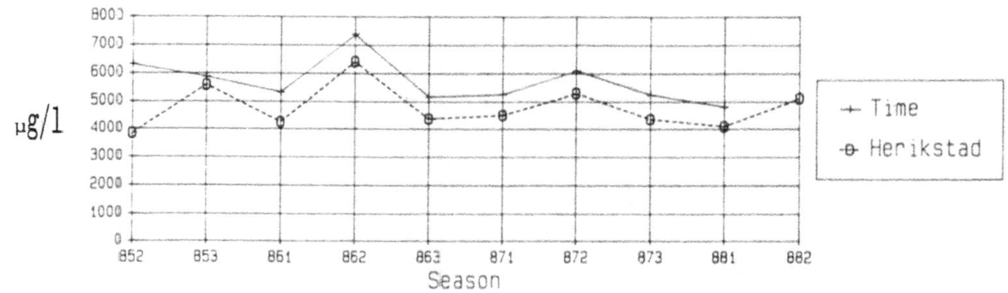

Fig. 7. Concentration of nitrite/nitrate by season.

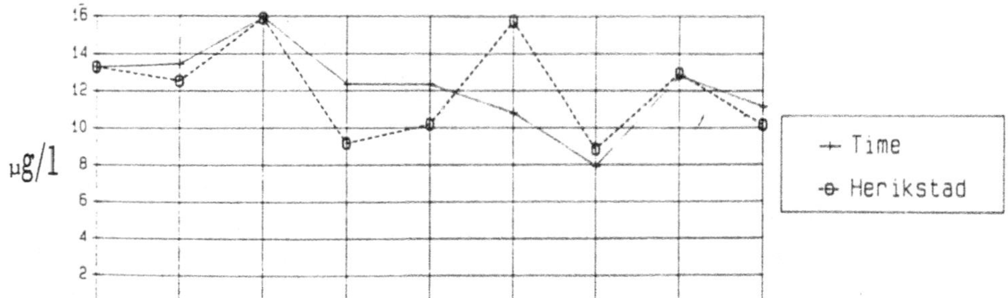

Fig. 8. Concentration of suspended matter by season.

TABLE II

Grouping of polluting chemical components according to principal components analysis (varimax rotation)

Principal component	Variable	Loadings: Coordinates in component space		
		I	II	III
Component I:	Dissolved, total P	0.93	-0.13	-0.07
	Dissolved phosphate	0.83	-0.15	-0.10
	Total organic C	0.81	0.02	0.02
	Total P	0.81	0.40	-0.04
	Potassium	0.80	0.27	0.13
Component II:	Suspended matter	-0.06	0.94	0.11
	Suspended glow rest	0.12	0.81	0.16
Component III:	Nitrate/nitrate	-0.20	0.08	0.94
	Total N	0.16	0.21	0.93

in the data material. The components correspond roughly to phosphorus, suspended matter and nitrogen (Table II), in order of decreasing significance.

The most typical representatives for these groups are dissolved, total phosphorus, suspended matter and nitrite/nitrate, and these variables have therefore been chosen in Figures 6-8 displaying concentrations by season.

It should be noted that there is a relationship between this grouping and the estimated regression coefficients between loadings and runoff (Table I). The five pollutants in the phosphorus group all have an elasticity with regard to runoff which is larger (or equal) to 1. This reflects that the concentration of these factors tends to increase with increasing runoff.

6. Conclusions

Natural variations in water quality are large, making it difficult to detect effects of changes in human activity. This is widely recognized by statisticians working with problems related to environmental monitoring.

In the Norwegian experiment aimed at estimating possible effects of measures against agricultural pollution, such effects are not immediately obvious in the current data. The problem calls for statistical techniques taking the most important sources of variation into account. Some effects can be shown using analysis of covariance. Periods with expected effects have been identified and the measures have been roughly quantified by agricultural expertise. Provided this is a valid coding, the effects have an order of magnitude corresponding to 25% reduction in the concentrations of phosphorus and nitrogen in the test water course in the county of Rogaland. However, the relative short measurement period (about 3 yr so far) makes it necessary to be careful in interpreting the results.

Many of the nine chemical parameters measured and analysed show a similar pattern and principal component analysis has been performed to study the correlation and group the parameters. It turns out that three components represent a total of 81% of the total variation in the data material. These components correspond roughly to phosphorus, suspended matter and nitrogen.

Use of graphics is necessary and has been central in the project, both to explore the data and to convey the results to the agricultural experts and others. The results have been displayed in a form supporting the statistical methods and analyses applied.

References

Clifford, R., Wilkinson, J. W., and Clesceri, N. L.: 1986, 'Statistical Assessment of a Limnological Data Set', A. H. El-Shaarawi and R. E. Kwiathowski. *Statistical Aspects of Water Quality Monitoring*, 363-380.

Norusis, M. J.: 1986, *SPSS/PC+ for the IBMPC/XT/AT and SPSS/PC+ Advanced Statistics*, Chicago: SPSS Inc.

Silver, C. A.: 1986, 'Statistical Approaches to Grounwater Monitoring', *Journal of Hazardous Materials* **13**, 207-216.

AIR POLLUTION AND DISEASES OF THE RESPIRATORY TRACTS IN PRE-SCHOOL CHILDREN: A TRANSFER FUNCTION MODEL

ULRICH HELFENSTEIN

Biostatistisches Zentrum, medizinische Fakultät, Universität Zürich, Plattenstrasse 54, CH-8032 Zurich Switzerland

URSULA ACKERMANN-LIEBRICH

Abteilung für Sozial- und Präventivmedizin, Universität Basel, St. Albanvorstadt 19, CH-4052 Basle Switzerland

CHARLOTTE BRAUN-FAHRLÄNDER

Abteilung für Sozial- und Präventivmedizin, Universität Basel, St. Albanvorstadt 19, CH-4052 Basle Switzerland

and

HANS URS WANNER

Institut für Hygiene und Arbeitsphysiologie, ETH Zürich ETH-Zentrum, CH-8092 Zurich, Switzerland

(Received March 1990)

Abstract. The purpose of the present statistical analysis was the assessment of the relation between time series of environmental factors and of frequencies of diseases of the respiratory system in pre-school children. During about one year, daily measurements of air pollutants and climatic variables were taken. During the same period of time two series of medical data were collected: (i) The daily relative number of pre-school children, exhibiting diseases of the respiratory tracts who either came to the outpatients' clinic of the children's hospital or were reported by paediatricians in Basle (ENTRIES). (ii) The daily relative frequency of symptoms of the respiratory tracts observed in a group of randomly selected pre-school children (SYMPTOMS).

By means of transfer function models the relation between the two target variables and the 'explaining' variables was analysed. Several practical problems did arise: Choice of the appropriate transformation of the different series, interpretation of the crosscorrelation function using different methods of 'prewhitening', time splitting and nonstationarity of the crosscorrelation structure. In particular, it was found that after prewhitening the crosscorrelation function between the explanatory series SO_2 and the response series SYMPTOMS changes with time. While during the 'winter period' an instantaneous relation between these two series (and to a lesser extent between NO_2 and SYMPTOMS) was identified, no such relation was found for the other seasons.

Introduction

Children are considered to be particularly sensitive to air pollutants, one of the reasons being their narrower respiratory tracts. It has been shown (Colleg and Brasser 1980; Love *et al.*, 1982) that in regions with higher concentrations of air pollutants higher frequencies of respiratory diseases in children are found. However, it is not clear if in a single region time varying environmental factors are associated with the frequencies of respiratory diseases in children taking into account the stochastic dependence of succesive observations. The purpose of the present paper is to explore such a possible relation by means of transfer function models (Box and Jenkins, 1976; Box and Tiao, 1975; Abraham and Ledolter, 1983).

Between 1st November 1985 and 23rd November 1986, daily measurements of several air pollutants and climatic variables were taken. During the same period of time two series of medical data were collected: (i) The daily relative number of pre-school children, exhibiting diseases of the respiratory tracts who either came to the outpatients's clinic of the children's hospital or were reported by paediatricians in Basle (abbreviated ENTRIES). (ii) The daily relative frequency of respiratory symptoms observed in a group of randomly selected pre-school children (SYMPTOMS).

The series ENTRIES was composed of the following respiratory diseases: Rhinopharyngitis, bronchitis, sinusitis, pneumonia, asthma, pseudocrupp, otitis and angina. The following symptoms were aggregated to the series SYMPTOMS: Cough, runny or stuffy nose, sore throat, earache and fever (more than 38 °C).

The following climatic variables and air pollutants were available as explanatory series: Daily means of temperature, relative humidity, athmospheric pressure and of the velocity of the wind; the daily duration of sunshine and the amount of rainfall, the regional mean of SO_2, NO_2, and the daily mean of particulate matter.

In the next section the transfer function model is shortly described. Thereafter the practical problems in the process of model identification and the corresponding results are presented. We concentrate on the two input series SO_2 and NO_2 and the output series SYMPTOMS since only here a significant relation between input and output was identified. Figure 1 shows the series of the two air pollutants SO_2 and NO_2 and of the target variable SYMPTOMS. The value of the series corresponding to 1st November 1985 is plotted above point 5 on the time axis etc.

Statistical Methods and Results

1. The transfer function model

Let $y_t, y_{t+1}, \ldots$ be the observations (entries, number of symptoms) at times $t, t+1, \ldots$ Then it is assumed that the variable y_t may be considered as beeing composed of two parts:

$$y_t = u_t + n_t,$$

where u_t is the part which may be explained in terms of the input variable x_t (air pollutant, temperature, e.t.c.). n_t is an error or noise process, which describes the unexplained part of y_t.

It is assumed that the explained part u_t is given by a weighted sum of the present and of past values of x_t:

$$u_t = v_0 x_t + v_1 x_{t-1} + \ldots,$$

and n_t is an ARIMA process:

$$\nabla^d n_t = z_t,$$

and

$$z_t = \phi_1 z_{t-1} + \ldots + \phi_p z_{t-p} + a_t - \theta_1 a_{t-1} - \ldots - \theta_q a_{t-q},$$

$$\nabla n_t = n_t - n_{t-1}, \quad \nabla^2 n_t = \nabla(\nabla n_t).$$

The a_t are i.i.d. random variables with expectation 0 and variance σ_a^2. $\phi_1, \ldots, \phi_p$ are the autoregressive parameters, $\theta_1, \ldots, \theta_q$ are the moving average parameters; ∇ is the differencing operator, d is the order of the differencing operator (usually $d = 0$, 1 or 2).

2. Choice of transformation

The first step in the analysis is to 'make' the series stationary. If it is found that the variance is related to the mean, then a variance-stabilizing transformation has to be applied. A useful class of transformations is given by (Jenkins, 1979):

$$x_t^{(\lambda)} = \begin{cases} x_t^\lambda, & \lambda \neq 0 \text{ (power transformation} \\ \ln x_t, & \lambda = 0 \text{ (logarithmic transformation).} \end{cases}$$

A practical tool for the choice of transformation was found to be the mean-range plot: The series is subdivided into subsets (of size say 7) and the mean and the range of each subset is calculated. Then the mean is plotted against the range for each subset (Jenkins, 1979). If the range is independent of the mean no transformation is needed ($\lambda = 1$). If the plot displays random scatter about a straight line the logarithmic transformation is suggested ($\lambda = 0$). The method does not provide a 'precise' value of λ but allows to distinguish between the most important cases, i.e. between say λ equal 1, 0.5 or 0. Figure 2 shows the mean-range plot for the series SO_2. One can see that the square root transformation ($\lambda = 0.5$) does not suffice, whereas the logarithmic transformation is appropriate. (For the series of the particulates e.g. the square root transformation was found to be appropriate).

Besides stabilizing the variance, the logarithmic transformation of the series SO_2 had another 'simplifying' consequence. Figure 3 (a) and (b) show the relation between the input series SO_2 and the output series SYMPTOMS. While the input-output relation between the nontransformed data shows a marked nonlinear behaviour (a), the relation between the transformed series is approximately linear.

3. Identification of univariate models

After choosing the appropriate transformation, an univariate model was identified for each input and output series. Since in Basle January and February are the months with the strongest winter heating (SO_2), a separate model has been identified for this 'winter period'. The univariate models of the input series are necessary in order to apply a procedure called prewhitening (compare next section). The univariate model of the output series serves two purposes: (i) It provides an initial guess of the noise structure. (ii) The residual variance of the univariate model may be used as a 'yard stick' when comparing different transfer function models.

For the three series ln (SO_2), ln (NO_2) and SYMPTOMS the iterative method of model

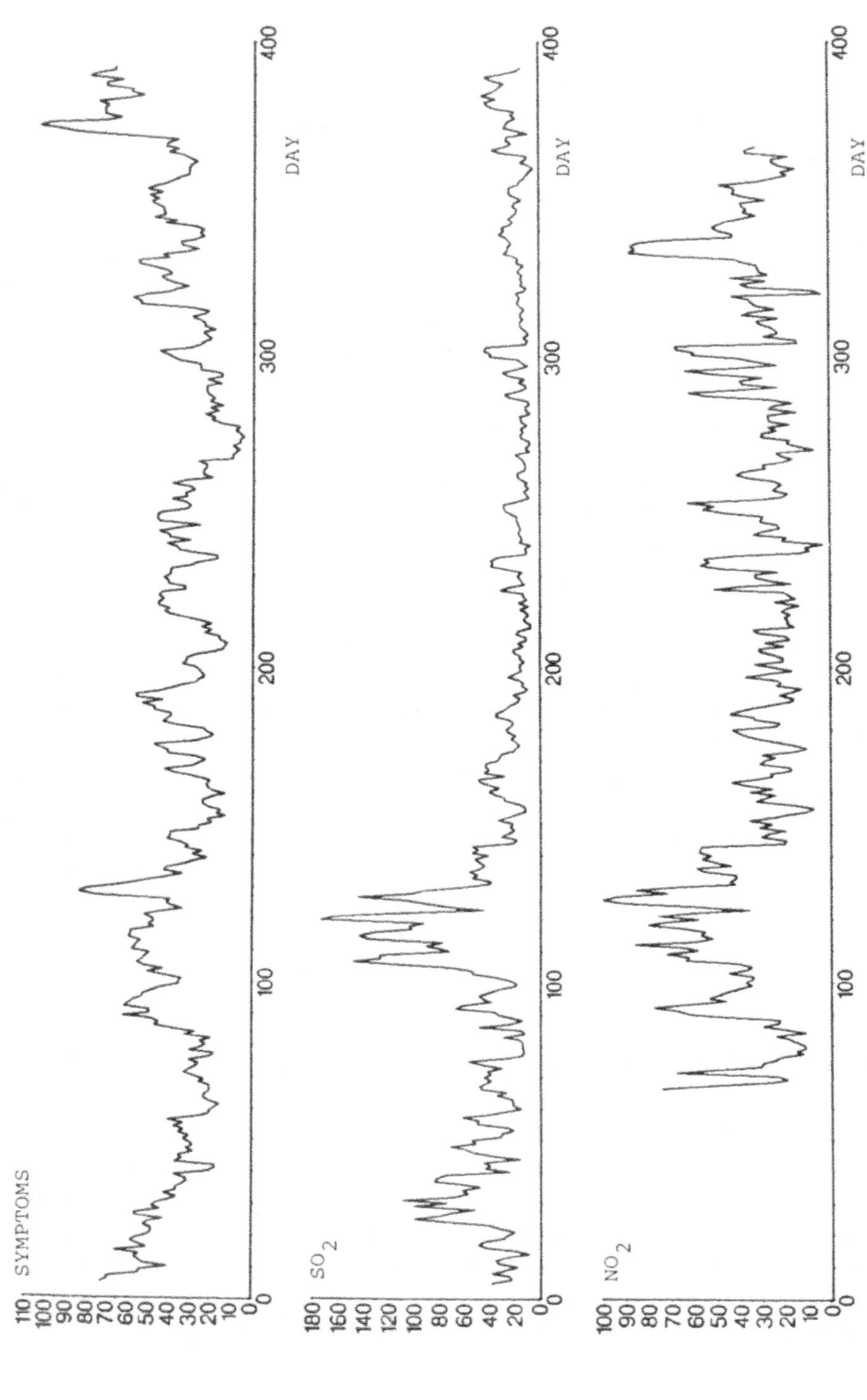

Fig. 1. Uppermost curve: Relative number of symptoms x 100 (SYMPTOMS). Second curve: SO_2 ($\mu g\ m^{-3}$). Lower part: NO_2 ($\mu g\ m^{-3}$).

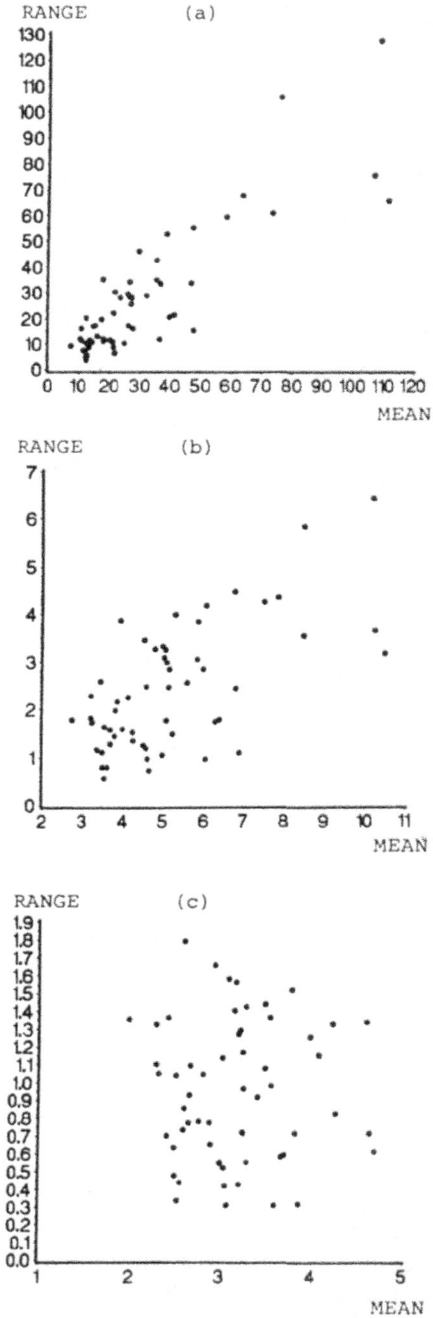

Fig. 2. Mean-range plot for SO_2. (a) Non-transformed data. (b) Applying the square root transformation. (c) Applying the logarithmic transformation.

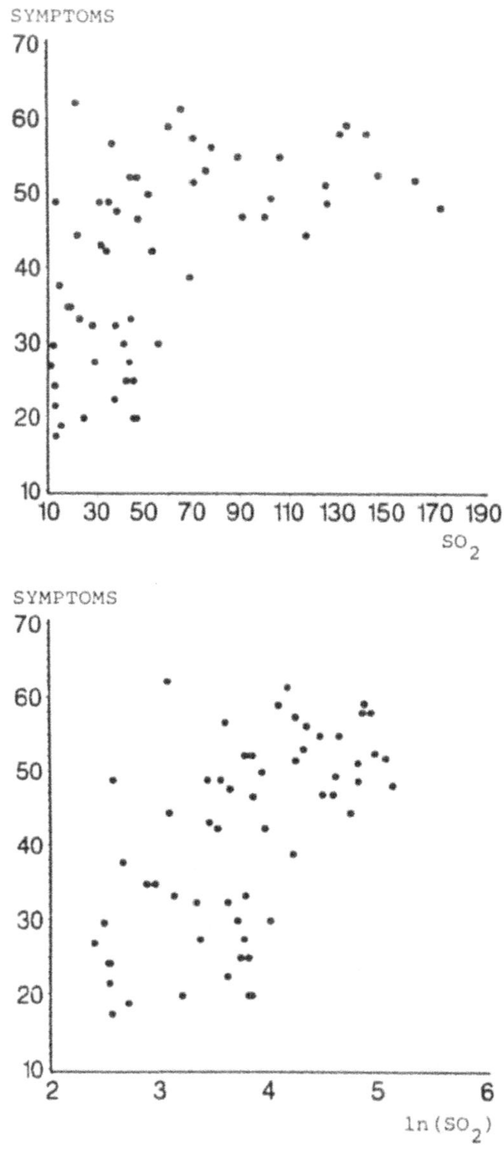

Fig. 3. (a) Non-transformed data. (b) Applying the logarithmic transformation to SO_2.

identification, fitting and diagnostic checking was straightforward. The autocorrelation function (ACF) of all series showed a slow decay and the partial autocorrelation function (PACF) showed a marked peak at lag 1. This behaviour is characteristic for an AR(1) process or for a non-stationary process. For the two input series an AR(1) model was tentatively identified. The ACF of the residuals showed no marked peaks. The goodness of fit tests (Ljung and Box, 1978) exhibited no sign of model inadequacy. For the output series SYMPTOMS the autoregressive coefficient was approximately 1. Therefore this nonstationary series had to be differenced. Fitting a MA(1) model to the series of differences showed no sign of model inadequacy. This univariate model is presented in the first line of Table I.

4. Prewhitening and Identification of Transfer Function Models

The relation between two time series x_t and y_t is determined by the crosscorrelation function (CCF):

$$\rho_{xy}(k) = \text{correlation } (x_t, y_{t+k}), \quad k = 0, \pm 1, \pm 2, \ldots$$

This function determines the correlation between the two series as a function of the timeshift k.

A main difficulty arises in the interpretation of the empirical crosscorrelation function. As shown by Bartlett (1935) and by Box and Newbold (1971) the empirical CCF between two completely unrelated time series which are themselves autocorrelated can be very large due to chance alone. Also, the crosscorrelation estimates at different lags may be correlated. This is due to the autocorrelation within each individual series.

Two ways out of this difficulty have been proposed: (i) The univariate model for the input series (SO_2, NO_2) converts the correlated series x_t into an approximately independent series α_t. Applying the identical operation to the output series y_t (SYMPTOMS) produces a new series β_t. The CCF between α_t and β_t (the prewhitened crosscorrelation function) shows at which lags input and output are related (Box and Jenkins, 1976).

An alternative way has been described by Haugh (1976) and by Haugh and Box (1977): (ii) For each series an individual model is identified. The CCF is then calculated for the two residual series.

Since it is not yet clear which of the two methods is superior (Haugh and Box, 1977) both procedures were tried. The CCF of the two original series in (SO_2) and SYMPTOMS is not interpretable (Figure 4a). Figure 4b shows the prewhitened CCF using method (i). There is a marked peak at lag 0 but not at other lags. The CCF using method (ii) showed approximately the same result. Figure 5 presents the two residual series from the corresponding univariate models. One may clearly recognize a synchronisation between the two series.

The above results suggested the following parsimonious transfer function model for x_t (ln (SO_2) respectively ln (NO_2)) and y_t (SYMPTOMS).

$$y_t = v_0 x_t + n_t,$$

$$\nabla n_t = (1 - \theta B) a_t,$$

or

$$\nabla y_t = v_0 \nabla x_t + (1 - \theta B) a_t.$$

The diagnostic checks on the residuals of the transfer function models showed no sign of

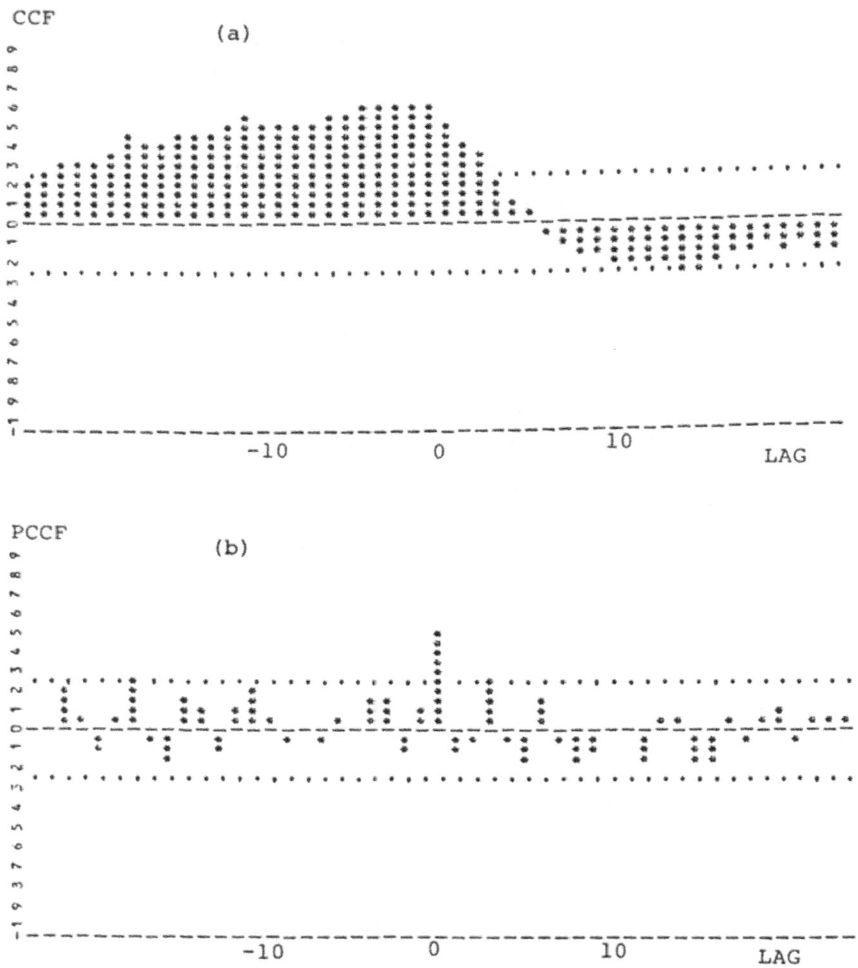

Fig. 4. Crosscorrelation function between ln (SO_2) and SYMPTOMS. (a) Before prewhitening (CCF). (b) After prewhitening (PCCF).

model inadequacy. A final model was fitted with both SO_2 and NO_2 as input variables. Table I summarizes all these models. One can see from the residual variances of the corresponding models that the series SO_2 contributes more to the explanation of the series SYMPTOMS than the series NO_2. The simultaneous introduction of the two input series SO_2 and NO_2 into the model shows no stronger reduction of the residual variance of SYMPTOMS than the model with SO_2 alone.

While the above results were found for the 'winter period' no relations between the input and output series were detected for the other seasons. This means that the crosscorrelation function changes with time.

Since the overall series SYMPTOMS is composed of several subseries (cough, etc.),

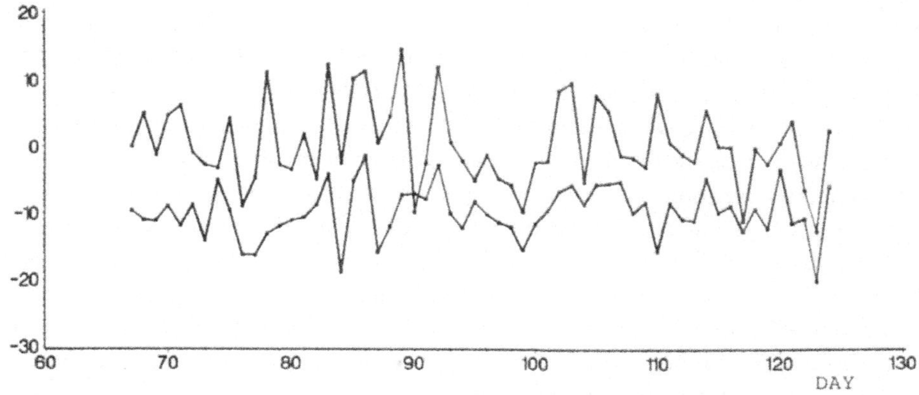

Fig. 5. Upper curve: Residuals of SYMPTOMS. Lower curve: Residuals of ln (SO_2) (rescaled).

compare the introduction), it seemed interesting to explore how the individual symptoms are related to the input variable ln (SO_2). The most pronounced instantaneous relation was found for the symptom 'runny or stuffy nose' ($v_0/SE(v_0) = 3.16$) and a somewhat weaker for the symptom 'cough' ($v_0/SE(v_0) = 2.54$). For the other symptoms no significant lag 0 transfer function weight was found.

No relations were found between the series ENTRIES and the series of explanatory variables.

Conclusion

After the choice of the appropriate transformations and after prewhitening, the CCF between the series ln (SO_2) and the series SYMPTOMS showed a peak at time lag zero but not at other time lags for the period January to February 1986. A similar result was found

TABLE I

Summary of transfer function models for SYMPTOMS

Model type	Estimated model	Residual variance (x10^4)
Univariate	$\nabla y_t = (1 - 0.24B)\, a_t$ ± 0.13	38.7
1 *Input*		
x_{1t}: ln (SO_2)	$\nabla y_t = 0.078\, \nabla x_{1t} + (1 - 0.26B)\, a_t$ $\pm 0.017 \quad \pm 0.13$	28.8
x_{2t}: ln (NO_2)	$\nabla y_t = 0.068\, \nabla x_{2t} + (1 - 0.19B)\, a_t$ $\pm 0.022 \quad \pm 0.13$	33.9
2 *Input*		
x_{1t}: ln (SO_2) x_{2t}: ln (NO_2)	$\nabla y_t = 0.067\, \nabla x_{1t} + 0.025\, \nabla x_{2t} + (1 - 0.26B)\, a_t$ $\pm 0.020 \quad \pm 0.024 \quad \pm 0.13$	28.8

for the CCF between in (NO_2) and SYMPTOMS. The transfer function models revealed that SO_2 contributes more than NO_2 to the explanation of the series SYMPTOMS. The simultaneous introduction of the two input series SO_2 and NO_2 into the model showed no stronger reduction of the residual variance of SYMPTOMS than for SO_2 alone.

While the above results were found for the 'winter period' no relations between input and output were detected for the other seasons. This means that the crosscorrelation function changes over time. In addition to that it is interesting to note that all 'climatic' variables (e.g. temperature) did not contribute to the explanation of the output series.

No relations were found between the series ENTRIES and the explanatory variables.

The overall series SYMPTOMS is composed of several subseries (cough, rhinitis etc.). Identification of transfer function models for each subseries revealed the strength in which each individual symptom was related to the input variable SO_2. The strongest instantaneous relation was found for the symptom 'runny or stuffy nose' and a somewhat weaker for the symptom 'cough'. For the other symptoms no significant lag 0 transfer function weight was found.

References

Abraham, B. and Ledolter, J.: 1983, *Statistical Methods for Forecasting*, John Wiley, New York.

Bartlett, M. S.: 1935, 'Some Aspects of the Time-Correlation Problem in Regard to Tests of Significance', *Journal of the Royal Statistical Society* **98**, 536-543.

Box, G. E. P. and Jenkins, G. M.: 1976, *Time Series Analysis, Forecasting and Control*, Revised Edition, Holden Day, San Francisco.

Box, G. E. P. and Newbold, P.: 1971, 'Some Commments on a Paper of Cohen, Gomme and Kendall', *Journal of the Royal Statistical Society A* **134**, 229-240.

Box, G. E. P. and Tiao, G. C.: 1975, 'Intervention Analysis with Applications to Economic and Environmental Problems', *Journal of the American Statistical Association* **70**, 70-79.

Colleg, J. R. T. and Brasser, L. J.: 1981, 'Study on Chronic Respiratory Diseases in Children in Relation to air Pollution', *WHO Regional Office for Europe*, Copenhagen.

Haugh, L. D.: 1976, 'Checking the Independence of Two Covariance-Stationary Time series: A Univariate Residual Cross-Correlation Approach', *Journal of the American Statistical Association* **71**, 378-385.

Haugh, L. D. and Box, G. E. P.: 1977, 'Identification of Dynamic Regression (Distrubuted Lag) Models Connecting Two Time Series', *Journal of the American Statistical Association* **72**, 121-130.

Jenkins, G. M.: 1979, *Practical Experiences with Modelling and Forecasting Time Series*, Gwilym Jenkins & Partners (Overseas) Ltd., St. Helier.

Ljung, G. M. and Box, G. E. P.: 1978, 'On a Measure of Lack of Fit in Time Series Models', *Biometrika* **65**, 297-303.

Love, G. J. *et al.*: 1982, 'Acute Respiratory Illness in Families Exposed to Nitrogen Dioxide Ambient Air Pollution in Chattanooga Tennessee', *Arch. Environ. Health* **37**, 75-80.

THE NIAGARA RIVER: A WATER QUALITY MANAGEMENT OVERVIEW

F. J. PHILBERT

Water Quality Branch, Inland Waters Directorate, Conservation and Protection Service, Ontario Region, Environment Canada, Burlington, Ontario, Canada, L7R 4A6

(Received February 1990)

Abstract. The Niagara River constitutes part of the Laurentian Great Lakes and St. Lawrence River system which represents approximately 80% of North America's supply of surface fresh water. The river is a major source of water for industry, municipalities, recreation and power generation and is the link between Lakes Erie and Ontario. The river forms part of the Canada-U.S. border and falls under the jurisdiction of both countries.

The massive industrialization of the region surrounding the river has led to a typical resource use conflict situation in which pollution of the river continues to be a major public concern.

A number of constitutional, institutional and jurisdictional factors make the management of the Niagara River an involved and complicated matter. The interests, intent, philosophies, laws and regulations are not necessarily the same among the numerous jurisdictions involved. Despite these differences, however, Canada and the United States have succeeded in developing and implementing a model cooperative international management plan for the river. An overview of the main international aspects relating to the development and implementation of this plan, the Niagara River Toxics Management Plan, is presented.

Introduction

The Niagara River constitutes part of the Laurentian Great Lakes and the St. Lawrence River system which represents one of the world's largest masses of surface fresh water and forms a waterway that stretches more than one third of the way across North America. The Great Lakes and interconnecting channels system is shared by Canada and the United States of America and represents approximately 80% of North America's, or about 18% of the world's, fresh liquid surface water. They sustain life and serve domestic, commercial, industrial, agricultural, transportation, tourism, fishery, recreational and waste assimilation needs for an estimated 7.5 million Canadians and 30 million Americans.

The area of the lakes is approximately 244 160 km^2, with a total land and water drainage basin area of about 765 990 km^2 (US EPA and Environment Canada, 1987). About 36%, i.e. 88 600 km^2, of the Great Lakes area lies within Canada.

The Niagara River serves as the main drainage system for the three Upper Lakes (Superior, Huron and Michigan) and Lake Erie into Lake Ontario (Figure 1). The 58 km river, flowing northward from Lake Erie to Lake Ontario, has an average flow of 5700 cubic metres per second. As a source of municipal drinking water, it serves a combined Canadian/United States population of more than 400 000 people. It is divided into the upper and lower reaches by the world famous Niagara Falls. It provides about 85% of the total tributary flow to Lake Ontario and has a significant impact on the quality of the Lake

Environmental Monitoring and Assessment **17**: 157–166, 1991.

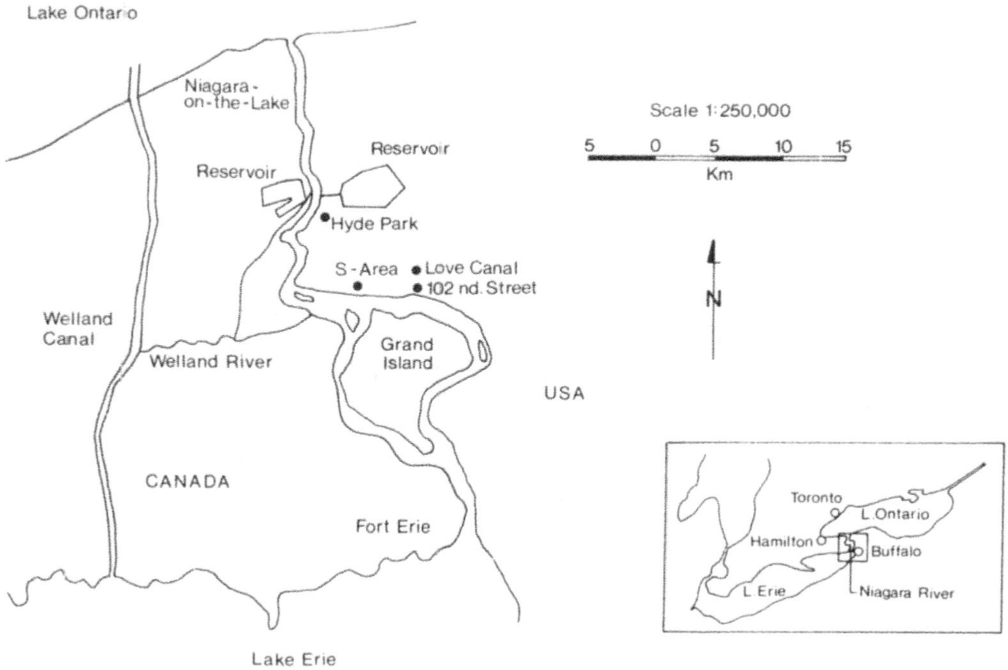

Fig. 1. The Niagara River.

which itself is the source of drinking water for approximately 3.8 million Canadians and about 805 000 Americans. The lake also supports a healthy tourist trade, sport and commercial fishery and a variety of recreational activities. Thus, the Niagara River, which itself is a major source of water for industry, municipalities, commerce, power generation, recreation, and tourism, impacts on a major population in the area.

The Great Lakes and their interconnecting channels are undoubtedly a precious resource of vital importance to all facets of life and activity in the Great Lakes Basin. Despite their grandeur and magnificence, however, they are indeed a fragile and vulnerable ecosystem susceptible to the damaging effects of the myriad of degrading influences to which they are exposed.

Institutional Framework

Over the years, the increasing pollution of natural water systems in Canada and the United States had led to action by both countries, mainly in the form of federal, provincial and state legislation and regulation, and international accords, treaties, and agreements, of which the Boundary Waters Treaty is among the most noteworthy. In keeping with their shared responsibility and the realization that the Great Lakes and other boundary waters need to be managed cooperatively, Canada and the US signed the Boundary Waters Treaty in 1909 which specified the rights and obligations of both countries concerning boundary waters and, in particular, the obligation of each country to not pollute boundary waters nor waters flowing across the boundary, to the injury of health or property of the other. A binational organizational – The International Joint Commission (IJC) was established and given authority to investigate and resolve disputes between the two nations over the use of water resources having transboundary implications. Thus, the treaty instituted a novel framework for international cooperation in the use and management of shared resources. Basically, the IJC serves as an investigatory and advisory body on matters relating to water quality and as a quasi-judiciary on matters pertaining to the regulation of boundary water levels and flows.

The intent and obligations of Canada and the US as set out in the Boundary Waters Treaty were reinforced when, in 1972, the Canada-United States Agreement on Great Lakes Water Quality was signed. A revised agreement signed in 1978 and an amending protocol signed in 1987, now provide an international framework for a coordinated binational ecosystem approach for clean-up efforts and management strategies for the Great Lakes ecosystem. For example, whereas in the mid-forties the emphasis was on bacterial contamination, discolouration and odour problems, and in the late sixties and early seventies concern was centred around eutrophication problems, the focus of attention shifted in the late seventies and early eighties to (i) controlling discharges of toxic substances into the Great Lakes and their connecting channels and (ii) fostering an 'ecosystem' approach in managing the system. The fundamental objective of the current agreement and its amending protocol is perhaps best expressed in Article II of the Agreement (International Joint Commission, 1988) which in part states that, 'The parties agree to make a maximum effort to develop programs, practices and technology necessary for a better understanding of the Great Lakes Basin Ecosystem and to eliminate or reduce to the maximum extent practicable the discharge of pollutants into the Great Lakes System'. In keeping with this objective, therefore, the parties have adopted the policy that, 'The discharge of toxic substances in toxic amounts be prohibited and the discharge of any or all persistent toxic chemicals be virtually eliminated'.

The Niagara River Pollution Problem

The abundance and availability of the Niagara River water for both municipal/domestic uses and as a source of inexpensive hydro-electric power, led to the extensive industrialization of the area surrounding the River. This in turn resulted in the classic case of resource use conflict in which the river became the receptor of inordinate amounts of pollutants originating primarily from a massive complex of chemical, steel and petrochemical plants and municipal outfalls along its banks. Point sources (municipal and industrial discharges) along the river and in the Upper Great Lakes region, non-point sources, including active and inactive hazardous waste disposal sites and, to a lesser extent, agricultural and urban run-off, constitute the major input sources of contaminants to the river. It is no wonder, therefore, that pollution of the Niagara River has been, and continues to be, a major public concern. The level of public awareness and concern has heightened considerably during the past decade.

The Niagara River water quality pollution problem has been recognized since around the mid 1940's. The IJC identified it in 1951 as a 'problem area', with the initial pollution concerns being related to bacterial contamination, phenols, oil, odour, excessive levels of iron and chloride, as well as general discolouration. In the past decade, the River has been identified as the predominant source of organic and inorganic contaminants to Lake Ontario. These include metals, PCBs and Mirex (Kauss, 1983; Thomas *et al.,* 1988), chlorinated benzenes (Fox *et al.,* 1983; Carey and Fox, 1986), volatile halocarbons (Comba and Kaiser, 1984) and metals such as mercury, lead, zinc, and copper (Thomas *et al.,* 1988; Whittle and Fitzsimmons, 1983; Stevens and Neilson, 1988). Present attention is focused on toxic chemicals in the River and Lake Ontario and their effects on human health and the ecosystem as a whole. The river is now designated by the IJC as one of the 42 'Areas of Concern' in the Great Lakes basin exhibiting severe water pollution problems and where beneficial uses of water or biota have been adversely affected or where specific water quality objectives, established by the IJC, or local standards are being continually exceeded.

Management Framework

A number of constitutional, institutional and jurisdictional factors make the management of the Niagara River an involved and complicated matter. Although the discharge of contaminants to the River is controlled through regulatory programs carried out by the New York State Department of Environmental Conservation and the Ontario Ministry of the Environment on their respective sides of the river, there are, nevertheless, numerous organizations, including Canadian and US federal, provincial, state and municipal governments, having some sort of interest, jurisdiction, or involvement in the use and management of the river. Unfortunately, however, the interests, intent, philosophies, laws and regulations are not necessarily the same among the agencies. Nevertheless, despite these differences, Canada and the United States have succeeded in developing and implementing a management plan for the Niagara River, which could only be described as an exemplary cooperative approach to a major international environmental problem.

Over the past several decades, and particularly since the mid 1940's, the Niagara River pollution problem has been intently studied and the prinicipal jurisdictions have been attempting, both unilaterally through their own regulatory programs and legislation, and on a joint international basis, to address the issue. In January 1981, the IJC issued a special report (International Joint Commission, 1981) on pollution in the Niagara River. The Commission made a number of specific recommendations which, in part, called for:

(1) the undertaking of a comprehensive and coordinated study of the River, including the identification of sources, concentrations, fate and probable effects of all detected organic compounds and metals so as to permit assessment of the problem and to implement the required remedial or preventative action on a common basis;

(2) that a comprehensive and continuing monitoring program for the entire Niagara River and western end of Lake Ontario be developed and maintained, coordinated and supported by all relevant jurisdictions.

The first comprehensive integrated assessment of toxic chemicals in the Niagara River was undertaken in 1981, by the Niagara River Toxics Committee (NRTC), a four party committee consisting of representatives from Environment Canada, the Ontario Ministry of Environment, the U.S. Environmental Protection Agency, and the New York State Department of Environmental Conservation. The NRTC goal was to determine what chemicals were in the River, identify their sources, recommend control measures and establish a procedure to monitor the effectiveness of those control measures.

The results of the NRTC three-year study were presented in a comprehensive report (Niagara River Toxics Committee, 1984) completed in October 1984. The report established that there was extensive toxic chemical contamination of the Niagara River. Significant findings included the following:

(i) The total quantified load of EPA priority pollutants from the 69 municipal and industrial point source discharges to the river and its tributaries sampled during the study, approximated 1400 kg d^{-1};

(ii) thirty-seven of those point sources accounted for 95% of the total quantified loading of EPA priority pollutants, with the Niagara Falls New York Wastewater Treatment Plant identified as the single largest source of priority organic pollutants to the river;

(iii) ninety-one percent (29) of those sources were on the US side of the river with the other 9% (8) on the Canadian side;

(iv) over 215 hazardous waste disposal sites were identified in the Niagara/Erie county area of New York State, 164 of which were within a 5 km band on the US side along the river;

(v) sixty-one of the 164 US sites (notably Love Canal, Hyde Park, S-Area and 102nd Street) and five of 17 closed and active landfill sites on the Canadian side were assessed as having a significant potential to contaminate the river;

(vi) a total of 261 chemicals were found at least once in the water, sediment or biota (fish, clams, algae) sampled from the study area including the eastern end of Lake Erie and the western end of Lake Ontario;

(vii) two hundred and twenty seven of these chemicals were considered capable of having potentially adverse environmental effects on the Niagara River and Lake Ontario, with 57 having been detected at levels which exceeded some environmental or public health criteria at least once;

(viii) there was evidence of widespread ground water contamination, specifically by metals and synthetic organic compounds, within the 5 km band along the river; and

(ix) loadings to the river from waste sites could not be calculated because of insufficient data.

The NRTC report contained 24 specific recommendations, directed at the agencies involved to, (i) improve control programs, (ii) address the clean-up of hazardous waste sites, (iii) identify further sources and characteristics of chemicals, and (iv) implement programs to monitor the effectiveness of control measures. One recommendation called for the formation of a binational committee to coordinate the implementation of the recommendations in the report. These recommendations, combined with a desire among the four parties to maintain the initiatives and momentum of the NRTC, led to the development of the Niagara River Toxics Management Plan and the institution of a formal four party committee structure which administers its implementation.

The Niagara River Toxics Management Plan

A work plan, which is updated annually, and a Declaration of Intent together constitute the Niagara River Toxics Management Plan (NRTMP, 1988). Several months of intensive work and negotiations led to the development and adoption of an initial work plan by the four principal jurisdictions, i.e. the United States Environmental Protection Agency, Environment Canada, the New York State Department of Environmental Conservation, and the Ontario Ministry of the Environment. This four-party work plan was released on October 30, 1986. The 'Declaration of Intent', a political agreement relating to the NRTMP, was signed by the heads of the four agencies on February 4, 1987.

The 'Declaration of Intent', which formalizes the plan, commits the four participating agencies to work together to fully implement the actions and programs outlined in the Toxics Management Plan by, (i) coordinating the existing chemical pollutant control activities on the River in both countries, (ii) establishing a common basis for assessing toxic chemical loadings to the River, (iii) identifying priorities for control measures to reduce loadings, and (iv) evaluating the success of these measures on an ongoing basis. It calls for the issue of a report to the public every six months on the progress being made to reduce persistent toxic chemicals of concern entering the river. It specifically calls for a target reduction level of 50 per cent of loadings of persistent toxic chemicals of concern, from sources on both sides of the Niagara River, by 1996. For point sources, this has been based on data collected between April 1, 1986, and March 31, 1987. For non-point sources it has been based on information collected between April 1, 1987 and March 31, 1988.

The fundamental goal of the Plan is to reduce the loadings of toxic chemicals to the Niagara River. The objectives of the plan are to: (i) control or eliminate discharges of priority toxic chemicals at their source, (ii) identify corrective action for clean-up of the river, (iii) measure progress continuously, and (iv) report publicly and regularly on progress. The strategy, organization, and activities necessary to ensure the timely and effective achievement of the goals and objectives of the Declaration of Intent are specified in the Work Plan segment of the plan. Thus, under the NRTMP, activities are identified for the coordination of pollution control programs, the establishment of a common basis for assessing pollutant loadings, the identification of priorities for control measures and the evaluation of control measures.

As was noted earlier, the NRTC could provide no indication of the magnitude of toxic loadings to the Niagara River from non-point sources. The NRTMP identifies two activities directed to the estimation of such contributions. Attempts are being made to derive an initial estimate using ambient river monitoring data from the head and mouth of the Niagara River (input-output differential monitoring) together with updated loading data from municipal and industrial sources. In addition, pending the development of further site-specific data and more direct measurements, estimates of potential contaminant loadings would be derived from existing hydrogeological and contaminant data at the various sites.

The NRTMP identifies an organization and implementation structure for the coordination and evaluation of pollution control measures which are directed to the systematic reduction of loadings of toxic chemicals to the Niagara River. The plan is comprised of the following eight basic components: (i) river monitoring, (ii) point sources, (iii) non-point sources, (iv) chemicals of concern, (v) technical and scientific cooperation, (vi) a communication plan, (vii) organization and implementation, and (viii) reporting. It calls for a senior level coordination committee to coordinate and oversee plan implementation. Under the umbrella of the Coordination Committee, a number of sub-committees have been formed to perform tasks specified in the plan (Figure 2).

A public involvement component of the Communication Plan allows for active participation by the public in matters pertaining to the NRTMP. Public meetings which have been held semi-annually since January 1987, serve to present current programs and report progress on specific plan activities.

The NRTMP formed the basis for the development of a similar Toxics Managament Plan for Lake Ontario which was adopted by the agencies on February 28, 1989.

The Niagara River Remedial Action Plan

As stated above, the IJC has designated the Niagara River as one of the 42 'Areas of Concern' in the Great Lakes Basin. The development and subsequent implementation of a 'Remedial Action Plan' (RAP) to restore and protect the River's beneficial uses is another noteworthy management initiative currently underway.

Canada and the United States, under the terms of the amended 1978 Great Lakes Water Quality Agreement are required, among other things, to (i) develop and implement systematic ecosystem based strategies to restore and protect beneficial uses in Areas of Concern (or in open lake waters), (ii) develop and implement RAPs for Areas of Concern in accordance with prescribed guidelines, (iii) consult with the public and all affected state, provincial and municipal governments in the RAP development process, (iv) submit RAPs to the IJC for review and comment at prescribed stages of their development, and (v) report bienially to the IJC on progress on the development and implenentation of the RAPs and on the restoration of beneficial uses. The first report is due December, 1989.

RAPs are the brain-child of the IJC and are intended to be comprehensive blueprints for pollution abatement and control measures required to address water quality and water use problems specific to a particular Area of Concern. The Plans include the conduct of surveillance and monitoring programs to measure the effectiveness of clean-up measures, warn of emerging problems and track down contaminant sources. The RAP development process includes a strong public participation component allowing for input and the active participation of all interested groups or individuals.

One of the provisions of the Declaration of Intent is for the parties to submit the

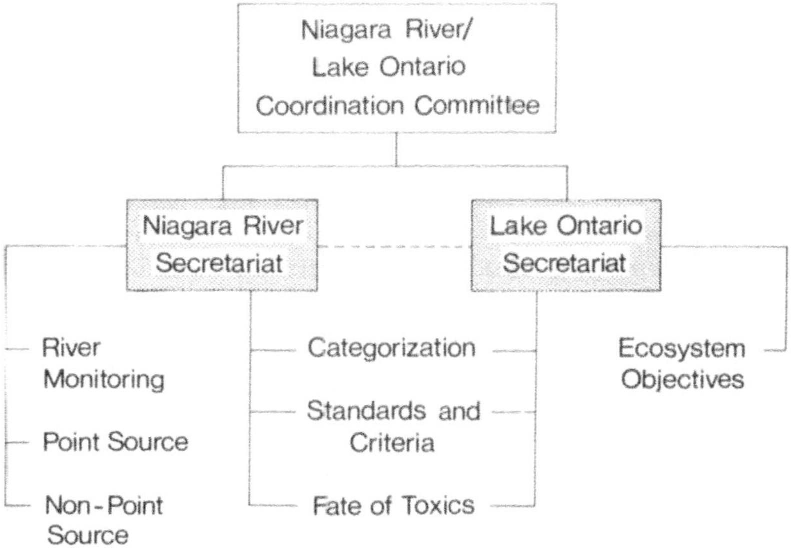

Fig. 2. NRTMP Management Structure (From NRTMP, 1988 Revision).

NRTMP and progress reports to the IJC as part of the IJC's Great Lakes Remedial Action Plan Program. Thus, the basic objectives of the RAP process and those of the NRTMP are mutually reinforcing. However, RAPs are broader in scope than the NRTMP, encompassing in addition to toxic chemicals, other issues such as aquatic habitat degradation, and contamination by conventional pollutants, nutrients, and bacteria.

In Canada, RAPs are being prepared jointly by the federal and provincial governments with the aim of cleaning up the 17 Areas of Concern which are on the Canadian side of the Great Lakes. The Niagara River is one of five Areas of Concern shared with the United States and, as such, has been the subject of discussion for the establishment of a binational RAP. The process will likely entail the development of RAPs by Canada and the US as separate jurisdictional responsibilities with each jurisdiction providing cooperative input as needed. The RAPs will include a description of how each jurisdiction intends to remedy, within its territory, the Niagara River pollution problems. The binational RAP will then evolve when, at an appropriate stage in the RAP development process, the jurisdictions jointly develop a common statement of environmental problems and goals for the River (NRTMP, 1988).

Major Accomplishments to Date

Overall, since its inception approximately two and a half years ago, the management framework established under the Niagara River Toxics Management Plan has proven to be a success. Since the institution of the Plan, the four agencies, acting individually and together, have initiated a variety of programs and activities in conformance with the Plan requirements. There has been a concerted, coordinated effort by the four principal jurisdictions in the planning and implementation of ambient, point source, and non-point source monitoring programs on the River. Agreement has been reached on sampling, analytical, and data interpretation and reporting protocols for the ambient monitoring program. A system of categorizing toxic chemicals has been developed to determine either that a chemical warrants corrective action on a priority basis, or that it can be controlled more routinely through the implementation of existing and developing programs that apply to the control of all toxics (NRTMP, 1988). Screening protocols have been or are being developed by the four agencies to identify candidate chemicals to be targeted for 50% reduction. They involve a consideration of chemicals showing a significant increase in concentration at the mouth of the river relative to the head, as well as a comparison of water quality and fish tissue data against established standards and criteria, together with a determination of the relative contribution of Niagara River sources for these chemicals. Already, using these screening protocols, the following ten chemicals from an initial group of 16 have been identified as the first set of pollutants targeted for a 50% loading reduction by 1996:

benz(a)anthracene	mirex
benzo(a)pyrene	hexachlorobenzene
benzo(b)fluoranthene	PCBs
benzo(k)fluoranthene	mercury
tetrachloroethylene	2,3,7,8 - TCDD

Further assessment of chemicals for a 50% loading reduction is continuing.

Of the 1400 kg day^{-1} identified by the NRTC, both New York State and Ontario have reported reductions in priority pollutant loadings associated with municipal and industrial sources amounting to 80% in 1985-1986 for New York and 60% in 1986-1987 for Ontario. Present estimates are in the order of 309 kg day^{-1}. The US EPA has attempted to estimate the potential loadings to the Niagara River from leaking hazardous waste sites using existing hydrogeological and contaminants data. The best estimate of current total actual loadings is 216 kg day^{-1} (178 kg day^{-1} organics) (Brooksbank, 1989).

Summary and Conclusion

The release of the Niagara River Toxics Committee report in 1984 was an important milestone in a series of investigations and reports, over the years, on toxic chemicals entering the Niagara River. The results of the NRTC report is an apt illustration of how much can be accomplished when governments work together to resolve their common environmental problems. The study was a landmark in advancing our understanding of the Niagara River pollution problem. The Report findings and recommendations provided a framework within which the Canadian and US governmens were able to establish priorities for cleaning up the Niagara River. This led to the immediate formulation and successful implementation of a joint management plan for the river.

The Niagara River Toxics Management Plan (and the evolving Niagara River Remedial Action Plan) has built on the precedent established over the past 18 yrs by the Canada-US

Great Lakes Water Quality Agreement with the primary objective to develop and implement coordinated programs to eliminate or reduce, to the maximum extent practicable, the discharge of pollutants into the Niagara River. The adoption and implementation of the NRTMP places a firm commitment on each of the four principal jurisdictions to follow an agreed upon management strategy for the effective coordination and evaluation of collective pollution abatement programs and acitivities aimed towards the achievement of significantly reduced loadings of toxic chemicals to the river. Implementation of the plan is considered to be progressing satisfactorily and significant progress has already been reported by the four jurisdictions involved.

Acknowledgement

The provision of some material for this paper and helpful review comments by M. G. Brooksbank are gratefully acknowledged.

References

Brooksbank, M. G.: 1989, Personal Communication, Conservation and Protection, Ontario Region, Environment Canada, Toronto, Canada.

Carey, J. H. and Fox, M. E.: 1987, 'Distribution of Chlorinated Benzenes in the Niagara River Plume', NWRI Contribution. #87-86.

Comba, M. E. and Kaiser, K. L. E.: 1984, 'Tracking River Plumes with Volatile Halocarbon Contaminants: The Niagara River-Lake Ontario Example', *J. Great Lakes Res.* **10(4)**, 375-382.

Fox, M. E., Carey, J. H. and Oliver, B. G.: 1983, 'Compartmental Distribution of Organochlorine Contaminants in the Niagara River and the Western Basin of Lake Ontario', *J. Great Lakes Res.* **9(2)**, 287-294.

International Joint Commission: 1981, 'Special Report (under the 1978 Great Lakes Water Quality Agreement) on Pollution in the Niagara River'.

International Joint Commission: 1988, 'Revised Great Lakes Water Quality Agreement of 1978'.

Kauss, P. B.: 1983, 'Studies of Trace Contaminants, Nutrients and Bacteria Levels in the Niagara River', *J. Great Lakes Res.* **9(2)**, 249-273.

Niagara River Toxics Committee: 1984, A joint publication of Environment Canada, United States Environmental Protection Agency, Ontario Ministry of the Environment, and New York State Department of Environmental Conservation.

Niagara River Toxics Management Plan: 1988 Revision, A document by Environment Canada, United States Environmental Protection Agency, New York State Department of Environmental Conservation, and Ontario Ministry of the Environment.

Stevens, R. J. J. and Neilson, M. A.: 1988, 'Inter- and Intralake Distributions of Trace Contaminants in Surface Waters of the Great Lakes', *J. Great Lakes Res.*

Thomas, R. L., Gannon, J. E., Hartig, J. H., Williams, D. J. and Whittle, D. M.: 1988, 'Contaminants in Lake Ontario – A Case Study', Proc. of World Conf. on Large Lakes, May 1986, Mackinac Is. Mich.

United States Environmental Protection Agency and Environment Canada: 1987, 'The Great Lakes – An Environmental Atlas and Resource Book', (ISBN No. 0-662-15189-5).

Whittle, D. M. and Fitzsimmons, J. D.: 1983, 'The Influence of the Niagara River on Contaminant Burdens of Lake Ontario Biota', *J. Great Lakes Res.* **9(2)**, 295-302.

TIME SERIES VALUED EXPERIMENTAL DESIGNS: A REVIEW

BRAJENDRA C. SUTRADHAR

Department of Mathematics and Statistics, Memorial University of Newfoundland, St. John's, Newfoundland, Canada A1C 5S7

and

IAN B. MACNEILL

Department of Statistical and Actuarial Sciences, The University of Western Ontario, London, Ontario, Canada N6A 5B9

(Received February 1990)

Abstract. A review is given of the literature on time-series valued experimental designs. Most of this literature is divided into two categories depending upon the factor status of the time variable. In one category, time is an experimental factor, and in the other it is a non-specific factor and enters the design in the context of replications. Analyses in both the time and frequency domain are reviewed. Signal detection models, Bayesian methods and optimal designs are surveyed. A discussion is also presented of application areas which include field trials and medical experiments. A main theme of the literature is that application of standard *F*-tests to highly correlated data can be misleading. A bibliography of relevant publications from 1949 onward is presented.

1. Introduction

In many situations where an investigation is repeated over time on physically independent material, and where external conditions can be treated as random, it may be sensible to treat time as a non-specific factor; that is, to consider time in the context of replications. For example, the use of automatic data acquisition equipment may make it possible to obtain many observations on the same treatment combination but with only a small time interval between consecutive observations. An example where such a time series valued experimental design and its concomitant analysis would apply is a process control problem in which it is expensive to change the process parameters, but in which it is possible to make observations in a short period of time for a fixed set of parameters. These observations form a time series characterized by a high degree of correlation among contiguous observations. In other experimental situations time is considered as one of the experimental factors, and not just as part of the replication process. The time series aspect of such designs may exhibit autocorrelation sufficiently high that time series methods are required for an appropriate analysis of time as a specific factor. In many cases where data are collected over time it may not be clear that the white noise assumption is valid; one suspects for the most part these cases are analyzed routinely by ANOVA methods without challenging the independence assumption.

This article is meant to serve two purposes: first, to acquiant the general reader with the

Environmental Monitoring and Assessment **17**: 167–180, 1991.

fact if the observations under a treatment in an experimental design form a time series or the data are collected over time, the classical ANOVA methods for testing the treatment effects without challenging the independence assumption, may be highly misleading; a modified analysis is required to adjust for correlation induced biases; and secondly, to give the researcher interested in this area of statistics access to the current and past literature. The discussion below surveys the various models for time series valued experimental designs and reviews the different areas where these designs are used; these areas range from field trials to clinical experiments.

2. The Factor Status for Time

To initiate the discussions we consider a general two-way analysis of variance model

$$y_{ij}(t) = \mu + \alpha_i + \mathrm{B}_j + z_{ij}(t) \qquad (2.1)$$

$$i = 1, \ldots, k; j = 1, \ldots, m; t = 1, \ldots, n_{ij},$$

where: $y_{ij}(t)$ is the t-th observation in the (i,j)-th cell, μ is the general effect, α_i is the i-th treatment or row effect, β_j is the j-th column or block effect, and $z_{ij}(t)$ is the error variable. Assume that the $z_{ij}(t)$ are autocorrelated and that they could follow as complex a model as the multiplicative seasonal ARMA process

$$\phi_p(B)\Phi_P(B^s)z_{ij}(t) = \theta_q(B)\Theta_Q(B^s)a_{ij}(t), \qquad (2.2)$$

where: $\phi_p(B)$, $\theta_q(B)$ are polynomials of degrees p and q in non-negative powers of B with zeros outside the unit circle; $\Phi p(B^s)$, $\Theta_Q(B^s)$ are polynomials of degrees P and Q in non-negative powers of B^s with zeros outside the unit circle; B is the backshift operator; and $a_{ij}(t)$ are components of white noise series that are independent for all $i = 1, \ldots, k$ and $j = 1, \ldots, m$. For example, if $p = 1, q = 0, P = 0, Q = 0$ in (2.2), the errors $z_{ij}(t)$ follow the AR(1) process; that is

$$z_{ij}(t) = \phi_1 z_{ij}(t-1) + a_{ij}(t).$$

The two-way ANOVA model (2.1) represents a time series valued experimental design. If a large number of observations are collected on the same treatment combination but with only a small time interval between consecutive observations, it may be sensible to treat time as a non-specific factor. Models similar to (2.1) with time as a non-specific factor have been studied by a number of authors. Among them, we mention Berndt and Savin (1975, 1977), Brillinger (1980), Ljung and Box (1980), Azzalini (1981), Yang and Carter (1983), Rothenburg (1984), Mansour *et al.* (1985), Pantula and Pollock (1985), and Sutradhar *et al.* (1987).

In certain experimental situations where data are collected at a few equally spaced time points, generally with a relatively large time interval between consecutive observations, it may be sensible to treat time as a specific factor. The two-way ANOVA model with time as a specific factor may be expressed as

$$y_i(t) = \mu + \alpha_i + \gamma_t + z_i(t), \quad (2.3)$$
$$i = 1, \ldots, k; t = 1, \ldots, n$$

where α_i is the i-th treatment or row effect, γ_t is the t-th time or column effect, and the $z_i(t)$'s are autocorrelated. Models similar to (2.3) have also been studied by many authors. For example, we refer to Box (1954), Shumway (1971), Brillinger (1973, 1980), Ljung and Box (1980), Andersen *et al.* (1981), Azzalini (1984), and Sutradhar and MacNeill (1989).

3. Experimental Designs with Time as a Specific Factor

Two of the earliest papers pertaining to experimental designs with correlated errors are those of Williams (1952), and Box (1954). Williams examined the efficiency of systematic designs compared to randomized schemes under the assumptions that the errors form a one-dimensional sequence and that they are correlated as in a stationary linear autoregressive process. Suppose that the n time points in (2.3) represent n blocks. Also suppose that $y\{(t-1)+i\}$ denotes the observation due to the i-th treatment under t-th block. Then under the assumption on the error variable that

$$a(h) = \phi a(h-1) + \xi(h),$$

for $h = 1,2, \ldots, kn$ and $\xi(h) \sim N(0, \sigma^2)$, it has been shown by Williams (1952) that for $k > 2$, systematic designs are more efficient than randomized designs for all $-1 < \phi < 1$. For $k = 2$, systematic designs are more efficient for positive ϕ but less efficient for negative ϕ. Since the work of Williams (1952), several articles have appeared on optimal designs with correlated errors. They will be discussed later in a separate section.

In a similar vein to that of Williams (1952), Box (1954) has also discussed the two-way ANOVA model (2.3). But unlike Williams, Box (1954) assumed that $z_i(t)$ are correlated errors within rows (treatments), rows being independent and identically distributed. Analogous to the classical analysis of variance, Box examined inferences about the column effects of the model. More specifically, following the approach suggested by Welch (1937, 1947), Box approximated the distribution of the usual F-test statistic for the hypothesis of constant γ_t values. Box has shown that the test of no column (time) effects is not seriously affected by the presence of serial correlations. However, Box did not consider the analogous approximation to the distribution of the F-statistic for testing the hypothesis of constant α_i values. The tests for no column effects as well as no row effects were studied in detail by Andersen *et al.* (1981). These test statistics are typically of the form $k\text{SSD}_1/\text{SSD}_2$, where k is a positive constant and the SSD's are sums of squares. The basic difference between the two tests is that in the former case, the two sums of squares involved are independent, but, in the latter case, the sums of squares are generally dependent. In the uncorrelated case, the SSD's exactly follow gamma distributions. In finding the approximation for the usual F-statistics, Andersen et al approximated the distributions of the SSD's by gamma distributions with first and second moments equal to the corresponding moments of the SSD's; see Section 4 in Andersen *et al.* They discussed the application of their tests with plasma citrate concentration data. Recently, Sutradhar and MacNeill (1989) extended Andersen et al's results to the case of two-way correlated data in a two-way ANOVA table. They proposed certain modified F-statistics for testing the presence of row and column effects. Sutradhar and MacNeill have used a Gaussian approximation to obtain the percentile values of the modified F-statistics. They applied their modified F-test to examine the adequacy of the multiplicative seasonal model with zero mean component for the airline data modelled by Box and Jenkins (1976).

Azzalini (1984) considers a model similar to, but different from, that considered in Andersen et al (1981). In the notation of (2.3), Azzalini considers α_i as a random effect such that $\alpha_i \sim N(x_i'\boldsymbol{\beta}, \sigma_\alpha^2)$, where x_i is the p-dimensional vector of covariates and $\boldsymbol{\beta}$ is a $p \times 1$ vector of unknown parameters. The time effect in Azzalini (1984) is the same as γ_t in (2.3). Azzalini developed a modified likelihood function and obtained algebraic expressions for the maximum likelihood estimators of $\boldsymbol{\beta} = (\beta_1, \ldots, \beta_p)'$, $\boldsymbol{\gamma} = (\gamma_1, \ldots, \gamma_n)'$, σ_α^2 and σ_a^2 as functions of ϕ, and suggested that an estimator of ϕ may be obtained by iteratively maximizing the likelihood. Azzalini also discussed tests of hypotheses concerning $\boldsymbol{\beta} = 0$, $\sigma_\alpha^2 = 0$, and $\gamma_1 = \ldots = \gamma_n = 0$.

4. Experimental Designs with Time as a Non-Specific Factor

There are many situations in practice where an investigation is repeated over time on physically independent material, and where external conditions may be treated as random. In such situations, it may be sensible to treat time as a non-specific factor. An appropriate model for the two-way ANOVA model with time as a non-specific factor is given by (2.1). In the case of one-way ANOVA, the model reduces to

$$y_i(t) = \mu + \alpha_i + z_i(t), \qquad (4.1)$$
$$i = 1, \ldots, k; \; t = 1, \ldots, n$$

where $y_i(t)$ is the observation at time t due to the i-th treatment, μ is the overall mean effect, α_i is the effect of the i-th treatment, and $z_i(t)$ is a component of a time series process.

Some attention has been paid to the use of the regression approach to analysis of model (4.1) with correlated errors. In an early paper, Andersen (1949) discussed effects of autocorrelation on the use of the least squares method of estimation. He demonstrated the presence of a large bias towards randomness in estimates of the autoregressive parameters of error terms which are based on calculated residuals. He discussed the loss of efficiency of the least squares method of estimation and prediction if the error terms are highly correlated. The effects of autocorrelation have also been studied by Wold (1949) and by Cochrane and Oreutt (1949).

Analogous to the usual analysis of variance approach, Shumway (1970) discussed frequency dependent estimation and tests of hypothesis for regression models with correlated error variables. Shumway also derived a frequency dependent goodness-of-fit criterion analogous to R^2.

Gallant *et al.* (1974) considered the study of experimental material which exhibits two characteristics: first, that it is possible to obtain replicates subject to various configurations of factors thought to affect the experimental material; and second, that the process of recording the phenomenon under study generates a long sequence of correlated observations. First they replaced the time series under a cell in the design by its Schuster periodogram. Secondly, they transformed the periodogram to obtain certain sequences indexed by frequency which correspond to analysis of variance statistics, such as treatment means and F-statistics, appropriate to the experimental design chosen. Finally, these sequences were used to compute ANOVA statistics.

Brillinger (1973), in a frequency domain set-up discussed the frequency-specific effects on certain F-statistics in connection with inferences about parameters in ANOVA models with autocorrelated errors. He illustrated the two way random effects model by an analysis of a sample of ten temperature series classified as of European or North American origin.

Azzalini (1981) considered the model

$$y_i(t) - \mu = \phi\{y_i(t-1) - \mu\} + a_i(t),$$

where: $y_i(t)$ is the t-th observation ($t = 1,2,\ldots,n_i$) of the i-th time series $i = 1,\ldots,k$; ϕ is the AR(1) parameter for each time series and $a_i(t)$ are the independent error variables having identical normal distributions with zero mean and variance σ^2. Azzalini dealt mainly with the estimation of the parameters μ, ϕ and σ^2 of the above model with special emphasis on the asymptotic results when $k \to \infty$ and n_i is fixed, say $n_i = n$.

Berndt and Savin (1975) have discussed Wald, likelihood ratio, Lagrange multiplier and max-root tests for testing the linear hypothesis in the multivariate linear regression model. They have shown that these tests based on exact distributions conflict with each other when applied to a given data set. In a later paper, Berndt and Savin (1977) showed that even in the asymptotic case, the Wald, likelihood ratio and Lagrange multiplier tests yield conflicting inferences. Rothenberg (1984) suggested that Edgeworth-corrected critical values may be used for the above three tests as they do not conflict in the case of a one-dimensional hypothesis; for example, in testing $\alpha_1 = \alpha_2$, where, for $k = 2$ in (4.1), α_1 and α_2 are the treatment effects. However, these size-adjusted tests fail to give unique inferences for multidimensional hypotheses; for example, in testing $\alpha_1 = \alpha_2 = \alpha_3$ or $\alpha_1 = \alpha_2 = \alpha_3 = \alpha_4$, where the α 's are the treatment effects.

Brillinger (1980), using the Fourier transform of the data, proposed modified F-statistics as a function of the frequency, λ, for testing the significance of a deterministic signal as a component of the basic linear model with correlated data. Recently, Sutradhar *et al.* (1987) use modified time domain F-statistics for testing treatment/group effects when observations under treatment groups are autocorrelated. The modifications proposed to the F-test both by Brillinger (1980) and Sutradhar *et al.* (1987) account for autocorrelation in the replications. Consider the model (4.1) with

$$z_i(t) = \phi z_i(t-1) + a_i(t),$$

where $a_i(t)$ are i.i.d. $\mathrm{N}(0, \sigma_\alpha^2)$. Then for testing the null hypothesis $H_0 : \alpha_1 = \alpha_2 = \ldots = \alpha_k = 0$, against the alternative $H_A : \alpha_i \neq 0$, for some i, Sutradhar, MacNeill and Sahrmann (1987) proposed the modified F-statistic F^* given by

$$F^* = \frac{dQ_1}{Q_2}, \tag{4.2}$$

where

$$Q_1 = \sum_{i=1}^{k} (\bar{y}_{i\bullet} - \bar{y}_{\bullet\bullet})^2,$$

$$Q_2 = \sum_{i=1}^{k} \sum_{t=1}^{n} (y_i(t) - \bar{y}_{i\bullet})^2,$$

and

$$d = \{kn(n-1)/(k-1)\}\{c_2(\phi)/c_1(\phi)\},$$

with

$$\bar{y}_{i\bullet} = \sum_{t=1}^{n} y_i(t)/n,$$

$$\bar{y}_{\bullet\bullet} = \sum_{i=1}^{k} \sum_{t=1}^{n} y_i(t)/kn,$$

$$c_1(\phi) = \frac{1}{(1-\phi)^2}\left[1 - \frac{2\phi(1-\phi)^n}{n(1-\phi^2)}\right],$$

and

$$c_2(\phi) = \frac{1}{n-1}\left[\frac{n}{1-\phi^2} - c_1(\phi)\right].$$

Since Q_1 and Q_2 are not independent, the distribution of F^* in (4.2) is complicated. Sutradhar *et al.* (1987) conducted a simulation study to obtain the 5% and 1% percentile points of the null distribution of F^*, for $k = 2$ and $n = 75$, 100. It is shown that the distribution of F^* is dramatically affected by ϕ, the AR(1) parameter.

The expressions $(k-1)c_1(\phi)$ and $k(n-1)c_2(\phi)$ may be thought of as the 'degrees of freedom' for Q_1 and Q_2 respectively. These expressions measure the amount of information for testing purposes in the correlated data; when ϕ is large and positive, the amount of information is much less then would be provided by the same number of uncorrelated observations. MacNeill *et al.* (1985) indicated that the effect of the ratio $c_2(\phi)/c_1(\phi)$ on the F-statistic may be approximated for a wide range of sample sizes by $(1-\phi)/(1+\phi)$ for $|\phi| < 0.9$.

Sutradhar and Bartlett (1989) approximate the distribution of a ratio of two general quadratic forms involving central or noncentral variables and definite or non-definite matrices. First, they constructed the moments of the distribution of the ratio of the two quadratic forms by using mixed cumulants of the quadratic forms up to sixth order. Then they approximated the distribution of the ratio of two quadratic forms by a Johnson (1949) curve which has the same first four moments. The approximation has been applied to calculate the percentile points of the modified F-statistic F^* in (4.2) for AR(1) data. This approximation is evaluated by a comparison with the percentile values obtained by simulations in Sutradhar *et al.* (1987). The simulation supports the percentile values of F^* obtained by moment approximation precedure; see also Table II in Sutradhar and Bartlett (1989). Furthermore, Sutradhar and Bartlett have derived a statistic similar to F^* in (4.2) for SARMA $(p,q)(P,Q)_s$ data and obtained 5% values of the distribution for selected θ and Θ for the $(0,1)(0,1)_{12}$ process, where θ and Θ are the parameters of the non-seasonal and seasonal moving average processes respectively.

Several articles have appeared which consider nested designs with correlated errors. Yang and Carter (1983) considered a one-way linear model for nested designs with time series data, 'time' being considered as a non-specific factor. In a slightly different notation, they consider the model

$$y_{ij}(t) = \alpha_i + \beta_{j(i)} + z_{ij}(t), \qquad (4.3)$$
$$i = 1, \ldots, k; j = 1, \ldots, m; t = 1, \ldots, n;$$

where $y_{ij(t)}$ is the observation at time t on the j-th individual under i-th treatment, α_i is the i-th treatment effect, $\beta_{i(i)}$ is the individual effect where the j-th individual is nested under the i-th treatment group, and $z_{ij}(t)$ are the residuals which follow an ARMA (p, q) process. Under the assumption that $\beta_{j(i)} \sim \mathrm{N}(0, \sigma_\beta^2)$ and $a_{ij}(t) \sim \mathrm{N}(0, \sigma_a^2)$ for all i, j, they proposed a test statistic W for testing the hypothesis $H_0 : \alpha_1 = \alpha_2 = \ldots = \alpha_k$ against the alternative $H_1 : H_0$ is not true, where W is given by:

$$W = \frac{\left[\sum_{i=1}^{k} \left(\sum_{j=1}^{m} y'_{ij}\, B\, 1_n \right)^2 - k^{-1} \left(\sum_{i} \sum_{i} y'_{ij}\, B\, 1_n \right)^2 \right] / \{(k-1)c\}}{\sum_{i=1}^{k} \left[\sum_{j=1}^{m} y'_{ij}\, B_{ij} - (kc)^{-1} \left(\sum_{j} y'_{ij}\, B\, 1_n \right)^2 \right] / \{k(mn-1)\}}, \qquad (4.4)$$

with

$$y'_{ij} = (y_{ij1}, \ldots, y_{ijn})',$$
$$\mathbf{1}'_n = (1, \ldots, 1)_{1\times n},$$
$$\gamma_\beta^2 = \sigma_\beta^2/\sigma_a^2,$$
$$B = A^{-1} - (\gamma_\beta^2 A^{-1} J_n A^{-1}) / (1 + \gamma_\beta^2 \mathbf{1}'_n A^{-1}\, \mathbf{1}_n),$$
$$J_n = \mathbf{1}_n \mathbf{1}'_n,$$
$$c = \mathbf{1}'_n B\, 1_n,$$

and A is the correlation matrix for the error vector with elements $z_{ij}(t), t = 1, \ldots, n$. To derive W, they applied well-known full and reduced model methodology based on the hypotheses. They demonstrate that when the time series parameters are known, under H_1, W has the non-central F-distribution with d.f. $k - 1$, $k(mn - 1)$ and non-centrality parameter

$$\delta = \left\{ mc \sum_{i=1}^{k} (\alpha_i - \bar{\alpha})^2 \right\} / \sigma_a^2 ,$$

where $\bar{\alpha} = \Sigma_{i=1}^{k} \alpha_i/k$, and c is as in (4.4). If γ_β^2 is estimated by using the same formula [Equation 2.3 in Yang and Carter (1983)] as in the white noise case, Yang and Carter's test statistic W has the apparent pedagogical virtue of reducing to the white noise case, but $\hat{W}$ then follows the F-distribution with $k-1, k(m-1)$ d.f. In turn, this apparent pedagogical property reveals that W in (4.4) should have $(k - 1), k(m - 1)$ d.f. instead of d.f. $k - 1$, $k(mm-1)$. Thus, the use of their test statistic W for highly correlated time series data is extremely doubtful. Moreover, the distributional discrepancy of their test statistic is quite serious for small m, and often m is so in practice.

Mansour *et al.* (1985) considered the same model (4.3) as did Yang and Carter (1983). They discussed mainly the estimation of the parameters of the model. Maximum likelihood techniques are employed to estimate the variane components $\sigma_\beta^2, \sigma_a^2$, and also ϕ, the parameter of the autoregression. They examined the biases of the m.l.e.'s of $\sigma_\beta^2, \sigma_a^2$ and ϕ through a Monte Carlo study. Their estimation procedure was illustrated with lactation data. Mansour *et al.* (1985) noted their experiences regarding analytical techniques for the AR(1) error model. The difficulties they discuss are likely to be more pronounced in more complex models.

Pantula and Pollock (1985) have studied model (4.3) in the context of radio-telemetry (bobcat telemetry) and of plant growth experiments. They estimated $\alpha = (\alpha_1, \ldots, \alpha_i, \ldots, \alpha_k)'$ by ordinary least square methods and used least squares residuals to estimate ϕ by the method of moments. Next, they estimated the variance components σ_β^2 and σ_a^2 as functions of ϕ. In estimating ϕ by the method of moments, residuals with time lags 1 and 2 were used, which seems arbitrary. For other choices of lags, this ϕ estimate would be different. Consequently, the estimates of variance components will be affected.

Recently, Sutradhar (1990) discussed the joint estimation of the parameters of the model, namely, α, $\sigma_\beta^2, \sigma_a^2$, and the autocorrelation parameter ϕ, by maximizing the exact likelihood function. It is well-known that under the null hypothesis $H_0 : \alpha_1 = \alpha_2 = \ldots = \alpha_k = 0$, the classical F-statistic, for testing treatment effects, still has the usual F-distribution with $(k-1)$ and $k(m-1)$ degrees of freedom, but the power function of the test is affected by the autoregression parameter ϕ. However, it is shown by Sutradhar (1990) that the classical F-test is inappropriate for testing the individual's variation σ_β^2. This is so because the quadratic forms involved in the F-ratio (for the latter test) are correlated as a consequence of the correlated errors. Following Sutradhar *et al.* (1987), Sutradhar discusses a modified F-test for testing the individual's variation.

Much of what has been discussed so far pertains to statistical inference for time series valued experimental designs with time as a specific or non-specific factor. Time series valued designs, however, are employed in several other areas such as signal detection, Bayesian inference, field experiment, biological experiment, optimal designs. We discuss each of these areas in turn and then discuss other papers which do not fit into any of these categories.

5. Signal Detection

Shumway (1971) dealt with the model

$$y_i(t) = s(t) + \eta_i(t),$$

with $E[\eta_i(t)\eta_h(t')] = R_{ih}(t - t')$, for $i = 1, \ldots, k$; $t = 1, \ldots, n$, where $R_{ih}(t - t')$ is the noise correlation at lag $t - t'$. In the model, $s(t)$ denotes the signal of the process at the t-th time point. He transformed the observations into the frequency domain where tests are approximately independent for adjacent frequencies. He then used a simple likelihood ratio approach to test the hypothesis that a common signal is present through all the process. This model is equivalent to the two-way ANOVA model with correlated errors, time being a specific factor.

As was mentioned above, Brillinger (1980) also discussed a model similar to that in Shumway (1971). Analogous to the classical ANOVA, Brillinger uses certain frequency dependent modified F-statistics to test hypotheses regarding the signal of the model.

6. Bayesian Inference

Zellner and Tiao (1964) utilized Bayesian methods to analyze the regression model with errors generated by a first order autoregressive scheme. They employed locally uniform prior distributions for the parameters in the model considered. For a simple regression model, they derived finite-sample joint, conditional and marginal posterior distributions of the parameters. These distributions can be used to make inferences about the parameters and to investigate how departures from independence affect inferences about the parameters.

Tiao and Tan (1966) used Bayesian methods to analyze the one-way random effects model

$$y_{it} = \mu + \alpha_i + z_{it},$$

where $\alpha_i \sim (0, \sigma_a^2)$, for $i = 1, \ldots, k$, and where the errors are assumed to follow a first order autoregressive process, i.e. $z_{it} = \phi z_{i(t-1)} + \epsilon_{it}$, with $\epsilon_{it} \sim (0, \sigma^2)$. They have shown that inferences about the variances σ_a^2 and σ^2 can be very sensitive to changes in the value assumed for ϕ. Then, they used the posterior distributions of ϕ to remove the uncertainty in the inferences about σ_a^2 and σ^2.

Larsen (1969) considered a Bayesian approach to the two-way ANOVA model when the error terms within rows are not independent, but have a covariance matrix dependent on at most two unknown parameters. More specifically, he considered the model

$$y_i(t) = \mu + \alpha_i + \beta_j + z_i(t),$$

$i = 1, \ldots, k; t = 1, \ldots, n$, with $\text{cov}(z_i(t), z_h(t')) = 0$ for $i \neq h$, and $\text{cov}(z_i(t), z_n(t')) = \sigma^2 v_{tt'}$ for $i = k$. For example, for the AR(1) error process, $v_{tt'} = \{\phi^{|t-t'|}\}/(1 - \phi^2)$. That is, the covariance matrix depends on two unknown parameters, σ^2 and ϕ. Let $A = V^{-1}$, where V is the covariance matrix of the error vector. Also let $t_i = \Sigma_{h=1}^{k} \alpha_h/\sqrt{k}$ be a linear contrast of treatment effects. Then Larsen obtained the posterior density of t_i,

$$g(t_i) = \int g_1(t_i; v|y) g_2(\phi)\, d\phi,$$

where

$$g(t_i, v|y) \propto \frac{a_{\bullet\bullet}^{-1/2(k-2)} |A|^{1/2(k-2)}}{\{SS_{res} + a_{\bullet\bullet}(t_i - \hat{t}_i)^2\}^{1/2\{(k-1)(n-1)+1\}}},$$

with $\hat{t}_i = \Sigma_{h=1}^{k} \hat{\alpha}_i/\sqrt{k}$, and $a_{\bullet\bullet} = \Sigma_t \Sigma_{t'} a_{tt'}/n^2$, $a_{tt'}$ being the (t, t')-th element of the matrix A. The above posterior density, $g(t_i)$, can be used for making inferences about the parameters of the model. Inferences based on the posterior density depend on the prior density $g_2(\phi)$. However, Larsen claims that the posterior density is rather robust against changes in the prior density of ϕ, provided that the prior density is rather flat over the possible range of the correlation parameter ϕ.

Ljung and Box (1980) also utilized Bayesian methods to make inferences about the parameters in analysis of variance models with autocorrelated error terms. They compared the row and column effects in a two-way classification model with correlation within rows. Unlike Tiao and Tan (1966), who examined the sensitivity of inferences about the variance components, they examined the sensitivity of inferences about the row and column effects, to assumptions about the correlation. More specifically, they showed that for the linear model

$$y = x\boldsymbol{\beta} + \epsilon,$$

inferences about $\boldsymbol{\beta}$ are very sensitive to assumptions about ϕ, where ϕ is the first order autocorrelation coefficient of the errors. They have shown this by plotting the conditional density $p(\boldsymbol{\beta}|\phi, y)$ of $\boldsymbol{\beta}$ for different values of ϕ. It was shown that the centre of the distribution is relatively insensitive to changes in ϕ, but the spread of the distribution increases considerably as ϕ increases. Thus inferences about β are very sensitive to assumptions about ϕ.

7. Field Experiments

Dubey *et al.* (1977) considered various experimental designs in a field experiment where the errors formed a stationary spatial process. They compared the efficiencies of different designs.

Björnsson (1978) discussed a regression model with correlated errors. The author used GLS regression techniques to analyse treatment contrast vectors. As an example of the application of time series models for residuals, he used hay yields for 18 y from five fertilizer experiments on permanent grassland at the Skriduklaustur experimental station in Iceland.

8. Medical Experiments

Wilson *et al.* (1981) discussed estimation of the parameters of the model:

$$y_t = \mu + \epsilon_i,$$

where $\epsilon_t = \phi\epsilon_{t-1} + \omega_t$, with $\omega_t \sim N(0, \sigma^2_\omega)$. In particular, for estimation of ϕ and σ^2_ω, they compared a variance-based estimation procedure with MLE procedure. The MLE procedure is much more difficult to compute and the algorithm may not necessarily converge. As a consequence, they suggest that in spite of the generally better behaviour of the MLE procedure, the variance-based procedure may be useful. They claim that the variance-based estimator is much easier to compute, does not require iteration, and does not depend on normality. They applied their methodology to a diastolic blood pressure study. The data collection procedure was as follows. Measurements were obtained every day for five days on a group of adults who were not under treatment for hypertension. For any given subject, measurements were obtained at the same time each day, and each subject was studied on five consecutive days of a single week. For each subject on any given day, three measurements were obtained, separated in time by five minutes, and the average of the three was used to represent the measurement for that day.

Taka and Armitage (1983) considered a regression model in which the response depends on a treatment effect, an error term and additionally on the previous response. The error terms have been assumed to be correlated. They estimated the parameters of the model by the likelihood approach and discussed certain hypothesis testing problems. They applied their methodology to parkinsonism data described by Hunter *et al.* (1970).

9. Optimal Designs

Jenkins and Chanmugan (1962) considered the problem of choosing 'best' experimental designs in the simplified situation where it is required to estimate the slope for a single independent variable. Thus, they discussed the determination of optimum block size when the errors constitute (1) a finite moving average process, or (2) an autoregressive process.

Berenblut and Webb (1974) considered design problems in the context of possible autocorrelations. They have suggested a design procedure which minimizes the generalized variance of the estimates of the parameters. More specifically, given that part of the design matrix appropriate to the non-treatment parameters, say X_0, and the alternatives concerning the error covariance matrix, namely, $H_0 : D(y) = \sigma^2 I$ and $H_1 : D(y) = \sigma^2 v(\phi)$, they constructed the matrix of treatment vectors X_1 which, together with X_0, provides optimum information concerning the minimum variance unbiased estimates of the parameters of the model whether the analysis appropriate to H_0 or to H_1 is used.

Sacks and Ylvisaker (1966, 1968) dealt with the problem of determing the points in time at which a given number of observations, subject to a specified stationary correlation structure of errors, should be made so as to satisfy an optimum design criterion. In contrast, Berenblut and Webb (1974) presumed that the observations are to be taken at equally spaced intervals of time throughout, a condition met frequently in practice.

Kiefer (1961) discussed the efficiencies of certain systematic designs.

Gasser (1975) studied the effects of correlation on the null and non-null distributions of the classical chi-square goodness-of-fit statistic. His findings are as follows. With low to moderate correlation, the distribution of χ^2 under H_0 does not change significantly. For processes strongly correlated there is a sharp contrast between high and low frequency patterns. Series with high frequency correlation have small to moderate deviation, whereas low frequency patterns lead to gross deviations from the χ^2-distribution. The powers of the tests are usually smaller than in the independence case. To study power loss, a set of distributions were chosen, namely, uniform, double exponential and χ^2 with various degrees. The effects of serial dependence on χ^2-statistics were also discussed by Brillinger (1974) and by Chanda (1975).

In survey sampling, the effects of serial correlation on systematic sampling have been discussed by Cochran (1946), and Quenouille (1949). Wu *et al.* (1988) discussed a regression model with an error structure that allows for intracluster correlation of the residual errors. They propose a modified F-test for testing the regression coefficient that takes into account the intraclass correlation coefficient. Wu *et al.'s* modification is similar to the modification suggested by Sutradhar *et al.* (1987) in the context of the one-way ANOVA model with correlated errors.

Tavare and Altham (1983) examined the effects of serially dependent data on standard tests, namely, the Pearson goodness-of-fit test and the test of independence in 2×2 contingency tables. They proposed a corrected chi-square statistic which adjusts for correlation induced biases. Barton *et al.* (1962), Goodman (1964), and Klotz (1978) studied the effect of dependence in Markov chains.

10. In Conclusion

The work that has been surveyed above shows that analyses that do not properly account for autocorrelations in time dependent data collected according to an experimental design may be highly misleading. Also, this survey reveals that while much has been achieved in the study of time series valued experimental designs, there are still many lacunae in the area's methodological development. This is particularly true for designs and time series models of a more complicated or specialized nature.

References

Andersen, A. H., Jensen, E. B., E. B., and Schou, G.: 1981, 'Two-Way Analysis of Variance with Correlated Errors', *International Statistical Review* **49**, 153-167.

Andersen, R. L.: 1949, 'The Problem of Autocorrelation in Regression Analysis', *Journal of the American Statistical Association* **44**, 113-129.

Azzalini, A.: 1981, 'Replicated Observations of Low Order Autoregression Time Series', *Journal of Time Series Analysis* **2**, 63-70.

Azzalini, A.: 1984, 'Estimation and Hypothesis Testing for Collections of Autoregressive Time Series', *Biometrika* **71**, 85-90.

Barton, D. E., David, F. N., and Fix, E.: 1962, 'Persistance in a Chain of Multiple Events When There is Simple Dependence', *Biometrika*, **49**, 351-357.

Berenblut, I. A. and Webb, G. J.: (1974), 'Experimental Design in the Presence of Autocorrelated Errors', *Biometrika* **61**, 427-437.

Berndt, E. R. and Savin, N. E.: 1975, 'Conflict Among Criteria for Testing Hypotheses in the Multivariate Linear Regression Model', *Discussion paper 75-21 (revised)*, Department of Economics, University of British Columbia.

Berndt, E. R. and Savin, N. E.: 1977, 'Conflict Among Criteria for Testing Hypotheses in the Multivariate Linear Regression Model', *Econometrika* **45**, 1263-1277.

Björnsson, H.: 1978, 'Analysis of Series of Long-Term Grassland Experiments with Autocorrelated Errors', *Biometrics* **34**, 645-651.

Box, G. E. P.: 1954, 'Some Theorems on Quadratic Forms Applied in the Study of Analysis of Variance Problems, II. Effect of Inequality of Variance and Correlation Between Errors in Two-Way Classification', *Annals of Mathematical Statistics* **25**, 484-498. Brillinger, D. R.: 1973, 'The Analysis of Time Series Collected in an Experimental Design', in P. R. Krishnaiah (ed.), *Multivariate Analysis III*, Academic Press, New York, pp. 241-256.

Brillinger, D. R.: 1974, 'The Asymptotic Distribution of the Whittaker Periodogram and a Related Chi-square Statistic for Stationary Processes', *Biometrika* **61**, 419-422.

Brillinger, D. R.: 1980, 'Analysis of Variance and Problems Under Time Series Models', in P. R. Krishnaiah (ed.), *Handbook of Statistics*, Vol. 1, North Holland Publ. Co., pp. 237-278.

Chanda, K. C.: 1975, 'Chi-square Goodness-of-fit Tests for Strong Mixing Stationary Processes', *Report ARL TR 75-0016 Aerospace Res. Labs*, Wright-Patterson A.F.B.

Cochran, W. G.: 1946, 'Relative Accuracy of Systematic and Stratified Random Samples for a Certain Class of Populations', *Annals of Mathematical Statistics* **17**, 164-177.

Cochrane, D. and Oreutt, G. H.: 1949, 'Application of Least Squares Regression to Relationships Containing Autocorrelated Error Terms', *Journal of the American Statistical Association* **44**, 32-61.

Devore, J. L.: 1976, 'A Note on the Estimation of Parameters in a Bernoulli Model with Dependence', *Annals of Statistics* **4**, 990-992.

Duby, C., Guyon, X. and Prum, B.: 1977, 'The Precision of Different Experimental Designs for a Random Field', *Biometrika* **64**, 59-66.

Gallant, A. R., Gerig, T. M. and Evans, J. W.: 1974, 'Time Series Realizations Obtained According to an Experimental Design', *Journal of the American Statistical Association* **69**, 639-645.

Gasser, T.: 1975, 'Goodness-of-fit Tests for Correlated Data', *Biometrika* **62**, 563-576.

Goodman, L. A.: 1964, 'The Analysis of Persistence in a Chain of Multiple Events', *Biometrika* **51**, 405-411.

Hunter, K. R., Stern, G. M., Laurence, D. R. and Armitage, P.: 1970, 'Amanatadine in Parkinsonism', *Lancet* **1**, 1127-1129.

Jenkins, G. M. and Chanmugan, J.: 1962, 'The Estimation of Slope When the Errors Are Autocorrelated', *Journal of the Royal Statistical Society, Series B* **24**, 199-214.

Johnson, N. L.: 1949, 'Systems of Frequency Curves Generated by the Method of Translation', *Biometrika* **36**, 149-176.

Kiefer, J.: 1961, 'Optimum Experimental Designs V, with Applications to Systematic and Rotable Designs', in *Proceedings of the Fourth Berkeley Symposium on Mathematical Statistics and Probability*, Vol. 1, California, Berkeley, pp 381-405.

Klotz, J.: 1972, 'Markov Chain Clustering of Births by Sex', *Proceedings of the Sixth Berkeley Symposium on Mathematical Statistics and Probability* **4**, 173-185.

Klotz, J.: 1973, 'Statistical Inference in Bernoulli Trials with Dependence', *Annals of Statistics* **1**, 373-379.

Ladd, D. W.: 1975, 'An Algorithm for the Binomial Distribution with Dependent Trials', *Journal of the American Statistical Association* **70**, 333-340.

Larsen, W. A.: 1969, 'The Analysis of Variance for the Two-Way Classification Fixed-Effects Model with Observations within a Row Serially Correlated', *Biometrika* **56**, 509-515.

Lindquist, B.: 1978, 'A Note on Bernoulli Trials with Dependence', *Scandinavian Journal of Statistics* **5**, 205-208.

Ljung, G. M. and Box, G. E. P.: 1980, 'Analysis of Variance with Autocorrelated Observations', *Scandinavian Journal of Statistics* **7**, 172-180.

MacNeill, I. B., Sahrmann, H. F. and Sutrudhar, B. C.: 1985, 'Time Series Valued Experimental Design', in *Conference volume for the 45th Session of the International Statistical Institute*, pp. 469-470.

Mansour, H., Nordheim, E. V. and Rutlegde, J. J.: 1965, 'Maximum Likelihood Estimation of Variance Components in Repeated Measures Designs Assuming Autoregressive Errors', *Biometrika* **41**, 287-294.

Pantula, S. G. and Pollock, K. H.: 1985, 'Nested Analysis of Variance with Autocorrelated Errors', *Biometrics* **41**, 909-920.

Quenouille, M. H.: 1949, 'On a Method of Trend Elimination', *Biometrika* **36**, 75-91.

Rotenburg, T. J.: 1984, 'Hypothesis Testing in Linear Models When the Error Covariance Matrix is

Nonscalar', *Econometrika* **52**, 827-842.

Sacks, J. and Ylvisaker, D.: 1966, 'Designs for Regression Problems with Correlated Errors', *Annals of Mathematical Statistics* **37**, 66-89.

Sacks, J. and Ylvisaker, D.: 1968, 'Designs for Regression Problems with Correlated Errors; Many Parameters, *Annals of Mathematical Statistics* **39**, 49-69.

Shumway, R. H.: 1970, 'Applied Regression and Analysis of Variance for Stationary Time Series', *Journal of the American Statistical Association* **65**, 1527-1546.

Shumway, R. H.: 1971, 'On Detecting a Signal in N Stationarily Correlated Noise Series', *Technometrics* **13**, 499-519.

Sutradhar, B. C.: 1990, 'Exact Maximum Likelihood Estimation for the Mixed Analysis of Variance Model with Autocorrelated Errors', *The Statistican* **39**.

Sutradhar, B. C. and Bartlett, R. F.: 1989, 'An Approximation to the Distribution of the Ratio of Two General Quadratic Forms with Applications to the Time Series Valued Designs', *Communications in Statistics, Theory and Methods* **18**, No. 4.

Sutradhar, B. C. and MacNeill, I. B.: 1989, 'Two-Way Analysis of Variance with Stationary Periodic Time Series', *International Statistical Review* (to appear).

Sutradhar, B. C., MacNeill, I. B. and Sahrmann, H. F.: 1987, 'Time Series Valued Experimental Designs: One-way Analysis of Variance with Autocorrelated Errors', in I. B. MacNeill and G. J. Umphrey (eds.), Kluwer Acad. Publ., *Time Series and Econometric Modelling,* Dordrecht: Holland, pp. 113-129.

Taka, M. T. and Armitage, P.: 1983, 'Autoregressive Models in Clinical Trials', *Communications in Statistics, Theory and Methods* **12**, 865-876.

Tavare, S. and Altham, M. E.: 1983, 'Serial Dependence of Observations Leading to Contingency Tables, and Corrections to Chi-Squared Statistics', *Biometrika* **70**, 139-144.

Tiao, G. C. and Tan, W. Y.: 1966, 'Bayesian Analysis of Random-Effects Models in the Analysis of Variance II. Effect of Autocorrelated Errors', *Biometrika* **53**, 477-495.

Welch, B. L.: 1937, 'The Significance of the Difference Between two means when the Population Variances are Unequal', *Biometrika* **29**, 350-362.

Welch, B. L.: 1947, 'The Generalization of 'Student's' Problem when Several Different Population Variances Are Involved', Biometrika **34**, 28-35.

Williams, R. M.: 1952, 'Experimental Designs for Serially Correlated Observations', *Biometrika* **39**, 151-167.

Wilson, P., Hebel, J. R. and Sherwin, R.: 1981, 'Screening and Diagnosis When Within-Individual Observations Are Markov-Dependent', *Biometrics* **37**, 553-565.

Wold, H. O. A.: 1949, 'On Least Squares Rgression with Autocorrelated Variables and Residuals', *Proceedings of the International Institute of Statistics* 277-289.

Wu, C. F. J., Holt, D. and Holmes, D. J.: 1988, 'The Effect of Two-stage Sampling on the *F*-Statistic', *Journal of the American Statistical Association* **83**, 150-159.

Yang, M. C. K. and Carter, R. L.: 1983, 'One-Way Analysis of Variance with Time-Series Data', *Biometrics* **39**, 747-751.

Zellner, A. and Tiao, G. C.: 1964, 'Bayesian Analysis of the Regression Model with Autocorrelated Errors', *Journal of the American Statistical Association* **59**, 763-778.

EXTENT OF ACIDIFICATION IN SOUTHWESTERN QUÉBEC LAKES

J. DUPONT

Ministère de l'Environnement du Québec, 3900, Marly, Ste-Foy (Québec), Canada, G1X 4E4

(Received February 1990)

Abstract. This paper presents the results from the first two areas covered by the statistically-oriented Québec Spatial Lake Acidity Monitoring Network (RESSALQ). It is used in combination with the existing LRTAP-Québec temporal network, the Québec precipitation sampling network (REPQ) and a dose-effect model (SIGMA/SLAM) in order to assess the global extent of damages related to acidity, to detect changes in water quality, and to measure the effects of wet sulphate deposition reduction and those of sulphate target loadings on lake acidity. Results obtained with this network were also used in combination with data issued from the Eastern Lake Survey, in order to establish the relative acidity and sensitivity status of lakes in Eastern North America. While Florida has the highest proportion of very acidic and very sensitive lakes, Québec has a higher overall proportion of sensitive and acidic lakes compared to other areas of the United States.

Introduction

The quantification of the water quality status of a group of lakes and its statistical extrapolation to all the lakes within a given area is a relatively new approach in acid precipitation studies. Since 1980, several studies have been conducted in order to assess the effects of long range transport of airborne pollutants (LRTAP) on Québec lake ecosystems. Before 1986, these studies were mostly performed on calibrated watersheds, temporal lake water quality networks and regional lake surveys. In fact, very few monitoring projects in the past have used a statistically-oriented approach for regional surveys. The best known project of this kind is perhaps the National Surface Water Survey initiated in 1984 by the Environmental Protection Agency (EPA). The first step of this monitoring project was to acquire representative lake water quality samples from major sensitive areas in the United States. The Eastern Lake Survey (ELS) was conducted in 1985 on lakes with an area over four hectares, while the Western Lake Survey (WLS) was initiated in 1986 on lakes with an area greater than one hectare.

The ELS has allowed scientists to extrapolate the acidity level of more than 1600 sensitive lakes, located in three major sectors of the Eastern United States (Northeast, Southeast and Upper Midwest), to a target lake population of 179 53 (Linthurst *et al.*, 1986). This exercise has generated, for each region, the number and percentage of lakes belonging to different levels of variables such as lake pH. This spatial monitoring network was also designed to obtain comparative surface water quality patterns for these regions (Landers *et al.*, 1988) and their sub-regions (Brakke *et al.*, 1988; Eilers *et al.*, 1988a, b).

Other spatial monitoring projects conducted around the world were also based on regional surveys such as the Adirondack Lake Surveys (ALS), conducted on 1247 lakes

Environmental Monitoring and Assessment **17**: 181–199, 1991.

between 1984 and 1986 (Quinn and Simonin, 1987). Similar surveys were also conducted in Scandinavia (Henriksen *et al.,* 1988). Other spatial lake surveys were conducted in the United States (Glass and Loucks, 1986) and in Canada (Jeffries *et al.,* 1986; Kelso *et al.* 1986; Dupont and Grimard, 1989). These lake inventories however, were not all statistically oriented in order to generate non-biased estimates of the actual lake acidity status.

The obtention of a reliable image of a population, based on a given sample, is submitted to very strict constraints. In fact, the objectives, the method of sampling and the type of data treatment have to be defined before the onset of a lake survey. It is very important to resort to the Sampling Theory in order to establish the optimal sampling method. In this paper, results from the statistically oriented Québec Spatial Lake Acidity Monitoring Network (RESSALQ) are presented.

Our main goal is to quantify the extent of damages caused by acid deposition to lakes in several regions of concern located on the Canadian Shield. We will discuss the results observed in the first two hydrographic regions, Outaouais and Mauricie, covered by the RESSALQ network and compare them with results obtained from the ELS survey for the Northeast, the Upper-Midwest and the Florida sub-regions. The two Québec areas are part of an overall territory covering five hydrographic regions, where wet sulphate depositions often exceed the 20 kg ha^{-1} yr^{-1} target loading proposed by the United States – Canada Work Group (United States-Canada, 1983) in order to protect moderately sensitive ecosystems.

Sampling Method

The Simple Random Sampling method was chosen for the lake selection in the RESSALQ network. According to Frontier (1983), the purpose of this method is to randomly and independently choose a '*n*' unit sample from a '*N*' unit population. Each element taken from the population has the same probability of being chosen, while each '*n*' unit sample has the same probability of being constituted. This approach has the advantage of being very simple to use. The only limitation lies in the need to establish a complete list of all the elements which constitute the target population.

We have assumed that one sample per lake was enough to chemically characterize a given lake. It is evident that sampling a lake in a given moment of the day, in a given season of the year, and in a given year, cannot fully describe the complex dynamics of the lake (Landers *et al.,* 1988). However, a single water sample per lake can be justified if we want to obtain an index of the lake characteristics, which is actually the case.

Lake sampling was performed in winter. This period was chosen for several reasons. The between-lake variability is at its lowest, and the hydrological and hydrobiological conditions are mostly stable. Lake selection was performed according to certain criteria. First of all, only the sensitive areas were considered as part of the RESSALQ network in order to detect potentially acidic lakes. Rock and soil sensitivity maps (Shilts *et al.,* 1981) were used to locate these areas. Most of the territory is considered to be sensitive with the exception of the areas near Maniwaki-Hull, the Abitibi low-lands and the St. Lawrence low-lands. Consideration of these non-sensitive lakes in the network would have increased the sample variance and would have decreased estimate precision.

Among other criteria, we have considered lakes with an area ranging from 10 to 2000 ha (0.1-20 km^2). Logistical and technical reasons ruled out the selection of smaller lakes because they tend to be very shallow. This fact alone implies that a great number of these small lakes would have been unable to meet the minimum depth requirement of 3 m. On the other hand, large water bodies (> 2000 ha) were also excluded because most of these are reservoirs which are difficult to characterize chemically from only one sample. In the same way, running waters, bog lakes and lakes influenced by local pollution (agriculture, mining operations and urban activities) were not considered.

Lake Sampling and Laboratory Analysis

Lake sampling was performed by helicopter in both areas. An integrated sample (0-5 m, or 2 m from the lake bottom) was collected under ice-cover at the center of each lake. Water samples were collected in four separate bottles: a 125 mL polyethylene bottle for nutrient analysis (NO_3, NH_4) with addition of 0.5 mL H_2SO_4 8N preservative, a 125 mL polyethylene bottle for cation analysis (Ca, Mg, Na, K, Fe, Mn) with 0.5 mL HNO_3 50:50 preservative, a fully filled 125 mL polyethylene bottle for the analysis of dissolved organic carbon (DOC) and dissolved inorganic carbon (DIC), and a 1000 mL polyethylene bottle for the analysis of other variables: conductivity, alkalinity, true color, pH, SO_4, Cl, strong acids (H^+), and filterable Al. Table I presents the analytical methods used for each of these variables. In the field, the samples were placed in an icebox, covered with snow and sent to the laboratory for analysis. pH, alkalinity, sulphate and true color were analyzed less than 48 hr after sampling.

A quality control plan was conducted to verify data variability due to sampling. This was done by use of sample triplicates. However, results have shown no significant difference between samples. A quality assurance plan was also conducted at our laboratory as part of the Federal inter-laboratory comparisons. Finally, statistical validation of the data was performed in order to detect abnormal values and transcription errors.

Statistical Framework

The RESSALQ network can be described as a two-step operation. The first step represents a 5-yr survey of the five southernmost hydrographic regions on the Canadian Shield: Outaouais, Mauricie, Saguenay/Lac-Saint-Jean, Côte-Nord and Abitibi. Each of these regions will be visited once between 1986 and 1990, during the month of March (Figure 1). In 1990, we will have an overall image of lake water quality and acidity in the sensitive areas of Southern Québec. A second step will take place between 1991 and 1995. The lakes sampled between 1986 and 1990 will be visited once again in order to detect a change (not a trend) in lake water quality. Figure 1 presents the location of the hydrographic regions and the limits of the area covered by the RESSALQ network. The

TABLE I

Analytical procedures used for the laboratory determination of chemical variables

Variable	Unit	Analytical method	Limit of detection
Lake water samples			
pH	unit	Electrometry	0.1
Strong acids	μeq L^{-1}	conductivimetric titration with sodium hydroxide	2
Alkalinity	mg L^{-1}	conductivimetric titration with nitric acid	0.1
SO_4	mg L^{-1}	Spectrophotometry	0.5
True color	Hazen	Spectrophotometry	0.1
Conductivity	μS cm^{-1}	Electrometry	0.1
$NO_3 + NO_2$	mg L^{-1}	Spectrophotometry	0.02
NH_4	mg L^{-1}	Spectrophotometry	0.02
Ca	mg L^{-1}	Plasma	0.1
Mg	mg L^{-1}	Plasma	0.1
Na	mg L^{-1}	Plasma	0.1
K	mg L^{-1}	Plasma	0.1
Cl	mg L^{-1}	conductivimetric titration with mercury nitrate	0.1
Fe	mg L^{-1}	Plasma	0.01
Mn	mg L^{-1}	Plasma	0.01
Filterable-Al	mg L^{-1}	Plasma	0.01
DIC	mg L^{-1}	UV irradiation and conductivimetry	0.1
DOC	mg L^{-1}	UV irradiation and conductivimetry	0.01
Precipitation water samples			
SO_4	μeq L^{-1}	Ion exchange resins with conductivimetric detection	2
Precipitation	mm	Precipitation collector	1

regions discussed in this article are shaded. This figure also presents the mean annual sulphate concentrations in wet deposition for 1986 (Jacques and Boulet, 1988).

The flow chart presented in Figure 2 shows the procedures used with the RESSALQ network. The Outaouais case study is shown as an example to illustrate the steps followed when conducting a study in a given area. The first steps consist of locating the extent of the sensitive areas and listing all lakes meeting the criteria (target lakes). From this target population (N_t), a sample (n) is randomly drawn using a computer.

The number of lakes to be sampled in each region is dependent of a precision to be maintained around the proportions and the mean. As for proportions, the number (n) is only dependent on the sample size, the desired confidence limit and the standardized value of the normal distribution for a given significance level (Cochran, 1977). This computation is calculated differently for the mean-based statistics, because n is dependent on the population mean and standard deviation, the standardized value of the normal distribution and a desired confidence limit. The standard deviation and the mean were computed from a 650 Outaouais lake dataset and a 500 Mauricie lake dataset where all of the lakes were sampled prior to 1986 (Dupont, 1988a, 1989). We have assumed that these data were reliable enough to give us an estimate of the water quality variability within each region of interest. The optimal number of lakes to be sampled was set in order to minimize

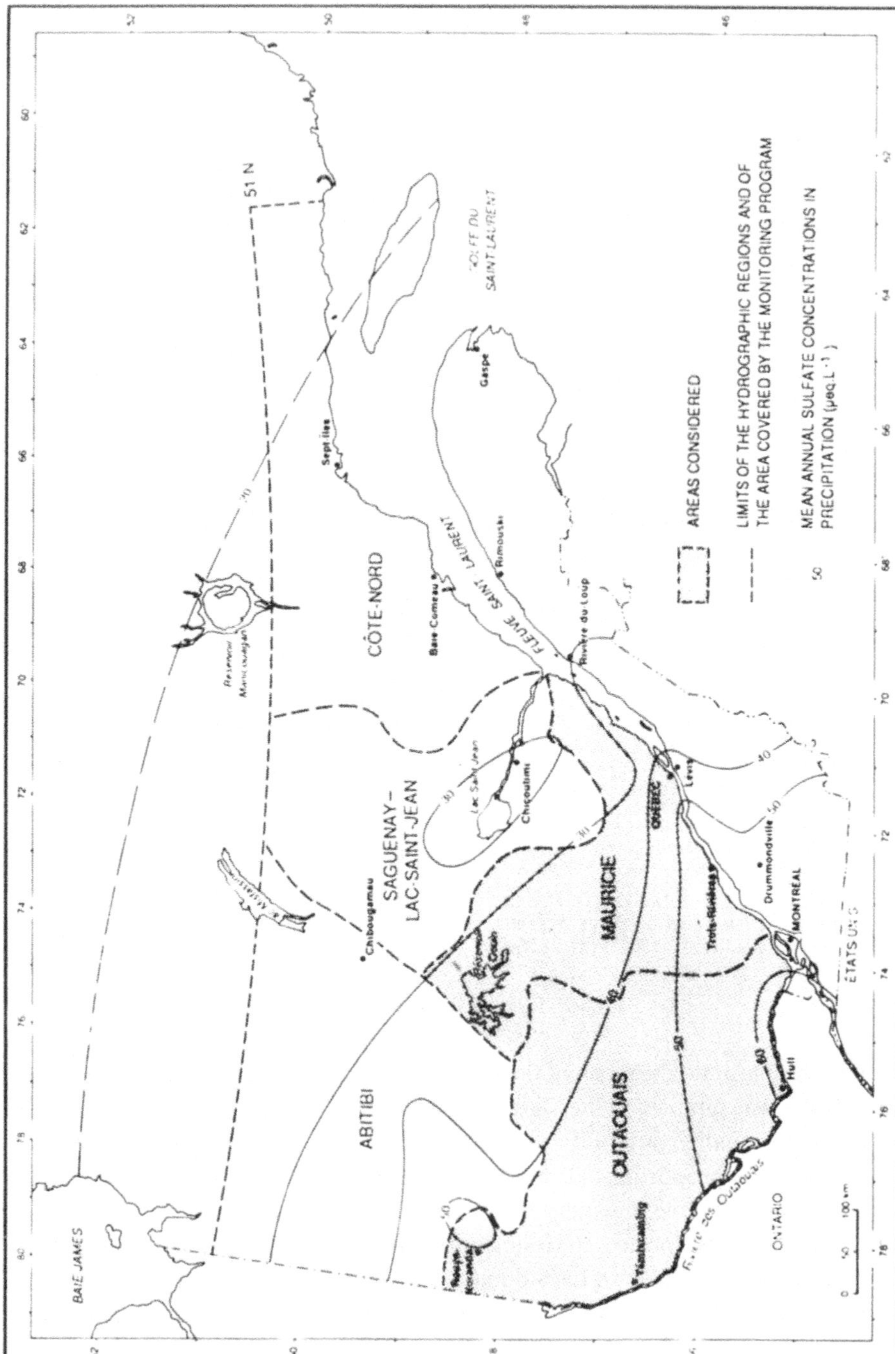

Fig. 1. Regions covered by the Québec Spatial Lake Acidity Monitoring Network (RESSALQ).

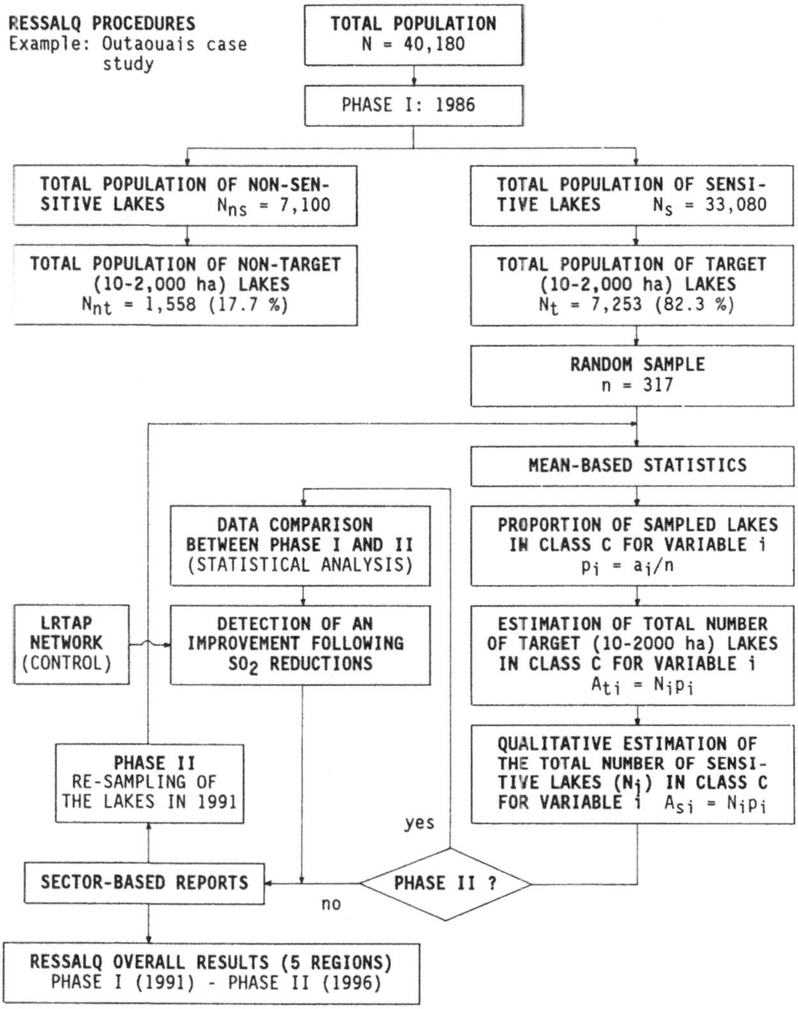

Fig. 2. RESSALQ procedures.

the absolute error around the mean of the most variable element, alkalinity. The greatest number of lakes was sampled in the Outaouais watershed because this is the area where the greatest water quality variability was found. A maximum relative error of 6% was achieved around the proportions. Table II presents the absolute error of 16 variables, along with other descriptive statistics.

Once analyses were performed, the following mean-based statistics were computed for each variable: mean ($\bar{x}$), standard deviation (s), standard error of the mean ($s_{\bar{x}}$), relative error the mean ($\epsilon_{\bar{x}}$), absolute error around the mean ($L_{\bar{x}}$) and the confidence intervals around the mean.

The same procedure was followed for the proportion-based statistics. According to the

TABLE II

Mean, standard deviation, minimum, maximum and absolute error of 16 chemical variables analyzed for the Mauricie and the Outaouais regions

	Mean	Standard deviation	Minimum	Maximum	Absolute Error
Mauricie					
pH (units)	6.0	0.4	4.8	7.1	0.1
Alkalinity (mg L^{-1})	3.6	3.4	0.1	23.4	0.4
True color (Hazen)	32.2	19.6	1.0	100.0	2.3
SO_4^{2-} (mg L^{-1})	3.7	1.4	1.5	10.0	0.2
$NO_3^- + NO_2^-$ (mg L^{-1})	0.11	0.10	0.02	0.70	0.01
NH_4^+ (mg L^{-1})	0.03	0.03	0.02	0.30	0.01
Cl^- (mg L^{-1})	0.8	2.7	0.2	44.0	0.3
Conductivity (μS cm^{-1})	22.4	13.5	10.4	202.0	1.6
DOC (mg L^{-1})	5.3	2.6	0.1	16.4	0.3
Ca^{2+} (mg L^{-1})	2.2	1.3	0.2	11.4	0.2
Mg^{2+} (mg L^{-1})	0.6	0.2	0.2	1.6	0.1
Na^+ (mg L^{-1})	0.9	1.4	0.2	23.4	0.2
K^+ (mg L^{-1})	0.4	0.2	0.1	1.1	0.1
Fe^{2+} (mg L^{-1})	0.31	0.24	0.01	1.86	0.03
Mn^{2+} (mg L^{-1})	0.02	0.02	0.01	0.24	0.01
Filterable-Al (mg L^{-1})	0.10	0.06	0.01	0.35	0.01
Outaouais					
pH (units)	5.9	0.5	4.2	7.4	0.1
Alkalinity (mg L^{-1})	4.0	5.4	0.1	49.9	0.6
True color (Hazen)	29.4	23.7	1.0	125.0	2.6
SO_4^{2-} (mg L^{-1})	5.6	1.5	2.0	10.0	0.2
$NO_3^- + NO_2^-$ (mg L^{-1})	0.09	0.07	0.01	0.39	0.01
NH_4^+ (mg L^{-1})	0.06	0.08	0.01	0.66	0.01
Cl^- (mg L^{-1})	0.6	0.6	0.2	9.0	0.1
Conductivity (μS cm^{-1})	28.7	12.3	14.0	115.0	1.3
DOC (mg L^{-1})	5.8	2.8	0.2	17.2	0.3
Ca^{2+} (mg L^{-1})	2.8	2.0	0.8	20.6	0.2
Mg^{2+} (mg L^{-1})	0.7	0.3	0.3	2.4	0.1
Na^+ (mg L^{-1})	0.8	0.4	0.3	5.4	0.1
K^+ (mg L^{-1})	0.5	0.2	0.1	1.4	0.1
Fe^{2+} (mg L^{-1})	0.25	0.23	0.01	1.54	0.02
Mn^{2+} (mg L^{-1})	0.03	0.03	0.01	0.34	0.01
Filterable-Al (mg L^{-1})	0.09	0.08	0.01	0.43	0.01

Sampling Theory, the study of proportions implies that each constituting element of a given population is part of either two classes C and C' (Cochran, 1977). The notation tied to this classification can be read as follows:

A_{ti} = number of elements from population in class C;
a_i = number of elements from sample in class C;
$P_i = A_{ti}/N_t$ = proportion of elements from population in class C;
$p_i = a_i/n$ = proportion of elements from sample in class C.

In this study, the previous notation applies to a target population N_t and a given variable i.

The value P_i is the estimator of P_i which is unknown, while $N_t p_i = \tilde{A}_{ti}$ is the estimator of A_{ti}. The parameter p_i follows a binomial distribution. However, when np_i is great enough, we can assume that p_i approximately follows a normal distribution. If p_i is lower or higher than a given threshold value (0.15/0.85 for the Outaouais region and 0.16/0.84 for the Mauricie region), then it is necessary to consider the binomial distribution (or even the Poisson distribution if the proportion is almost 0 or 1) in order to compute the confidence intervals. Taking into account these statistical considerations, we have computed, for each physical-chemical variable at a 1 - α significance level, the proportion (p_i), the standard deviation of p_i (s_{pi}), the total number of elements from the population in class C ($\tilde{A}_{ti}$, the confidence intervals around p_i and $\tilde{A}_{ti}$, the relative error ($\epsilon\tilde{A}_i$), and the absolute error ($L_{\tilde{A}i}$). When the binomial distribution applies, s_{pi}, $\epsilon_{\tilde{A}c}$, $L_{\tilde{A}c}$, and the confidence limits become asymmetric. In such a situation, we have referred to the Rohlf and Sokal (1969, 208) table. Mathematical formulation of the equations is presented in Dupont (1988a, 1989).

According to Figure 2, these computations will have to be calculated again after the second sampling. The resulting mean-based and proportion-based statistics will then be compared with those from the first survey in order to detect a potential change in lake acidity and water quality following scheduled reduction of SO_2 emissions. Statistical tools like the Student t-test and the analysis of variance ANOVA) will be used in order to test if there is a difference between Phase I and Phase II data. The LRTAP-Québec temporal network (Haemmerli, 1987) will be used as a control for the hydrometeorological variability.

It is possible to compare ELS and RESSALQ because both surveys were constructed around a statistical framework. However, some differences exist between them. The target lakes in the ELS survey have an area ranging from 4 to 2000 ha, which is a little wider than our range of 10 to 2000 ha, thus providing in theory a greater estimate of acidic or sensitive lakes than what would be observed with our range of lake areas. Another major difference existing between the networks is the fact that our lake sampling was performed in winter, while the ELS was sampled in the fall. The reasons for our choice were described earlier in this article. A final difference comes from the fact that our ANC measurements cannot present negative values. To solve this problem, we used the strong acid concentrations as a surrogate for negative alkalinity. These strong acids are measured when alkalinity is less than 0.1 mg L^{-1}.

Results

Table II presents the descriptive statistics ($\bar{x}$, s, minimum, maximum, and absolute error) concerning each of 16 physical-chemical variables analyzed at our laboratory. According to this table, Mauricie and Outaouais lakes appear to exhibit similarities in water quality. For example, mean values of pH, alkalinity, color, NO_3, DOC, Mg, Na, K, Fe, Mn and Al are similar. In most cases, this similarity also exists for the standard deviation, the minimum, the maximum and the absolute error. Such a similarity can imply that the physiographical factors are relatively homogeneous between both regions. However, the water quality similarity is not absolute. A greater mean value of SO_4, NH_4, conductivity and Ca is observed for the Outaouais lakes. The standard deviations for NH_4, alkalinity and Ca are also higher in this area. On the other hand, Mauricie lakes present a higher standard deviation for Cl and Na, which can be explained by the leaching of road salts into some lakes.

Figure 3 presents the pH measured on the 317 Outaouais lakes and the 266 Mauricie lakes according to three classes of acidity: ≤ 5.5 (acidic), 5.5 to 6.0 (transition) and > 6.0 units (non-acidic). On the same figure, we present the main patterns of lake acidity. The darker shade represents the areas where the lakes are the most acidic. Within these areas, 40 to 67% of the lakes are acidic, while 85 to 100% of the lakes show a pH ≤ 6.0. These areas are situated just southeast of Rouyn-Noranda, north of Québec City and in some sectors located to the north of both watersheds. The next class represents the areas where most lakes have a pH ranging from 5.5 to 6.0. There are also acidic (13%) and non-acidic lakes in these areas (18%). This pH class covers all of the northern portion of both watersheds where the altitude is high. The clearer shade represents the areas where most of the lakes are non-acidic. Some transition lakes are also present in these areas. The non-acidic lakes are mostly found on the southern border of the Canadian Shield, along the Outaouais and St. Lawrence Rivers. Finally, the unshaded areas are sectors known to be mostly non-sensitive.

Figure 4 presents lake alkalinity values according to three classes: ≤ 2.0, 2.0 to 5.0 and $>$ 5.0 mg L^{-1} of $CaCO_3$. This figure also presents the pattern observed in lake sensitivity based on the total alkalinity. The darkest shade indicates areas of extreme sensitivity where most lakes show values less than 2.0 mg L^{-1}. These areas are mostly found in the upper parts of the Outaouais and Mauricie regions or in high altitude locations. The next shade represents areas where lake sensitivity is high. In this class, most lakes have an alkalinity ≤ 5.0 mg L^{-1}. The extent of this sensitivity class is a little wider than the extreme sensitivity class, but its location remains mostly to the north. The clearer shade represents the areas where the sensitivity ranges from intermediate to low. These areas are mostly observed along the St. Lawrence River or near known non-sensitive areas (no shade).

Table III presents the lake proportions associated to classes of nine variables for the two regions of interest. This table also presents the relative error and the estimated number of 10-2000 ha sensitive lakes in each variable class. The estimates were obtained by extrapolating the water quality data measured on the 266 Mauricie lakes and the 317 Outaouais lakes to the 5667 and 7253 respective target lakes located in these two hydrographic regions.

According to Table III, the lake proportions computed for each hydrographic region differ at several levels. For example, the Outaouais region has a greater proportion of lakes with pH lower than or equal to 5.0, 5.5 and 6.0. The lake proportions are also greater for the Outaouais region in the lower classes of alkalinity and color, the upper classes of NO_3 and Al, and all classes of SO_4 and HCO_3/SO_4 ratio. On the other hand, the Mauricie region presents higher lake proportions in the upper classes of alkalinity and DOC, and all classes of Ca + Mg. The lake proportions in the other variable classes are quite similar.

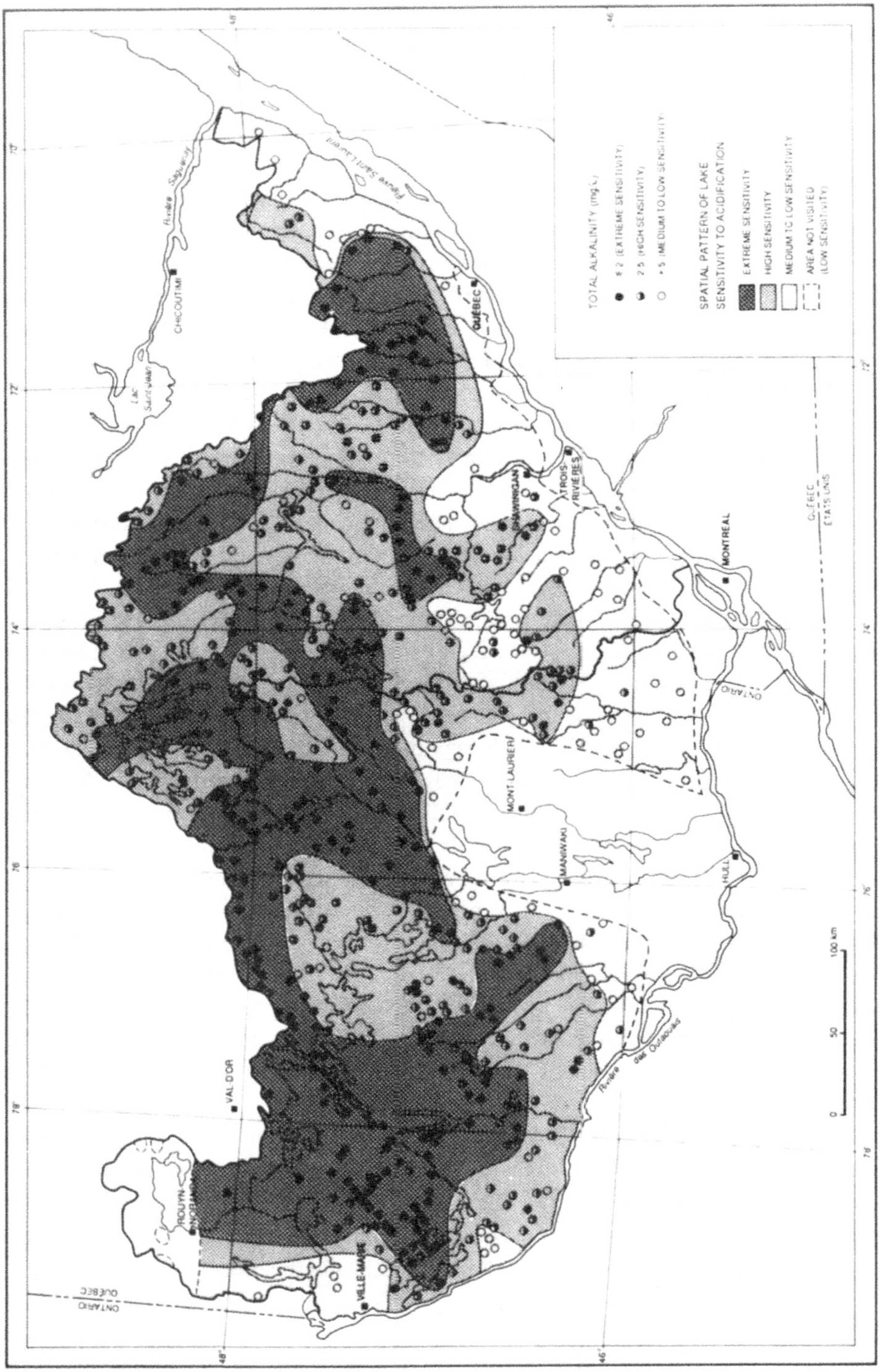

Fig. 3. pH variability within the Mauricie and the Outaouais hydrographic regions.

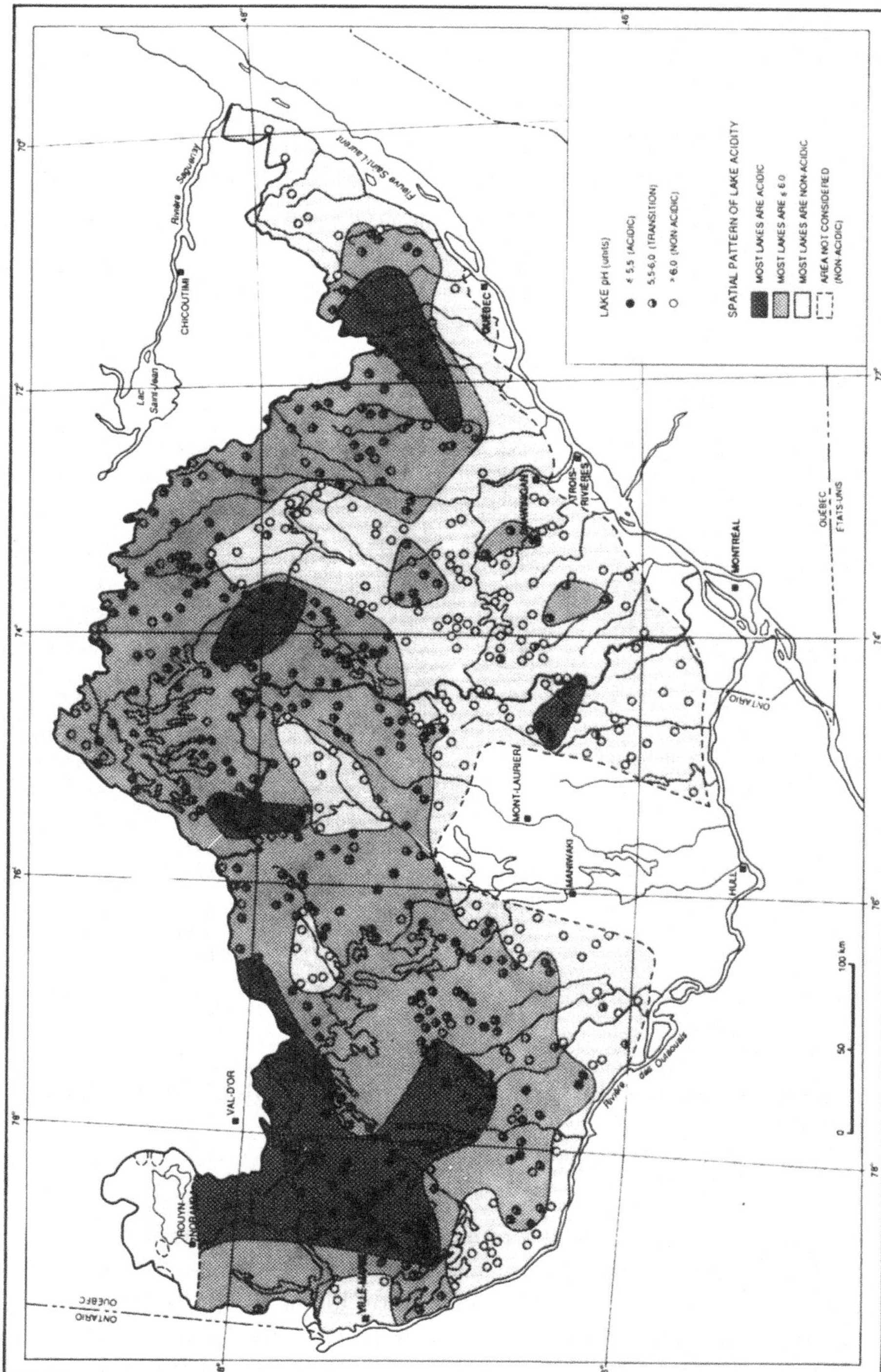

Fig. 4. Alkalinity and sensitivity variability within the Mauricie and the Outaouais hydrographic regions.

TABLE III

Proportions and absolute estimates of the number of lakes associated to classes of nine variables for the Mauricie and the Outaouais hydrographic regions

		Mauricie Region			Outaouais Region		
Variable	Class	%	Number	Error (%)	%	Number	Error (%)
pH (units)	$\leq$ 5.0	2.6	149	1.2-3.1*	4.1	297	1.9-2.5
	$\leq$ 5.5	11.7	660	3.2-4.8	23.3	1690	± 4.6
	$\leq$ 6.0	58.3	3302	± 6.0	62.5	4533	± 5.2
	$\leq$ 6.5	94.0	5326	3.5-2.5	92.7	6723	3.0-2.8
Alkalinity (mg L^{-1})	$\leq$ 0.1	0.8	43	0.8-2.1	0.9	65	0.7-1.9
	$\leq$ 2.0	33.8	1917	± 5.7	43.2	3133	± 5.4
	$\leq$ 5.0	80.1	4538	± 4.9	81.1	5882	± 4.2
	$\leq$ 10.0	95.1	5390	3.4-2.1	92.1	6680	4.7-3.2
SO_4^{2-} (mg L^{-1})	> 3.0	57.9	2152	± 6.0	94.6	6864	2.6-2.4
	> 4.0	29.7	852	± 5.6	80.8	5857	2.5-4.3
	> 6.0	4.5	320	3.8-1.7	30.9	2242	± 5.4
True color (Hazen units)	$\leq$ 10	11.6	660	3.2-4.8	23.7	1719	± 4.6
	$\leq$ 30	53.0	3004	± 6.1	60.3	4374	± 5.4
	$\leq$ 50	82.3	4666	± 4.7	84.2	6107	± 3.9
$Ca^{2+} + Mg^{2+}$ (μeq L^{-1})	$\leq$ 75	4.9	277	2.1-3.4	0.6	46	0.8-1.8
	$\leq$ 100	15.0	852	4.0-4.7	3.2	229	1.6-2.3
	$\leq$ 150	58.3	3302	± 6.0	33.1	2402	± 5.2
	$\leq$ 200	82.7	4687	± 4.6	70.7	5125	± 5.1
DOC (mg L^{-1})	$\leq$ 4.0	37.2	2103	± 5.9	32.2	2334	± 5.2
	$\leq$ 6.0	66.9	3782	± 5.7	63.7	4622	± 5.3
	$\leq$ 8.0	85.3	4836	5.1-3.6	84.2	6109	± 4.1
$[HCO_3]/[SO_4]$	$\leq$ 0.2	3.0	170	1.6-2.7	9.5	689	1.6-2.7
	$\leq$ 0.5	25.6	1449	± 5.3	54.1	3924	± 5.2
	$\leq$ 1.0	64.7	3664	± 5.8	84.2	6107	± 5.5
$NO_3^- + NO_2^-$ (mg L^{-1})	$\leq$ 0.05	32.0	1811	± 5.7	34.1	2471	± 5.3
	$\leq$ 0.10	62.0	3515	± 5.9	71.6	5194	± 5.0
	$\leq$ 0.15	78.9	4474	± 5.0	83.6	6063	± 4.2
Filterable-Al (μg L^{-1})	> 100	38.0	2152	± 5.9	36.6	2654	± 5.4
	> 150	15.0	852	4.7-4.0	17.7	1281	± 4.3
	> 200	5.6	320	3.8-2.1	10.1	732	3.1-3.7

Number : Statistical estimate of the number of target lakes in each region which is associated to given classes of variables
Error : Relative error around the proportion
* : Asymmetric distribution (lower limit – higher limit)

In some cases, the differences in lake proportions between both areas of concern are very large. For example, the percentage of lakes with a pH $\leq$ 5.0 and $\leq$ 5.5 is almost two times greater for the Outaouais region. The same is true for the lake proportion in which color is under 10 Hazen units. On the other hand, the proportion of lakes with a Ca + Mg concentration $\leq$ 100 μeq L^{-1} is five to eight times more important in the Mauricie area.

The extrapolation of the observed water quality to all of the sensitive target lakes (N_t) is important when assessing the number of acidic lakes or the number of lakes affected by acid deposition. Overall, the proportion of lakes with a pH ≤ 5.0 is relatively small, 2.6% in the Mauricie area (149 out of 5667) and 4.1% in the Outaouais region (297 out of 7253). However, the proportion of lakes with a pH ≤ 5.5 is far greater. In the Outaouais and Mauricie areas, 1690 lakes out of 7253 target lakes (23.3%) and 660 lakes out of 5667 target lakes (11.7%), respectively have a pH ≤ 5.5. When considering both areas together, 2350 lakes out of 12 920 were acidic at the time of sampling. The proportion of lakes with a pH of 6.0 and lower is even greater. There are 4533 Outaouais target lakes out of 7253 (62.5%) and 3303 Mauricie target lakes out of 5667 (58.3%) with a pH of 6.0 and lower, for a total of 7836 out of 12 920 target lakes.

Figure 5 presents the Outaouais and Mauricie cumulative frequency distributions (F(x) and 1 - F(x)) for pH, alkalinity, Ca, DOC, SO_4 and filterable Al. This figure also presents the cumulative frequency distributions of lake water quality in Florida, in the Northeast and in the Upper Midwest of the United States.

According to Figure 5, the Outaouais and Mauricie regions have higher proportions of highly-sensitive and low-pH lakes. The cumulative frequency distributions are very similar for pH, alkalinity and calcium. Nearly 90% of the Québec lakes have values of pH, alkalinity and Ca lower than 6.5 units, 100 μeq L^{-1} and 200 μeq L^{-1} respectively. This proportion is lower in the ELS regions. It varies from 20 to 50% depending upon the variable and the region considered. On the other hand, Florida shows higher proportions of very acidic and very sensitive lakes than elsewhere. Approximately 20% of Florida lakes have a pH under 5.0, a negative ANC and a Ca value less than 50 μeq L^{-1}, while the proportions are under 10% for the other areas. The Québec regions also differ markedly from their American counterparts when frequency distributions are compared, pH, alkalinity and Ca frequency distributions from Québec regions cover smaller ranges of values compared to those from the American sub-regions. The frequency distributions for the Outaouais and the Mauricie areas are very similar, with the exception that the Outaouais pH frequency distribution and the Mauricie calcium frequency distribution are shifted slightly towards lower values.

The sulphate inverse frequency distributions differ greatly from the cumulative frequency distributions discussed earlier, since Québec regions cannot be distinguished as easily from American sub-regions. With the exception of Florida, every distribution curves have the typical S shape of the normal distribution. The Florida inverse frequency distribution is more uniform. This region is also characterized by a high proportion of lakes with high sulphate concentrations (30% of the lakes have sulphate concentrations above 200 μeq L^{-1}). Except for Florida, the Outaouais region and the Northeast sub-region have the highest proportion of high sulphate concentrations, with the exception that the latter area has a greater proportion of lakes with low and high sulphate concentrations. The Mauricie area has a high proportion of lakes with low sulphate concentrations. The Upper Midwest is the only American sub-region with a higher proportion of low sulphate lakes.

The DOC cumulative frequency distributions are interesting in more than one way. First of all, the distribution curves of the Outaouais area, the Mauricie area and the

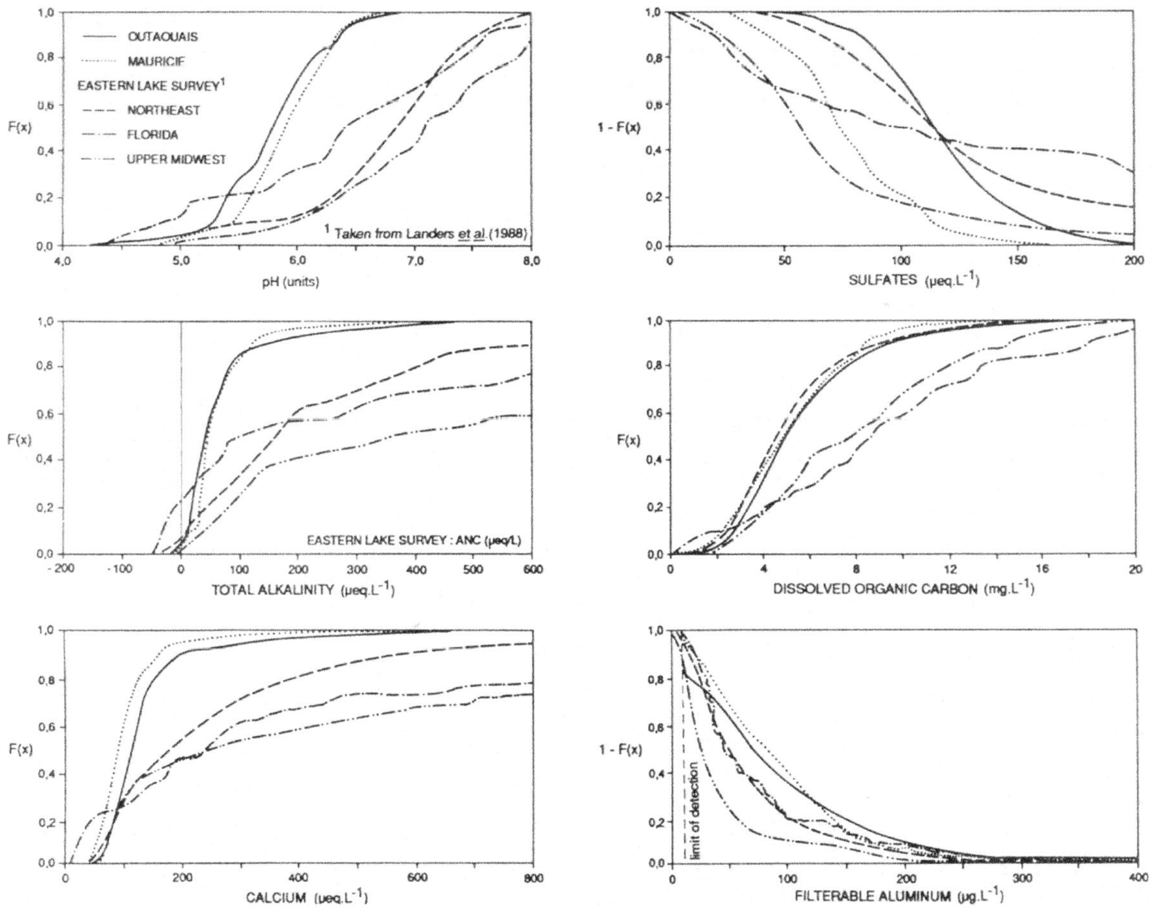

Fig. 5. Frequency distribution of pH, alkalinity, calcium, sulphate, dissolved organic carbon and filterable aluminium for the Mauricie and Outaouais regions, with reference to other surveyed areas of Eastern United States.

Northeast sub-region are similar; the Northeast sub-region contains a slightly higher number of lakes with low DOC. These curves seem to account for some homogeneity in the production of organic matter. The distribution patterns of the two other American sub-regions are not similar. The proportion of lakes with high DOC concentrations is greater in these areas, particularly in Florida.

Filterable aluminium is the last variable to be presented in Figure 5. In general, the highest concentrations of filterable Al (40-150 $\mu g\ L^{-1}$) were observed in the Outaouais and the Mauricie areas. Above 150 $\mu g\ L^{-1}$, the proportions of lakes in each area tend to become more similar.

Discussion

The first two steps of the RESSALQ network, achieved by the random sampling of the lake surface water quality in the Mauricie and the Outaouais hydrographic regions, have brought into light, the high sensitivity of lake water to acidification in the southern portion of the Canadian Shield. As shown previously, a high proportion of lakes are extremely sensitive to acidification ($\leq$ 2 mg L^{-1} $CaCO_3$), while more than 80% of the lakes have alkalinities under 5 mg L^{-1}. These low alkalinity values are also associated with low concentrations of calcium and magnesium, low values of conductivity, and a relatively high proportion of low-pH lakes.

This high sensitivity is related to the low acid neutralizing capacity generated by the chemical alteration of igneous rocks (granite, gneiss and quartzite) and its derived soils. The thinness of the soils also add to the high sensitivity (April and Newton, 1985). This last factor may be one of the major reasons why lakes located to the north of both regions (sharing a similar geology), are more sensitive to acidification than their southern neighbors. Altitude tends to increase along a south-north gradient and the thinnest soils are generally observed in higher places. The areas of high lake sensitivity presented in Figure 4 are almost all associated with an altitude exceeding 300 meters, while the less sensitive lakes are generally located at lower altitudes. The altitude and soil thickness are not the only factors explaining the presence of lakes with medium to high alkalinities in these watersheds. Presence of rocks rich in carbonates (marble present in the Baskatong-Maniwaki-Hull areas), of rocks more subject to chemical alteration (volcanic rocks near Rouyn-Noranda) and lacustrine clays near the Abitibi low-lands are also responsible for the lower sensitivity of some lakes.

The area showing the highest density of acidic lakes is located to the south and southeast of Rouyn-Noranda, where lake sensitivity is very high and wet deposition of sulphate exceeds 40 $\mu eq\ L^{-1}$. The same lake area was found to be under direct influence of SO_2 emissions from the Noranda smelter (Dupont, 1988b). Other areas with a high proportion of acidic lakes are observed north of Québec City and along the border of the Outaouais and Mauricie regions. In these areas, lakes are acidic even if the wet deposition of sulphate is close to or under 20 kg ha^{-1} yr^{-1}, suggesting that this criterion may be too high to adequately protect such ecosysyems. In the same way, other authors have shown the necessity of adopting a lower deposition criterion in order to adequately protect lakes similar to those found on the Canadian Shield (Gorham, 1983; Dickson, 1986, 145; Henriksen and Brakke, 1988). There are some differences between lakes from the Outaouais and the Mauricie regions. The proportion of acidic and transition lakes is higher in the Outaouais area, even if the Mauricie lakes are generally more sensitive to acidification. This can be explained by the fact that the Outaouais area has received greater amounts of acid deposition. In fact, the highest values of wet sulphate deposition were observed in this area. Moreover, the greater part of this region has received a deposition higher than 40 μeq L^{-1}, while such inputs were limited to the southern portion of the Mauricie hydrographic region. The fact that the Outaouais regions shows a higher mean concentration of lake sulphate (almost 2 mg L^{-1} more) is probably related to the higher sulphate wet deposition received in this area.

In comparison with the results obtained from the Eastern Lake Survey, the relative proportion of acidic lakes is of the same order and, in some cases, is worse than what was observed in the Eastern United States (11.7-23.3% in Québec *vs* 0.0-20.6% for ELS (Linthurst *et al.*, 1986)). However, the values become dramatic when considering absolute estimates. Mauricie and Outaouais regions alone have a combined number of 2350 acidic lakes out of the 12 920 target lakes (18.2%). This number is greater than the combined estimates of acidic lakes ($\leq$ 5.5 units) taken from all ELS sub-regions (1348 out of 17 953: 7.5%). The situation is probably even more dramatic because small lakes (< 10 ha) were omitted from our study. However, the higher partial pressure of CO_2 occuring in winter may explain in part, the greater proportion of low-pH lakes in Québec, but it cannot explain the large differences observed for alkalinity and calcium. The overall image becomes even more dramatic when considering that damages to lake organisms can occur at a pH of 6.0 and lower (Schindler, 1988). The pH of such lakes may be highly depressed at the time of spring snowmelt. This means that the organisms from 7836 target lakes out of the overall 12 920 (58.3-62.5%) could be affected in one way or another by acidity at some time of the year.

Most of the lakes not considered as target are presenting an area smaller than 10 ha (more than 93% of all non-target lakes). Less than 1% of the non-target lakes are influenced locally by man-made pollution, less than 5% are bog lakes, while less than 30 lakes have an area greater than 2000 ha. It is impossible to statistically extrapolate the results obtained from the sampled lakes to all of these sensitive lakes because the water quality of the former group is not necessarily representative of that of the latter. Nevertheless, it would be interesting to qualitatively extrapolate these results to all of the 33 080 sensitive Outaouais lakes or all of the 26 602 sensitive Mauricie lakes, while making the assumption that lake water quality is independent of the lake area. However, this assumption is not very realistic because it is well known that smaller lakes tend to be more sensitive and more acidic than larger ones (Quinn and Simonin, 1987). This implies that the qualitative extrapolation would generate at best, a minimal estimate of the overall number of acidic lakes in each region. In this manner, there would be a minimum of 7708 acidic lakes in the Outaouais area and 3112 acidic lakes in the Mauricie region, implying that there would be at least 11 000 acidic lakes in the southwestern part of the Canadian Shield in Québec. When considering a pH of 6.0 and lower, it is a minimum combined total of 36 500 out of almost 60 000 sensitive lakes which could be potentially affected by acidity.

With the exception of Florida, the proportions of very acidic lakes (≤ 5.0 units) remain comparable between all regions. In Florida, the proportion of very acidic lakes is at least twice that observed elsewhere. Except for those extremely sensitive and very acidic lakes, the Outaouais and Mauricie lakes are generally more acidic, more sensitive to acidification, have higher values of filterable aluminium, are equally (Northeast) or less affected (Florida and Upper Midwest) by organic matter, and are as affected by sulphate, compared to the lakes from the ELS sub-regions. All of these comparisons can be summarized by stating that overall, Southwestern Québec lakes are more susceptible to acid precipitation than American lakes.

The influence of airborne sulphate on lake acidity is becoming more and more evident. It has been shown that lake acidity is greatly dependent upon the ecosystem sensitivity, the presence of organic acids and the input of airborne sulphate (Dupont and Grimard, 1986, 1989; Gorham *et al.*, 1986; Neary and Dillon, 1988). Airborne NO_3 may also play a major role, but it was found to be less important in Québec lakes (Dupont and Grimard, 1989). The application of the SIGMA/SLAM model designed by these authors has shown that the Outaouais lakes may have had their pH reduced by a mean of 0.75 ± 0.34 unit through the influence of sulphate deposition alone (Dupont, 1988a). Mauricie lakes may also have been affected, but to a lesser degree (mean pH decrease of 0.59 ± 0.28 unit (Dupont, 1989). Since background sulphate wet deposition could acount for 4-7 μeq L^{-1} (Galloway *et al.*, 1987), the anthropogenic contribution would represent 75 to 93% of the overall lake acidification. The same model has also shown that a 15 kg ha^{-1} yr^{-1} target loading would be enough to increase the pH of most acidic and transition lakes above the pH 6.0 threshold (Dupont, 1988a, 1989). The achievement of such a target loading would necessitate a 50% reduction of wet sulphate deposition for 1986.

Conclusion

The lakes located in the Outaouais and Mauricie hydrographic regions in Southwestern Québec can be classified as part of the most sensitive and most acidic lake group in Eastern North America. More than 80% of these lakes show a high sensitivity to acid deposition. This high sensitivity and the high inputs of airborne pollutants, such as sulphates and nitrates, are responsible for the acidity of 23.3% of the Outaouais lakes and 11.7% of Mauricie lakes. The higher proportion of acidic lakes in the Outaouais area is caused by greater inputs of pollutants. These proportions imply that there are 2350 acidic lakes out of the 12 920 combined target lakes from both regions. The same percentages extended to all sensitive lakes of both hydrographic regions suggest that there would be a minimum number of 11 000 acidic lakes within the Southwestern portion of the Canadian Shield alone. Recovery could be possible for most of these acidic lakes with the application of a 15 kg ha^{-1} yr^{-1} target loading of sulphate in wet deposition. Such a target loading would require a 50% reduction in the wet sulphate deposition received in 1986.

Acknowledgement

I would like to thanks Messrs Yves Grimard, Marc Simoneau, Serge Tremblay and Lévis Talbot from the Direction de la qualité du milieu aquatique of the Ministère de l'Environnement du Québec for their precious comments while writing this paper. I would also like to give special thanks to Dr. Abdul El-Shaarawi for reviewing this paper.

References

April, R. and Newton, R.: 1985, 'Influence of Geology on Lake Acidification in the ILWAS Watersheds', *Water, Air, and Soil Poll.* **26**, 373-386.

Brakke, D. F., Landers, D. H. and Eilers, J. M.: 1988, 'Chemical and Physical Characteristics of Lakes in the Northeastern United States', *Environ. Sci. Technol.* **22**, 155-163.

Cochran, W. G.: 1977, *Sampling Techniques*, 3rd edition, Toronto, John Wiley and Sons, 428 pp.

Dickson, W. W.: 1986, 'Some Data on Critical Loads for Sulphur on Surface Waters' in J. Nilssen (ed.), *Critical Loads for Sulphur and Nitrogen*, Copenhagen, Nordik Ministerad, Miljo rapport 1986/11, pp. 143-158.

Dupont, J.: 1988a, *État de l'acidité des lacs de la région hydrographique de l'Outaouais*, Ministère de l'Environnement du Québec, Report No. PA-29, 99 pp.

Dupont, J.: 1988b, 'Influence des rejets atmosphériques d'anhydride sulfureux sur la qualité de l'eau des lacs de la région de Rouyn-Noranda', *Atmosphere-Ocean* **26**, 449-466.

Dupont, J.: 1989, *État de l'acidité des lacs de la région hydrographique de la Mauricie*, Ministère de l'Environnement du Québec, Report No. PA-33, 84 pp.

Dupont, J. and Grimard, Y.: 1986, 'Systematic Study of Lake Acidity in Québec', *Water, Air, and Soil Poll.* **31**, 223-230.

Dupont, J. and Grimard, Y.: 1989, 'A Simple Dose-Effect Model of Lake Acidity in Québec (Canada)', *Water, Air, and Soil Poll.* **44**, 259-272.

Eilers, J. M., Brakke, D. F. and Landers, D. H.: 1988a, 'Chemical and Physical Characteristics of Lakes in the Upper Midwest', *Environ. Sci. Technol.* **22**, 164-172.

Eilers, J. M., Landers, D. H. and Brakke, D. F.: 1988b, 'Chemical and Physical Characteristics of Lakes in the Southeastern United States', *Environ. Sci. Technol.* **22**, 172-177.

Frontier, S.: 1983, *Stratégies d'échantillonnage en écologie*, Masson, Les Presses de l'Université Laval, 494 pp.

Galloway, J. N., Dianwu, Z., Jiling, X. and Likens, J.: 1987, 'Acid Rain: China, United States, and a Remote Area', *Science* **236**, 1559-1562.

Glass, G. E. and Loucks, O. L.: 1986, 'Implication of a Gradient in Acid and Ion Deposition Across the Northern Great Lakes States', *Environ. Sci. Technol.* **20**, 35-43.

Gorham, E.: 1983, 'Acid Rain: What We Must Do', *The American Biology Teacher* **45**, 203-210.

Gorham, E., Underwood, J. K., Martin, F. B. and Ogden III, J. G.: 1986, 'Natural and Anthropogenic Causes of Lake Acidification in Nova Scotia', *Nature* **324**, 451-453.

Haemmerli, J.: 1987, 'Evolution temporelle de la qualité des eaux des lacs du réseau TADPA-Québec', *Le Naturaliste Canadien* **114**, 247-259.

Henriksen, A., Lien, L., Traaen, T. S., Sevaldrud, I. S. and Brakke, D. F.: 1988, 'Lake Acidification in Norway – Present and Predicted Chemical Status', *Ambio* **17**, 259-266.

Henriksen, A. and Brakke, D. F.: 1988, 'Sulphate Deposition to Surface Waters', *Environ. Sci. Technol.* **22**, 8-14.

Jacques, G. and Boulet, G.: 1988, *Réseau d'échantillonnage des précipitations du Québec: Sommaire des données de la qualité des précipitations 1986*, Ministère de l'Environnement du Québec, Report No. PA-31, 77 pp.

Jeffries, D. S., Wales, D. L., Kelso, J. R. M. and Linthurst, R. A.: 1986, 'Regional Chemical Characteristics of Lakes in North America: Part I-Eastern Canada', *Water, Air, and Soil Poll.* **31**, 551-567.

Kelso, J. R. M., Minns, C. K., Gray, J. E. and Jones, M. L.: 1986, 'Acidification of Surface Waters in Canada and its Relationship to Aquatic Biota', *Can. J. Fish. Aquat. Sci.* Special Publication No. 87.

Landers, D. H., Overton, W. S., Linthurst, R. A. and Brakke, D. F.: 1988, 'Eastern Lake Survey', *Environ. Sci Technol.* **22**, 128-135.

Linthurst, R.A., Landers, D. H., Eilers, J. M., Kellar, P. E., Brakke, D. F., Overton, W. S., Crowe, R., Meier, E. P., Kanciruk, P. and Jeffries, D. S.: 1986, 'Regional Chemical Characteristics of Lakes in North America, Part II: Eastern United States', *Water, Air, and Soil Poll.* **31**, 577-591.

Neary, B. P. and Dillon, P. J.: 1988, 'Effects of Sulphur Deposition on Lake Chemistry in Ontario, Canada', *Nature* **333**, 340-343.

Quinn, S. and Simonin, H.: 1987, 'The Bitter Taste of Acid Lakes', *Clearwaters* Winter **87**, 16-21.

Rohlf, F. J. and Sokal, R. R. 1969, *Statistical Tables,* San Francisco: W. H. Freeman and Company.

Schindler, D. W.: 1988, 'Effects of Acid Rain on Freshwater Ecosystems', *Science* **239**, 149-157.

Shilts, W. W., Card, K. D., Poole, W. H. and Sandford, B. V.: 1981, *Sensitivity of Bedrock to Acid Precipitation: Modification by Glacial Processes,* Geological Survey of Canada, Paper 81-14.

United States – Canada: 1983, *Memorandum of Intent on Transboundary Air Pollution,* Final Report, Impact Assessment Work Group I.

A STATISTICAL APPROACH TO FIELD MEASUREMENTS OF THE CHEMICAL EVOLUTION OF COLD (< 0 °C) SNOW COVER

CLAUDE LABERGE*

INRS-Eau, 2800 Einstein, Sainte-Foy, P.Q., Canada, G1V 4C7

and

GERALD JONES

INRS-Eau, 2800 Einstein, Sainte-Foy, P.Q., Canada, G1V 4C7

(Received February 1990)

Abstract. Two statistical methods for the analysis of data on the evolution of the chemical composition of cold snow (< 0 °C) in the field (Lac Laflamme, Quebec) were compared. The methods used on the data were regression analysis (One sample per sampling date over a long cold period) and ANOVA (replicate samples on a restricted number of sampling dates over shorter periods). The relative power of the tests to determine the detectable amplitude of chemical changes was derived from the theoretical power of the tests under comparable conditions of sampling (number of observations) and from the estimated error variances of the measured data.

The results of the study on the evolution of sulfates (SO_4) concentrations in discretely identified snow strata clearly showed that for six of the eight strata, significant losses of SO_4 occurred in snow during cold periods. The relative amplitude of the significant losses varied between 1% per day and 4% per day depending on the initial concentrations in the snow and the prevailing meteorological conditions.

The analysis of the data also demonstrated that for the same number of samples, the regression analysis is more efficient in detecting the chemical changes in snow than the alternative ANOVA method. The use of this information to plan sampling programs of cold snow under both field and laboratory conditions is discussed.

1. Introduction

Snow cover is a major reservoir of atmospheric pollutants in the hydrologic cycle of boreal ecosystems. In the past decade the acidification of surface waters in northern regions has resulted in an increase in the study of snow chemistry. The main emphasis has been placed on the temporal changes in the concentrations of strong-acid anions in snow during the cold (< 0 °C) accumulation period when the water equivalent of the pack increases and in the melt season when the phenomenon of 'acid shock' due to high hydrologic flux occurs.

One of the main difficulties in monitoring the chemical evolution of snow cover is the heterogeneous distribution of the chemical species in the pack (Tranter *et al.,* 1986). Although the chemical composition of individual snowfalls may be relatively homogeneous over a catchment area (10-100 km^2), subsequent wind redistribution (Delmas and Jones, 1987), small melt and freeze cycles (Colbeck, 1981), and local dust and forest canopy fallout (Jones and Sochanska, 1985) will provoke small-scale disparities (1-100 m^2) in the chemical characteristics of the snow strata laid down by the original precipitation events.

* To whom all correspondence should be adressed.

Environmental Monitoring and Assessment **17**: 201–216, 1991.

To follow chemical changes in cold snow, samples of the snow cover are taken over a period of time during which the temperature of the snow and the air are below 0 °C. The sampling method is destructive to the snow cover and successive samples have to be taken as close to the original point as possible. This, however, can be a cause for concern as the removal of snow from the original sampling point can influence the remaining snow in the immediate environment by modifying temperature gradients and air-snow interchange. As the sampling period evolves, the spatial requirements for the collection of the samples increases. This in turn enhances the probability that the sampling 'point' will cover areas of snow with different chemical characteristics. Thus, any attempt to study the evolution in the chemical composition of cold snow covers due to real processes of transformation or translocation 'in situ' has to take into consideration the apparent chemical changes that result from the spatial inequalities in snow quality.

To distinguish real chemical changes in snow from any apparent changes, the sampling strategy should be based on a statistical analysis of the spatial chemical characteristics of the snow cover at the beginning of the sampling period. In the absence of well defined spatial characteristics valid information on real in-pack (within the snow cover) changes can, however, still be obtained even if the spatial variability is less known. This would be true in the case where the in-pack chemical processes are consistent and result in concentration changes of species in the same direction (i.e. losses or gains) over time. The detection of a trend in time is then related to real changes large enough to overcome any masking effect by the chemical heterogeneity of the snow. In previous studies on the chemical evolution of cold snowpacks (Jeffries and Snyder, 1981; Jones and Bisson, 1984; Cadle *et al.*, 1984) the sampling methods did not give rise to sufficient data from which definite conclusions on real changes in in-pack chemistry of snow could be drawn. The optimization in the quality and quantity of field data from which unambiguous results may be obtained should thus be the major priority of the sampling methodology. Field operations are expensive. By reducing the cost in the sampling program much needed funds may be diverted elsewhere.

In order to optimize the field sampling programs, research workers in the different disciplines of snow science should carefully consider the statistical methods which are available for the treatment of data. Although these methods may be well known to statisticians, utilization by research workers for data analysis is often not applied in an efficient manner. The following article, primarly intended for non statisticians, describes two experiments which were carried out to study real changes in the chemistry of cold snow in the packs at Lac Laflamme, Quebec. The data from these experiments will be used to illustrate the comparative power of two different statistical methods.

During both study periods, discrete in-pack strata of snow were sampled. Each study had a different sampling strategy. In the first experiment one snow sample per stratum was taken on each sampling date over a relatively long cold period. In the second study many replicate samples of each stratum were taken per sampling date over short cold periods. The two data sets were treated by two simple but different statistical methods. The set from the long period was subjected to a trend detection analysis in which spatial variability is not a factor. The second set was analysed by analysis of variance (ANOVA) where the variance of replicates on successive sampling dates estimates the spatial variability of snow quality. The powers of the tests involved in each method are compared and the results from the two field studies are used to estimate the chemical changes that would be detectable by each method for a reasonable sample size.

2. Site Characteristics and Sampling Methodology

Two snow courses were laid out in the watershed of Lac Laflamme, a small headwater lake (47°19′ N, 71°07′ W) in the Parc des Laurentides, Quebec (Figure 1). The mean annual temperature is 0.2 °C (–15 °C January; 15 °C July). The total snowfall at Lac Laflamme is approximately 400 mm (Snow Water Equivalent, SWE) and the pack reaches a depth of 120-150 cm. The courses, approximately 25-30 m long and 1.5-2 m wide, were prepared before the winter season by removing low brush and debris from open areas of the boreal forest. The length of each snow course was staked out with two parallel lines of fence posts 2 m apart (Figure 2).

The first course was sampled in 1985; four adjacent discrete strata (1, 2, 3 and 4), identified by means of threads, were sampled (one sample per stratum) once a week during a cold period that ran from January 10 to March 27. The sampling dates were January 10, 16, 23 and 30, February 6, 13, 20 and 27, March, 7, 13, 20 and 27. During the last two sampling dates, however, some meltwater had began to penetrate the pack as the daytime air temperature increased in early spring.

The second snow course was sampled (5 samples per stratum) during intermittent short cold periods in 1988 when the January to March period experienced rain episodes. Rain-on-snow episodes displace chemical species in the pack (Jones *et al.*, 1989) and the experiment has to be started up again with new snow strata laid down subsequent to the rain events. Four strata (1, 2, 3 and 4) were sampled on two different sampling dates (Figure 2); Strata 1 and 2 were sampled on January 20 and 27, stratum 3 on February 5 and 15, and stratum 4 on February 23 and March 4.

The sampling technique for 1985 and 1988 was identical; it consisted in removing a core of constant cross-sectional area over the depth of each stratum by means of a small square corer of plastic. The samples were conserved at –20 °C until melted for analyses. A complete analysis for major ions was carried out on each sample (Jones, 1987); for purposes of discussion only the analyses for SO_4 are reported in this paper.

3. Statistical Approach: Background and Intercomparison between the Methods

Trend Analysis

The data from the 1985 study had to be treated by a time series analysis that would detect a trend in SO_4 concentrations. Of the many methods available for time series analysis, those of Box and Jenkins (1976) are the most widely used. In general, the treatment of data by the methods of these authors is restricted to modelling and prediction using autoregressive-moving average (ARMA) models; very little attention is given to the practical determination of changes in the value of the location parameter in a definite series.

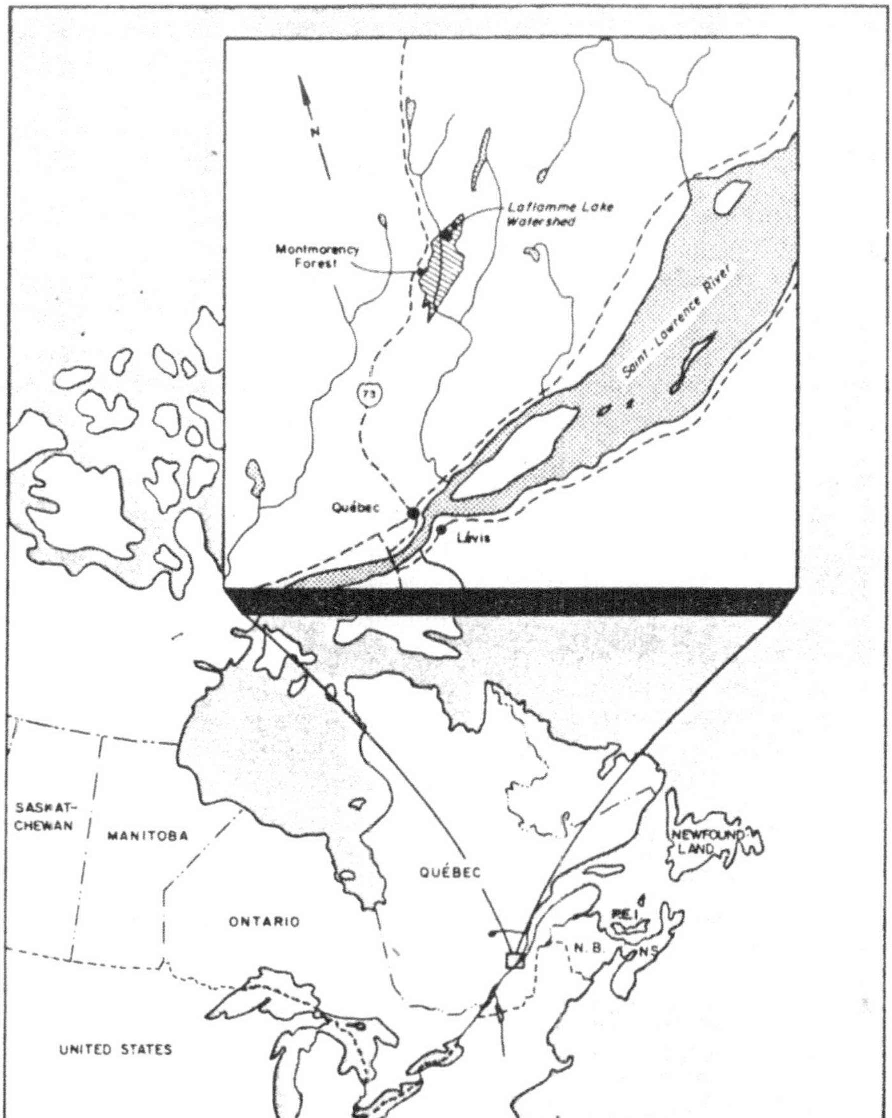

Fig. 1. Location of the Lac Laflamme watershed.

Furthermore, an efficient use of the Box and Jenkins methods requires a relatively large amount of data, i.e. at least 50 observations equidistant in time. These characteristics of the Box and Jenkins techniques make them unsuitable for the treatment of the data obtained in 1985 (10 observations over the period) and inappropriate for the majority of field experiments on snow cover. A more fitting method for such a small data set is the detection of trends over time. Two main types of trends are generally studied. First there

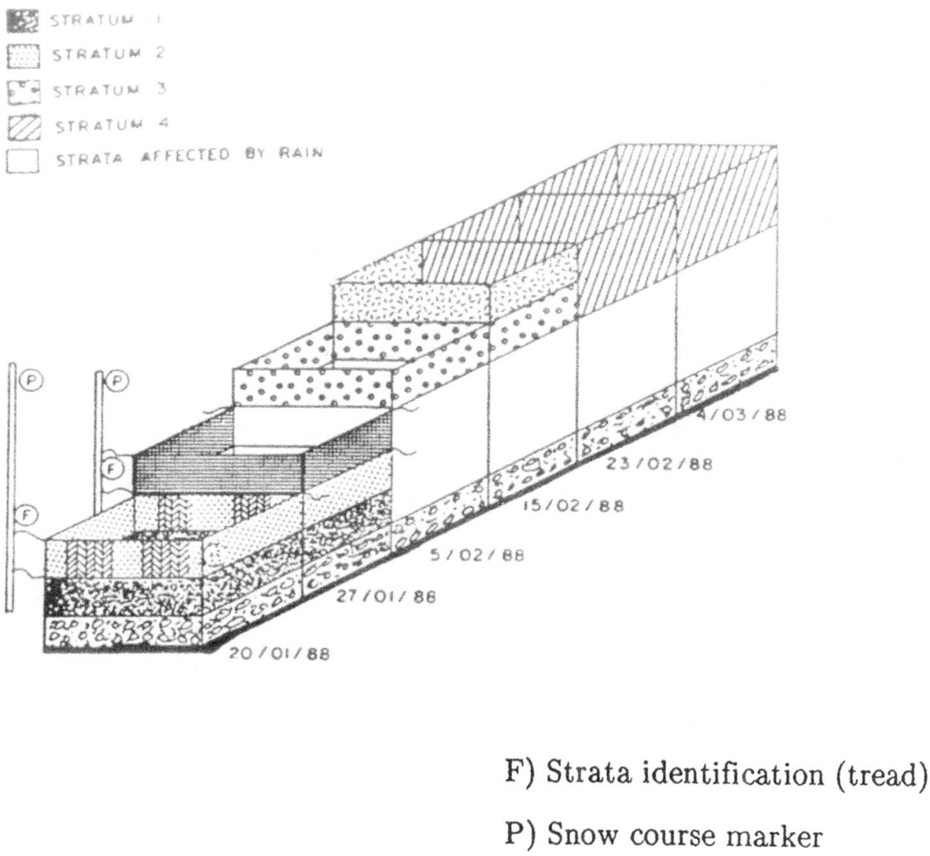

Fig. 2. Snow course lay out for the experiment of 1988, Lac Laflamme, Quebec.

are monotonic trends; these are gradual changes in time often associated with natural phenomena (e.g. precipitation, watershed runoff). On the other hand there are step trends due to stepwise changes in system conditions; these are often the result of forces external to the system (e.g. wastewater discharges). In the case of the 1985 experiment, the nature of the changes that were to be expected for in-pack chemistry lead us to believe that tests designed to detect monotonic trends should be used.

Tests for detecting monotonic trends may be of two types. These are parametric tests and nonparametric tests. Both types of test presuppose that successive observations are completely independent except for the possible trend. If some short-term dependence exists, i.e. autocorrelation or seasonality, then the tests have to be modified to compensate for these types of dependence (Lettenmaier, 1976; Hirsch and Slack, 1984). Parametric tests which are best adapted to the detection of monotonic trends are those of simple linear regression over time. In the case of the analysis of the change in the concentration of any chemical species over time the regression model is:

$$C_i = C_0 + \Delta t_i + e_i \tag{1}$$

where C_i is the concentration of the chemical at time t_i, C_0 is the original concentration, Δ is the slope of the regression and e_i is a random error component. The values of e_i are assumed to be independent and identically distributed $N(0, \sigma^2)$. Under these conditions the estimates for the slope of the regression and for its standard deviation indicate the amplitude and the significance of the trend. The null hypothesis, $H_0 : \Delta = 0$, is accepted if Δ is not significantly different from 0 and the presence of a trend is then rejected. The alternative hypothesis, $H_1 : \Delta \neq 0$, is then associated with the presence of a trend. In some cases the linear regression model will not adequately fit the data and a nonlinear regression may be necessary. Nonlinear trends can be transformed into linear form or can be fitted to nonlinear models (Ratkowsky, 1983).

The disadvantage of parametric tests is that they are sensitive to aberrant values; one outlier in the data set may considerably influence the overall result of these tests. To overcome this deficiency one can resort to nonparametric tests for the detection of monotonic trends. Although nonparametric tests like the Spearman test and the Kendall test (Conover, 1971) are adequate for the detection of such trends, they are not sensitive to aberrant values as they do not presuppose normal distribution of the data and use ranks of values rather than absolute numbers. On the other hand, these tests do not yield values for regression coefficients which indicate the amplitude of the trend. Neither can the level of the series be established. Instead they will test correlation coefficients (ρ) between ranks of concentrations and time. In an analogous manner to the parametric tests, the null hypothesis, H_0 is accepted if ρ is not significantly different from 0. If ρ is significantly different from 0 then the null hypothesis, H_0, is rejected and the presence of a trend, is accepted.

POWER OF THE TESTS AND NUMBER OF SAMPLES REQUIRED

The power of a test is the probability of correctly rejecting H_0 when H_1 is true, i.e. that Δ or ρ are different from 0. This is expressed by $1 - \beta$, where β is the type II error which qualifies the case where H_0 is incorrectly accepted. The power of a test allows the calculation of the minimum number of samples that are required to detect trends of predetermined values with a known probability. The power of the parametric test used in the linear regression model has been described by Bickel and Doksum (1977). Under H_1 we can obtain the power of this test from charts or tables of the noncentral Student distribution with noncentrality parameter δ defined by:

$$\delta^2 = \frac{\Delta^2}{\sigma^2} \sum_{i=1}^{N} (t_i - \bar{t})^2, \tag{2}$$

where N is the number of observations, t_i is the sampling time of observation i and $\bar{t} = 1/N \sum_{i=1}^{N} t_i$. Assuming equidistant observations (i.e. $t_i = i$) Equation (2) can be reduced to:

$$\delta^2 = \frac{\Delta^2}{\sigma^2} \frac{N(N+1)(N-1)}{12}. \tag{3}$$

Neter and Wasserman (1974, Table A-5) published the power function curves for the linear regression test. Table I presents the relationship between the power ($1 - \beta$), the number of observations (N), the noncentrality parameter (δ), and the relative amplitude of the slope with respect to the standard deviation of the error component (Δ/σ) for some values of these parameters. As an example, the table shows that with eight observations one can correctly detect a trend of amplitude, $\Delta = .31\sigma$, four times out of ten. As the number of observations increases, the value of Δ/σ decreases and if the criteria for an experimental program required the correct detection of a trend with an amplitude of less than 0.018σ at a success rate of 70% then more than 62 observations would be required. It should be noted that in the above examples the supposition that δ remains constant with the changes in δ is taken.

When observations are normally distributed, the power of nonparametric tests (Spearman, Kendall) is approximately of the same value of parametric tests (regression) when $N > 20$; for $N < 20$ the power of nonparametric tests is inferior to parametric tests. The former tests, however, should be used if outliers occur in the data sets or if the observations are not normally distributed.

ANALYSIS OF VARIANCE (ANOVA)

Trend analysis is less suited than ANOVA to replicate sampling over short time periods. The 1988 data set was thus treated by ANOVA (Montgomery, 1984). ANOVA is a statistical tool that permits the testing of the equality of several means ($\mu_1, \mu_2, \ldots, \mu_a$) and is thus a generalization of the Student t-test which can be used to test the equality of two means. The test presupposes a random sampling of the designated snow strata within the snow

TABLE I

Power of the regression test for the detection of trend

N[a]	δ[b]	Δ/σ[c]	$1 - \beta$[d]
8	1.5	0.23	0.25
8	2.0	0.31	0.40
8	2.5	0.39	0.55
22	1.5	0.050	0.28
22	2.0	0.067	0.45
22	2.5	0.084	0.64
62	1.5	0.011	0.30
62	2.0	0.014	0.50
62	2.5	0.018	0.70

[a] N, number of samples.
[b] δ, noncentrality parameter.
[c] Δ/σ, ratio of slope (Δ) to the standard deviation of error component (σ).
[d] $1 - \beta$, power of the test.

course on each sampling date. Although this was not done over the *whole* snow course on *each* date, the spatial heterogeneity in the chemical composition of the whole snow course was established on the first sampling date, and then a systematic and progressive sampling of the strata in the course was carried out (Figure 2). In an ANOVA the acceptance of the null hypothesis, $H_0: \mu_1 = \mu_2 = \ldots = \mu_a$, rejects any changes in the chemical composition of snow. These changes include systematic and non systematic differences in the level of the concentrations. Multiple comparison tests (Montgomery, 1984, pp. 64-71) should be used to determine the possibility of systematic evolution of the differences.

Power of the ANOVA tests and number of samples required

The power of the F test used by the ANOVA is based on the noncentral F distribution with noncentrality parameter (δ') defined by:

$$\delta'^2 = \frac{n \sum_{i=1}^{a} \tau_i^2}{a\sigma^2}, \tag{4}$$

where a is the number of sampling dates, n is the number of replicates, σ^2 is the variance of the error component and τ_i is the difference between the mean of the ith sampling date (μ_i) and the general mean ($\mu = {}^1/_a \Sigma_{i=1}^{a} \mu_i$). Montgomery (1984) gives the power curves for this test. Table II presents values of the power ($1 - \beta$) derived from some values of n, a, δ' and $[\Sigma\, \tau_i^2/\sigma^2]^{0.5}$. The table shows that the power increases (for any specified value of δ') as the number of replicates increases for the same number of sampling dates; if the number of sampling dates is not the same it is difficult to compare the powers of the respective sampling strategies as the sum, $\Sigma\, \tau_i^2$, does not contain the same number of terms. The direct comparison of power of trend analysis tests and ANOVA tests by comparing Tables I and II cannot be made. However, under the following circumstances a comparison of power between the statistical tests can be obtained.

Comparison of power: trend analysis versus ANOVA

In order to compare the power of trend analysis tests and ANOVA tests it is necessary to relate the amplitude of the trend Δ and the term $\Sigma\, \tau_i^2$ associated with temporal changes of the mean. This is straightforward if time intervals for successive samplings is the same and if there is a change of Δ between the sampling intervals for both trend analysis and ANOVA. Even if the actual time intervals are different, Δ and $\Sigma\, \tau_i^2$ can be related. In the case of Lac Laflamme the time interval between succesive samplings of the snow cover was approximately one week (6-10 days) in both 1985 (regression analysis) and 1988 (ANOVA).

Between two successive sampling dates, ANOVA is reduced to the standard Student t-test which leads to the following relationship between Δ and $\Sigma\, \tau_i^2$:

TABLE II

Power of the *F* test utilized in the analysis of variance

N^a	n^b	a^c	δ'^d	$[\Sigma\, \tau_i^2/\sigma^2]^{0.5^e}$	$1 - \beta^f$
8	4	2	1.5	1.06	0.42
8	4	2	2.0	1.41	0.65
8	4	2	2.5	1.77	0.84
9	3	3	1.5	1.50	0.42
9	3	3	2.0	2.00	0.65
9	3	3	2.5	2.50	0.84
12	4	3	1.5	1.30	0.49
12	4	3	2.0	1.73	0.73
12	4	3	2.5	2.17	0.92

[a] N, total number of samples.
[b] n, number of replicates per sampling dates.
[c] a, number of factor levels (i.e. number of sampling dates).
[d] δ, noncentrality parameter.
[e] τ_i, mean value at time i (μ_i) minus general mean ($\mu = 1/a\ \Sigma\ \mu_i$); σ^2, variance of error component.
[f] $1 - \beta$, power of the test.

$$\sum_{i=1}^{2} \tau_i^2 = \sum_{i=1}^{2} \left[\mu_i - \frac{\mu_1 + \mu_2}{2} \right]^2 = \sum_{i=1}^{2} \left[\frac{\Delta}{2} \right]^2 = \frac{\Delta^2}{2} \tag{5}$$

with the general mean $\mu = (\mu_1 + \mu_2)/2$ and $\Delta = \mu_2 - \mu_1$ since there is a change of Δ units between two sampling dates. In a similar way it can be shown that for more than two successive sampling dates ($a > 2$) the general relationship between Δ and $\Sigma\ \tau_i^2$ is given by Equation (6):

$$\sum_{i=1}^{2} \tau_i^2 = \frac{\sum_{i=1}^{a} (a - i) i^2\ \Delta^2}{a} . \tag{6}$$

By using Tables I and II, and Equations (5) and (6) one can now compare the relative efficiency of the two statistical methods to detect changes in a dynamic system (e.g. the cold snow cover) for the same number of observations. For example it can be seen from Table I that for a power value of 0.40 ($N = 8$, $\delta = 2.0$) the regression analysis will detect a slope of $\Delta = 0.31\sigma$. For the same number of observations ($N = 8$) and approximately the same value for the power of the ANOVA test of 0.42 (Table II, $n = 4$, $a = 2$, $\delta' = 1.5$), the term ($[\Sigma\ \tau_i^2/\sigma^2]^{0.5}$) is 1.06. Substituting $(1.06)^2\sigma^2$ for $\Sigma\ \tau_i^2$ in Equation (5) leads to a value of $\Delta = 1.5\sigma$. In this case, the ANOVA test can only detect a slope of 1.5σ from the same sample size. The comparison is, however, only valid if the variance of the error component (σ^2) is the same for both methods.

In the above example the data for the ANOVA tests were obtained on two successive samplings of four replicates per sampling date ($n = 4$, $a = 2$, Table II). Equation (6) permits the comparison between the ANOVA test, for which the successive sampling dates are more than two, and the trend analysis. Table II shows that the value of the power term for the condition of $N = 9$, $n = 3$, $a = 3$ and $\delta' = 1.5$ is very similar (0.42) to the regression test (0.40, $N = 8$) and the ANOVA test (0.42, $N = 8, n = 4, a = 2$, and $\delta' = 1.5$) cited above. If we consider that the number of observations are approximately the same ($N = 9$ v. $N = 8$), then the value of $[\Sigma\, \tau_j^2/\sigma^2]^{0.5}$ of 1.5 and Equation (6) show that the efficiency of the ANOVA can be improved to detect trends of amplitude, $\Delta = 1.1\sigma$, four times out of ten with the ANOVA ($n = 3, a = 3$). The trend analysis however, is still the most efficient method for the detection of a change in the concentrations.

4. Results and Discussion

1985: Trend Detection

Table III shows the SWE, concentrations of SO_4 at each sampling date, and the mean concentration for the study period of four adjacent strata in the snowcover at the Lac Laflamme site between January 10 and March 27, 1985. The table also reproduces the weighted concentrations of SO_4 for the strata combined as one stratum. The study of the composite stratum smooths out the irregularities that occur between individual strata and represents a better picture of the overall evolution of that part of the snowcover where the strata are found. Although there may be a consistent monotonic trend over the whole time period the different strata may be exposed to different phenomena at different times. Thus one stratum may indicate a decrease in the concentrations of SO_4 due to emigration of aerosols during snow metamorphism which is reflected by an increase in the SO_4 concentrations of the adjacent stratum. In addition dry deposition will increase the SO_4 particularly in those strata which comprise the surface of the pack early on in their existance even though the net dominant overall phenomenom may result in SO_4 losses for the whole pack.

The graphical representation of the results of each individual stratum are presented in Figure 3; Figure 4 records the overall evolution of the strata combined as one stratum. Strata 1, 3 and 4 (Figure 3) appear to show a decrease between January 23 and March 13 while the behaviour of stratum 2 is more erratic. Figure 4 also indicates that the overall trend between January 23 and March 13 is a decrease in SO_4 of approximately 7 μeq L^{-1}; as there was no loss or gain in SWE for this specific cold period the loss represents 39% of the original SO_4 load in the pack.

Regression analyses, however, on all the data for the strata show that the trends are significant only in the case of the first stratum and the composite stratum (Table IV). The overall trend of SO_4 loss is confirmed for the composite stratum. Spearman tests give exactly the same conclusions, thus showing that no aberrant data affected the regressions. If only the data between January 23 and March 13 are subjected to the same analysis then strata 3 and 4 also show a significant downward trend in SO_4 concentrations. The results of the tests lead us to conclude that there is a significant and progressive loss of SO_4

TABLE III

SO_4 concentrations (μeq L^{-1}) of four adjacent strata and of the composite stratum (strata 1-4) in the snow cover at Lac Laflamme, January to March, 1985

Date	Stratum 1	Stratum 2	Stratum 3	Stratum 4	Strata 1 - 4
10 Jan	25.6	10.4	8.1	21.0	13.9
16 Jan	29.6	9.8	15.4	29.4	17.1
23 Jan	39.8	9.0	12.3	20.8	17.8
30 Jan	17.9	11.5	13.8	26.3	14.5
6 Feb	17.9	9.4	12.7	24.6	13.2
13 Feb	15.2	9.8	9.8	22.3	11.8
20 Feb	17.3	9.0	8.3	21.0	11.5
27 Feb	11.5	9.6	10.6	19.8	10.9
7 Mar	11.9	10.4	9.0	17.3	10.8
13 Mar	11.3	10.8	8.5	15.8	10.6
20 Mar	12.3	10.2	10.6	24.2	11.6
27 Mar	20.4	12.3	12.3	23.8	14.9
Mean	19.2	10.2	11.0	22.2	13.2
SWE[a]	30.5	56.1	36.7	7.8	

[a] SWE, Snow water equivalent measured in millimeters.

from the snow strata during the cold period. The increase in SO_4 concentrations at the beginning of the period may have been due to dry deposition at the pack surface (Cadle *et al.,* 1985); the pack became deeper as the winter progressed, the lower strata (1 to 4) became isolated from the atmosphere, and dry deposition ceased to have a direct influence on the chemical evolution of these strata. On the other hand, the general increase of SO_4 at the end of the period is due to the percolation of meltwater from the upper part of the pack during the start of the springmelt season.

1988: ANOVA ANALYSIS

Table V shows the replicate values, the mean value, and the standard deviation of SO_4 concentrations for the four strata at the Lac Laflamme site in 1988. The weather in that year consisted of melt and rain-on-snow episodes and a prolonged period for the study of changes in cold snow did not occur. The longest period of persistent cold weather was experienced between February 5 and February 15. These unfavorable weather conditions did not allow the sampling of a stratum for more than two sampling dates. The Student *t*-test showed that there was a significant change over time of SO_4 concentrations for strata 2, 3 and 4 during their respective cold periods. In each case, losses (18%, 2; 39%, 3; 27%, 4) of SO_4 were registered; the amplitude of the losses averages at between 2.5% and 4% per day. The tests also permitted the estimation of standard deviations of the error component: the values are 0.26 μeq L^{-1}, stratum 1; 1.74 μeq L^{-1}, stratum 2; 0.22 μeq L^{-1}, stratum 3; and 0.46 μeq L^{-1}, stratum 4.

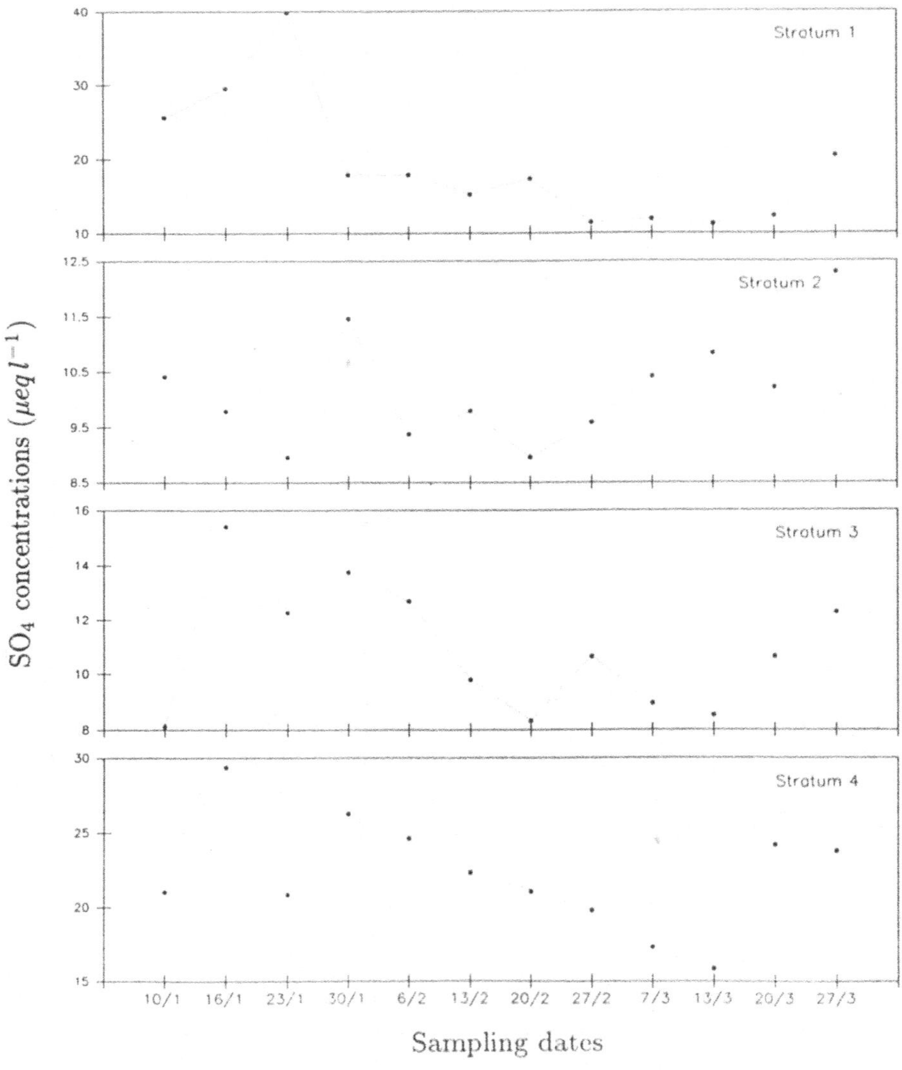

Fig. 3. Evolution of SO_4 concentrations (μeq L^{-1}) in four adjacent strata in the snowcover, Lac Laflamme, Quebec, January to March, 1985 (sampling dates are schematic only).

COMPARISON OF THE TWO METHODS, 1985, 1988

From the above estimations of σ for t-tests and regression analysis we can now compare the power of the tests to detect absolute changes in the concentrations of SO_4 over time. Table IV shows that the values of $\hat{\sigma}$ for SO_4 concentrations varied between 0.51 μeq L^{-1} and 3.57 μeq L^{-1} for the regression analysis of snow strata sampled in 1985. In general the values of $\hat{\sigma}$ are proportional to the absolute values for the mean concentrations showing that the coefficient of variation is relatively stable. This is also true in the case of the

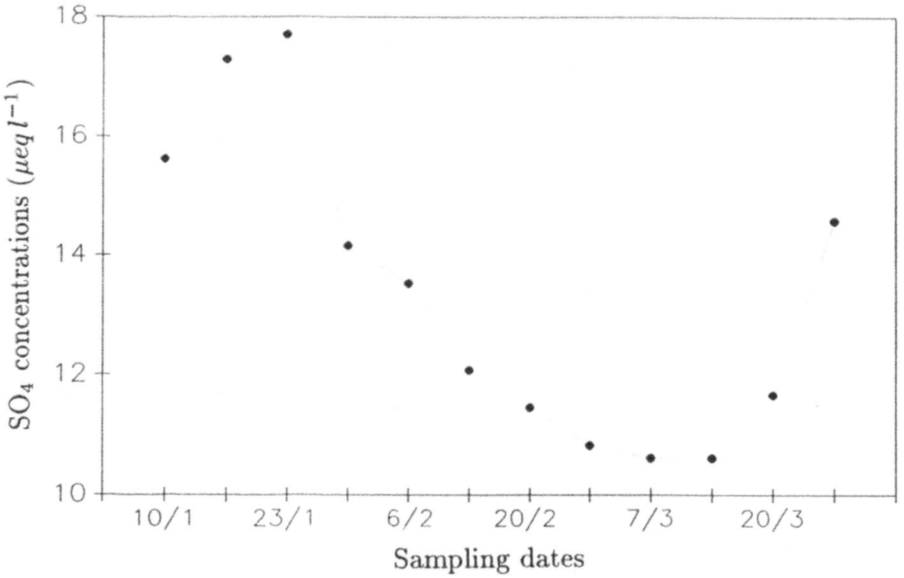

Fig. 4. Evolution of weighted SO_4 concentrations (μeq L^{-1}) in the composite stratum (strata 1 to 4, Figure 3), Lac Laflamme, Quebec, January to March, 1985 (sampling dates are schematic only).

TABLE IV

Regression tests and trend detection results of SO_4 concentrations in four adjacent strata and the composite stratum (1-4) in the snow cover at Lac Laflamme for the period January 10 to March 27, 1985

Stratum	Significant Trend (Yes/No)	$\hat{\Delta}^a$	$\hat{\sigma}^b$
1	Yes, negative	- 0.86	3.57
2	No	+0.06	0.51
3	No	- 0.11	1.25
4	No	- 0.21	1.94
Composite 1-4	Yes, negative	- 0.26	1.03

[a] $\hat{\Delta}$, estimated slope, measured in μeq L^{-1} $week^{-1}$.
[b] $\hat{\sigma}$, estimated standard deviations of error component, measured μeq L^{-1}.

ANOVA for the snow strata in 1988 (Table V). Thus, we can compare, on the one hand, the power of the two methods for strata which have low concentrations of SO_4, and, on the other hand, for strata which are more polluted in SO_4.

In the first case, stratum 2 in the 1985 series ($SO_4 \simeq 10$ μeq L^{-1}, Table III) can be considered as being equivalent in mean to stratum 4 ($SO_4 \simeq 10$ μeq L^{-1}) in the 1988 ANOVA study. To estimate the powers of the different tests used, the estimated values of the standard deviations of error component are $\hat{\sigma} = 0.51$ μeq L^{-1}, 1985, (Table IV) and $\hat{\sigma} =$ 0.46 μeq L^{-1}, 1988.

TABLE V

SO_4 concentrations (μeq L^{-1}) of four distinct strata in the snow cover of Lac Laflamme for different time periods, January 20 to March 4, 1988

		Stratum 1	Stratum 2	Stratum 3	Stratum 4
		20-1-88	20-1-88	5-2-88	23-2-88
Replicate # 1		11.98	46.92	7.25	10.00
Replicate # 2		11.40	50.65	7.29	10.42
Replicate # 3		11.42	51.33	7.31	11.67
Replicate # 4		11.52	50.98	7.33	10.83
Replicate # 5		11.81	50.46	7.46	10.72
	Mean	11.63	50.06	7.33	10.73
	$\hat{\sigma}^a$	0.26	1.80	0.08	0.62
		27-1-88	27-1-88	15-2-88	4-3-88
Replicate # 1		11.77	43.73	4.92	8.33
Replicate # 2		11.90	41.13	4.52	7.92
Replicate # 3		11.79	40.31	4.13	7.92
Replicate # 4		11.42	40.38	4.33	7.92
Replicate # 5		12.13	39.33	4.33	8.13
	Mean	11.79	40.99	4.44	8.04
	$\hat{\sigma}^a$	0.26	1.67	0.30	0.19

[a] $\hat{\sigma}$, standard deviation of replicates, measured in μeq L^{-1}.

In the second case, stratum 4, 1985 ($SO_4 \simeq 20$ μeq L^{-1}), the most consistently concentrated stratum of the regression analysis, can be compared with some limitations to stratum 2, 1988 ($SO_4 \simeq 40$ μeq L^{-1}), the most polluted stratum in SO_4 used for the ANOVA. Estimated values of σ in this case are 1.9 μeq L^{-1} in 1985 and 1.7 μeq L^{-1} in 1988.

For the sake of comparison, the number of total samples taken (N) is set at 8. This represents one sample per date for eight successive samplings in the regression analysis and four replicate samples per date for two samplings in ANOVA within the same time period. Substituting the respective estimations of σ for strata 2 and 4, 1985, into Table I (regression analysis), and strata 4 and 2, 1988, into Table II (ANOVA) permits the calculation of the amplitude of detectable changes over time of SO_4 concentrations at comparable powers. Thus Table VI shows that at a power of 0.40, the regression analysis can detect Δ values of more than 0.16 μeq L^{-1} per sampling interval in the case of the relatively dilute snow. The ANOVA test, however, detects values of Δ of only more than 0.69 μeq L^{-1} per sampling interval for similarly dilute snow at the same power. For polluted snow strata the regression analysis at a power of 0.40 can detect a trend amplitude of 0.59 μeq L^{-1} per sampling interval compared to 2.61 μeq L^{-1} per sampling interval by the ANOVA test at the same power.

TABLE VI

Comparison of the power ($1 - \beta$) of regression analysis tests and of ANOVA tests to detect trends of absolute amplitude (μeq L^{-1}) in SO_4 concentrations

N	δ	Regression Δ^a	Regression $1 - \beta$	ANOVA Δ^a	ANOVA $1 - \beta$
(a) Low concentrations of SO_4 ($\simeq$ 10 μeq L^{-1})					
8	1.5	0.12	0.25	0.69	0.42
8	2.0	0.16	0.40	0.92	0.65
8	2.5	0.20	0.55	1.15	0.84
(b) High concentrations of SO_4 (> 20 μeq L^{-1})					
8	1.5	0.44	0.25	2.61	0.42
8	2.0	0.59	0.40	3.47	0.65
8	2.5	0.74	0.55	4.36	0.84

[a] Δ, amplitude detected in μeq L^{-1} per sampling interval.

5. Conclusion

The results of the 1985 and the 1988 studies clearly show that losses of SO_4 can occur in snow during cold periods. Analysis of the data also demonstrates that for the same number of samples, the maximum distribution of the total number of samples over time (i.e. one sample per sampling date), and regression analysis, is more efficient in detecting the chemical changes in snow than the alternative method of regrouping the number of samples for a lesser number of successive sampling dates and using ANOVA.

This information can be used to plan future sampling programs of cold snow. Two scenarios for sampling can be envisaged. The first senario is that in which the cost of the field sampling is the major financial burden of the study. If one has a prior knowledge of the variability of concentrations of SO_4 in the snow, a minimum value for the amplitude of the changes that are detectable for the particular study in question may be set. By constructing tables similar to Table VI, the relationship between the number of samples, the maximum number of sampling trips that the budget will permit, the amplitude of the trend that is desired and the probable success rate of detecting the trend (power of the test) for ANOVA may be found. Conversely, if the number of samples is restricted by the budget but the field sampling is not, then a table of the power to detect the required amplitude of chemical change by regression analysis may be drawn up. This methodology, however, is only of value in simple systems, e.g. in regions where the probability of prolonged cold periods is high (Arctic, Antarctic). In the Lac Laflamme area the probability of accurately forecasting cold periods of more than one week or so is very low; in addition, the budget costs for analysis of samples and field sampling are comparable. In general, the program of snow sampling at this site relies more extensively on the ANOVA approach; the loss of power is then offset by fewer field samplings and lower probability of unfavorable weather conditions. On the other hand, the study of chemical changes in cold snow in the laboratory where experimental conditions may be easily controlled (Jones and Deblois, 1987) is more amenable to regression analysis.

Acknowledgement

This research was made possible with the financial aid of Environment Canada and the Natural Sciences and Engineering Research Council of Canada.

References

Bickel, P. J. and Doksum, K. A.: 1977, 'Mathematical Statistics: Basic Ideas and Selected Topics', Holden-Day, San Francisco.

Box, G. E. P. and Jenkins, G. M.: 1976, 'Time Series Analysis: Forecasting and Control', Revised Edition. Holden-Day, San Francisco.

Cadle, S. H., Dash, J. M. and Grossnickle, N. E.: 1984, 'Retention and Release of Chemical Species by a Northern Michigan Snowpack', *Water, Air, and Soil Pollut.* **22**, 303-319.

Cadle, S. H., Dash, J. M. and Mulawa, P. A.: 1985, 'Atmospheric Concentrations and the Deposition Velocity to Snow of Nitric Acid, Sulfur Dioxide and Various Species', *Atmospheric Environment* **19**, 1819-1827.

Colbeck, S. C.: 1981, 'A Simulation of the Enrichment of Atmospheric Pollutants in Snow Cover Runoff', *Water Resources Research* **17(5)**, 1383-1388.

Conover, W. J.: 1971, 'Practical Non-Parametric Statistics', 2nd Edition. John Wiley, New York.

Delmas, V. and Jones, H. G.: 1987, 'Wind as a Factor in the Direct Measurement of the Dry Deposition of Acid Pollutants to Snowcovers', in H. G. Jones and W. J. Orville-Thomas (eds.), *Seasonal Snowcovers: Physics, Chemistry, Hydrology,* NATO ASI Series C, Vol. 211, pp. 321-335.

Hirsch, R. M. and Slack, J. R.: 1984, 'A Non-parametric Trend Test for Seasonal Data with Serial Dependence', *Water Resources Research* **20**, 727-732.

Jeffries, D. S. and Snyder, W. R.: 1981, 'Variations in the Chemical Composition of the Snowpack and Associated Meltwaters in Central Ontario', in Proceedings of 38th Eastern Snow Conference. Syracuse, N. Y.: B. E. Goodison, pp. 11-22.

Jones, H. G.: 1987, 'Chemical Dynamics of Snowcover and Snowmelt in a Boreal Forest', in H. G. Jones and W. J. Orville-Thomas (eds.), *Seasonal Snowcovers: Physics, Chemistry, Hydrology,* NATO ASI Series C, vol. 211, pp. 531-574.

Jones, H. G. and Bisson, M.: 1984, 'Physical and Chemical Evolution of Snowpacks on the Canadian Shield (Winter 1979-1980), *Verh. Internat. Verein. Limnol.* **22**, 1786-1792.

Jones, H. G. and Deblois, C.: 1987, 'Chemical Dynamics of N-Containing Ionic Species in a Boreal Forest Snowcover During the Spring Melt Period', *Hydrological Processes* **1**, 271-282.

Jones, H. G. and Sochanska, W.: 1985, 'The Chemical Characteristics of Snowcover in a Northern Boreal Forest During the Spring Run-Off Period', *Annals of Glaciology* **7**, 167-174.

Jones, H. G., Tranter, M. and Davies, T. D.: 1989, 'The Leaching of Strong Acid Anions from Snow During Rain-on-snow Events: Evidence for Two Component Mixing', Atmospheric Deposition (Proceedings of the Baltimore Symposium, May 1989). IAHS, Publ. No. 179. 239-250.

Lettenmaier, D. P.: 1976, 'Detection of Trends in Water Quality Data from Records with Dependent Observations', *Water Resources Research* **12**, 1037-1046.

Montgomery, D. C.: 1984, *Design and Analysis of Experiments*, Second Edition. John Wiley, New York.

Neter, J. and Wasserman, W.: 1974, *Applied Linear Statistical Models*, Homewood: Richard D. Irwin.

Ratkowsky, D. A.: 1983, *Nonlinear Regression Modeling*, Marcel Dekker, New York.

Tranter, M. Brimblecome, P., Davies, T. D., Vincent, C. E., Abrahams, P. W. and Blackwood, I.: 1986, 'The Composition of Snowfall, Snowpack and Meltwater in the Scottish Highlands: Evidence for Preferential Elution', *Atmospheric Environment* **20(3)**, 517-525.

STATISTICAL CONTROL OF HYGIENIC QUALITY OF BATHING WATER

PER SETTERGREN SØRENSEN, JES LA COUR JANSEN
The Water Quality Institute, 11, Agern Allé, DK-2970 Hørsholm, Denmark
and
HENRIK SPLIID
Technical University of Denmark, DK-2800 Lyngby, Denmark

(Received May 1990)

Abstract. In Denmark the hygienic quality of the bathing water has been controlled, based on general guidelines since 1978. Today more than 1100 control sites in marine areas have been established to ensure safe bathing water quality. According to EEC directives and Danish tradition, the control is usually performed by measuring the content of the indicator bacteria Escherichia coli (*E. coli*) in 5 to 20 test samples each bathing season.

In Denmark, control programmes and data are evaluated using basic statistical quality control principles. This paper presents general guidelines for bacterial control, their statistical background and practical application. Furthermore, the evaluation and application of a specific programme for control of Staphylococcus aureus (*S. aureus*) are presented. This programme was used in a Danish bay where the authorities prescribe direct control of these potentially harmful bacteria.

Introduction

Monitoring of bathing water quality includes inspection at bathing sites of colour, smell, foam and other aesthetic properties, along with checks for possible signs of eutrophication or chemical pollution. However, control of the hygienic properties is the most important issue of bathing water quality control due to the risk of infectious diseases being transmitted through the water.

Hygienic properties related to contents of micro organisms cannot be controlled by mere qualitative inspection. Thus measurements of the microbiological condition of the water are necessary. However, a total examination of all pathogenic organisms, i.e. bacteria, viruses, parasites (eggs) etc. is too laborious in daily routine. Also, their excretion is expected to be intermittent, and hence the degree of contamination is assessed instead on the basis of the content of *indicator bacteria.*

Indicator bacteria are groups of bacteria characterized by properties such as being present in larger amounts than pathogenic bacteria whenever these occur, and being at least as resistant as the pathogenic bacteria in the aquatic environment. According to EEC directives and Danish tradition the species *Escherichia coli* (*E. coli*) has been chosen as the indicator bacteria. This group of indicator bacteria is often referred to as faecal coliforms.

For routine control of bathing water the general quality control criterion is stipulated in the guideline (Miljøstyrelsen, 1985): 'In at most 5% of the time during the bathing season the content of *E. coli* is allowed to exceed 1000 per 100 mL bathing water, based on statistical evaluation.'

Environmental Monitoring and Assessment **17**: 217–226, 1991.

Data Model and Basic Theory for General Control

The content of E. coli in the bathing water at a control station is assumed to be independent with a common log-normal distribution within the bathing season, hence the logarithms of contents of E. coli at bathing sites are modelled with normal distributions. These assumptions are based on theoretical considerations and on practical experience from measurements recorded during the period 1978 to 1985. Figure 1 supports the application of the log-normal distribution. A χ^2-goodness of fit test results in a $\chi^2(4) = 6.99$, which is not significant at a 10% level. This justifies the application of the log-normal distribution.

The critical fraction is defined as the relative part of the time the *E. coli* content exceeds 1000 bacteria per 100 mL water. The above control formulation demands that the critical fraction does not exceed 5% with a certain degree of certainty.

The basic inequality is

$$X \leq U,$$

where X is a random variable representing the logarithm of the *E. coli* concentration (per 100 mL) in the bathing water, and U is the logarithm of the control limit of 1000 per 100 mL, that is U equals $\log_e(1000) = 6.91$.

The critical fraction p_u is defined as the probability that $X \geq U$. Thus the objective of the procedure is to test the hypothesis

$$H_0\colon p_u \leq 0.05 \quad \text{against} \quad H_1\colon p_u > 0.05.$$

Under the assumption that X is normally distributed, these hypotheses can be rephrased (Schilling, 1982) yielding the following expressions:

$$H_0\colon \mu + z_{0.95} \cdot \sigma \leq U \quad \textit{against} \quad H_1\colon \mu + z_{0.95} \cdot \sigma > U,$$

where $z_{0.95}$ is the 95%-fractile in the standardized normal distribution, having a value of about 1.65. These quality criteria are illustrated in the diagram in Figure 2, showing the region of distributions that should pass the test (accept H_0) under ideal conditions (infinite number of samples).

Design of Sampling Plan

A sampling plan for operative quality control is designed from the stipulated data model using statistical quality control theory. Historically, the design is based on an Acceptable Quality Level (AQL) point and a fixed, predefined number of samples.

Based on practical considerations concerning sample size and quality objectives, an AQL point (0.05, 0.67) was chosen. This means that two thirds (67%) of all samples from a distribution with a true critical fraction of 0.05 should be accepted by the quality control scheme. The specified fixed number of samples is 5, 10 or 20, and 10 is used under usual conditions.

Based on the assumption that we have a sample $\{X_1, \ldots, X_n\}$ of size n from the random

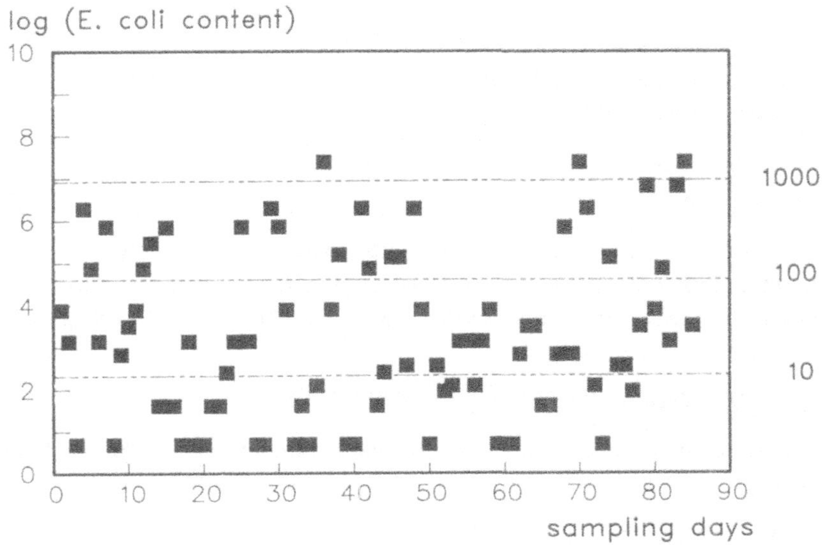

Fig. 1. Illustration of stationarity and lognormal distribution properties of the *E. coli* content during a bathing season at a Danish control station.

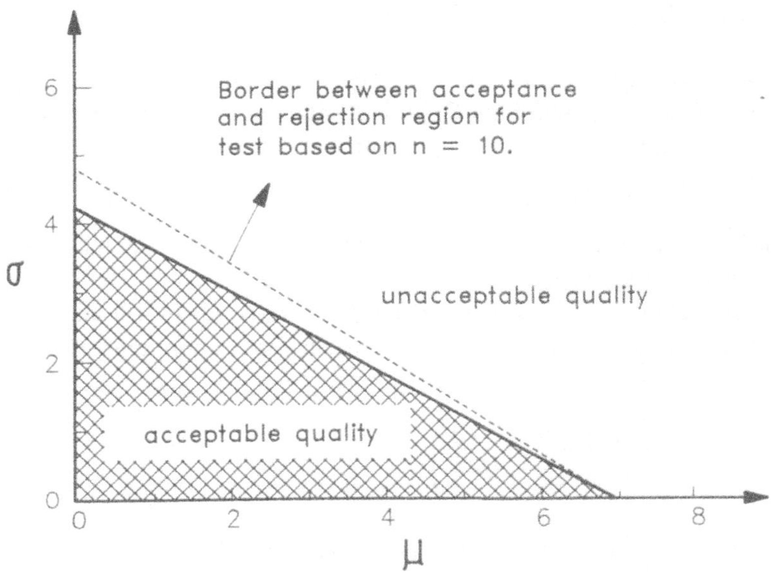

Fig. 2. Phase diagram showing acceptable quality and unacceptable quality regions for the logarithms of the *E. coli* content in bathing water.

variable X, and assuming that the X_i's are independent and normally distributed, consider the test statistic (Schilling, 1982):

$$Z = \frac{U - \bar{X}}{S},$$

where $\bar{X}$ and S are the sample mean and standard deviation, respectively.

Having set up this control strategy, the AQL relation and the given fixed sample sizes can be used to find reasonable values of the quality acceptance limit C. Using the AQL relation we have:

$$\text{AQL-criterion: } P\{\text{reject } H_0 | \; P_u = 0.05\} \leq 0.33.$$

Based on this relation the critical value C for the test quantity is determined from the inequality

$$C \cdot \sqrt{n} < t(n-1, -\sqrt{n} \cdot z_{0.05})_{0.33},$$

where $t(n-1, -\sqrt{n} \cdot z_{0.05})_{0.33}$ represents the 33%-fractile of the noncentral t-distribution with n–1 degrees of freedom and noncentrality $-\sqrt{n} \cdot z_{0.05}$. Although exact values for C and n satisfying this inequality can be computed quite easily we shall for illustrative purposes use the following approximation which is based on the normal distribution:

$$z_{0.33}\,(1 + \frac{C^2}{2}) \approx \sqrt{n}\;(C + z_{0.33})$$

from which appropriate values of C can be found corresponding to specified values of n. Solving for $n = 10$, an acceptance limit $C \approx 1.45$ is obtained, and hence the bathing water quality controlled on the basis of 10 samples during a bathing season is accepted, if

$$\bar{x} + 1.45 \cdot s \leq 6.91.$$

Substituting $\bar{x}$ and s for μ and σ, the borderline between acceptance and rejection regions depicted in Figure 2 is obtained.

The operating characteristic of the stipulated control strategy will analogously be determined from the noncentral t-distribution, but can also be approximated by a normal distribution:

$$\text{OC}(p) \approx 1 - \Phi\left(\frac{\sqrt{n}\;(C + z_p)}{\sqrt{1 + \frac{C^2}{2}}}\right)$$

In Figure 3, operating characteristic curves for sample sizes $n = 5$, 10 and 20 are illustrated.

As illustrated in Table I, the probability rate of acceptance of the bathing water quality amounts to only 20%, if the true critical fraction is exactly 0.05 and the sample size is $n = 5$. The control programme is intentionally designed like this due to the wish that only very

TABLE I

Control strategies as a function of selected sizes n and prescribed AQL points

Specifications				
AQL fraction	p	0.05	0.05	0.05
type I error	α	0.80	0.33	0.33
sample size	n	5	10	20
Results				
accept. limit	C	2.46	1.45	1.50

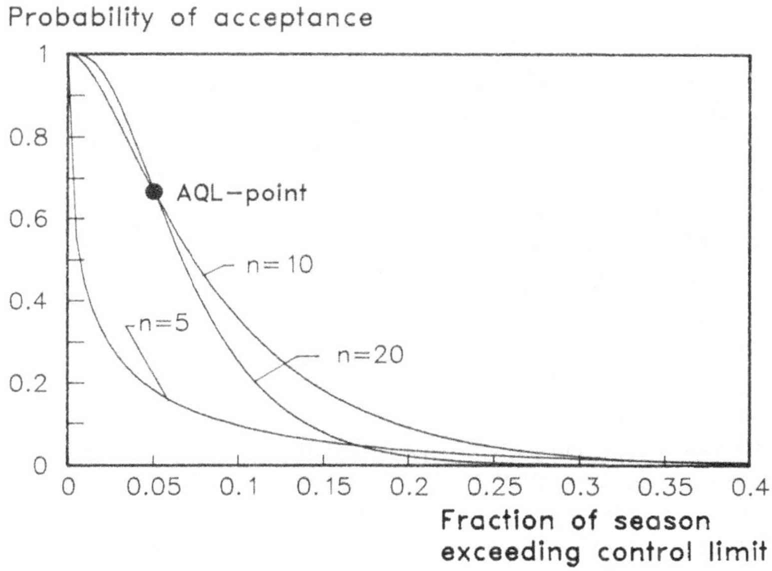

Fig. 3. Operating characteristic curves for sample sizes of 5, 10, and 20.

good quality water should be accepted at a sample size of 5. Sampling in these cases is merely carried out to monitor presumably uncontaminated water, i.e. principles of Limiting Quality (LQ) are applied in this case.

In the case of dubious bathing water quality, the sample size is increased to 20 samples per season in order to examine whether the water quality is in fact worse than the required level or if the rejection was caused by statistical coincidence at a quality better than, but near, the required level. Henceforth, the OC-curve is fixed at the specified AQL point, using the increased sample size to ensure a better discrimination between good and bad water quality.

Application of the Control Programme

Since 1978, control of bathing water has been required by Danish law in all local communities. Results from the period 1978 to 1984 gave rise to a revision yielding the above described control programme based on *E. coli.*

The revised control programme was implemented in 1985 as a result of a guide published by the National Agency of Environmental Protection. The guide contains data sheets and brief calculation guides which are very simple and easy to use. Using the guide allows local community administrations to conduct the statistical control of hygienic quality of their bathing areas themselves.

An example of a sample data set is shown in Figure 4.

When the control of bathing water quality in a season is finished, i.e. measurements are documented, statistics calculated and the bathing water quality rejected or accepted, a list of actions can be implemented. Possible actions include

- prohibition of bathing
- revision of the sample size
- strenghtening of the control limit
- close down of control station.

The sample sizes are to be regulated with regard to a rule based on general considerations concerning the 95% confidence limit of the critical fraction, as shown in Table II.

The proposition in Table II must be true for two consecutive bathing seasons before the sample size can be reduced to 5. On the other hand, when only one season has poor bathing water quality, the sample size must be increased to 20.

Control of *Staphylococcus aureus*

The possible need for direct control of Staphylococcus arose from two cases of infections which were suspected to stem from hospital outlets in the sewage and later on, in the water at the beaches. Preliminary discussions led to the concept of a control programme for these pathogenic bacteria based on the general statistical principles used in the control programme for *E. coli* in bathing water. No general guidelines existed in advance.

The species *Staphylococcus aureus (S. aureus)* was chosen as indicator of this bacteria group, and a control criterion very much like the one launched for *E. coli* was stipulated for *S. aureus* (VKI 1988): 'In at most 5% of the time during the bathing season the *S. aureus* content is allowed to exceed a count of 10 organisms per 100 mL of bathing water based on statistical principles.'

Data model and hypothesis for quality control

Due to limited analytical techniques, bathing water samples can only be analysed to yield a detection of whether the sample contains more or less than 10 organisms. Assuming stationarity during one bathing season, i.e. a fixed probability rate p of a count of 10 or more *S. aureus* per 100 mL, each observation X_i can be modelled by a Bernoulli distribution, i.e. a binomial distribution with $n = 1$:

$$X_i \in B(1,p),\ X_i = \begin{cases} 1 \text{ if count} \geq 10/100 \text{ mL} \\ 0 \text{ if count} < 10/100 \text{ mL.} \end{cases}$$

Hence the sum of observations $Z = \Sigma X_i \in B(n, p)$ follows a binomial distribution. On the

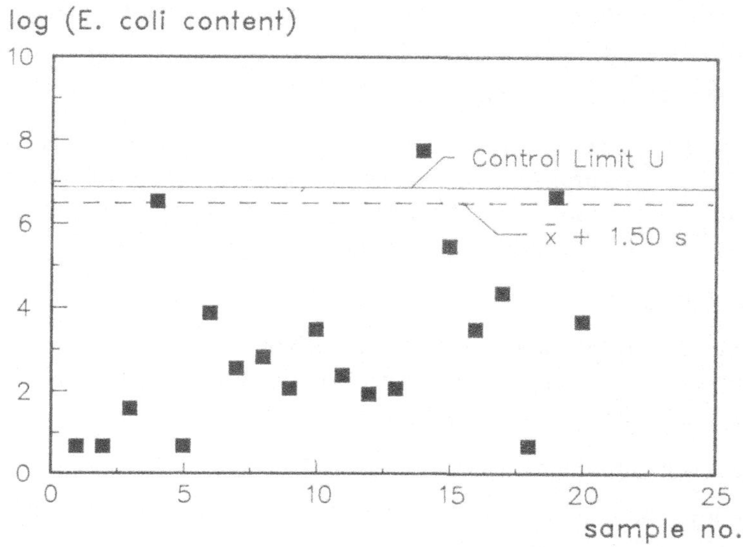

Fig. 4. E. coli results from a control station during one season (20 samples were taken).

TABLE II

Rules for revision of sample size when the bathing water quality is acceptable

$y = \dfrac{U - \bar{X}}{S} - \dfrac{2}{\sqrt{n}}$	New sample size		
	Current sample size		
	5	10	20
$y < 1.65$	10	10	20
$1.65 < y < 2.89$	10	10	10
$2.89 < y$	5	5	5

basis of Z it is possible to test a set of hypotheses very similar to the *E. coli* control programme:

$$H_0\text{: } p \leq 0.05 \quad \text{against} \quad H_1\text{: } p > 0.05.$$

DESIGN OF SAMPLING PLAN

The sum of detections of *S. aureus* above 10/100 mL is known to be a central estimator of the mean value np in the corresponding binomial distribution. The test of H_0 is designed on the basis of a usual AQL/LQ-scheme, with an LQ (Limiting Quality) of $p = 0.20$, i.e.

$$\begin{aligned} &\text{AQL:} \quad P\{\text{reject } H_0 | \, p = 0.05\} \leq \alpha \\ &\text{LQ:} \quad P\{\text{accept } H_0 | \, p = 0.20\} \leq \beta. \end{aligned}$$

Similar to the *E. coli* case, the control strategy is to reject H_0 if $Z > C$, otherwise, it is accepted. As C is an integer between 0 and n, a pragmatic approach was used to search for pairs of (n, C) such that

$$B(n, 0.05)_{1-\alpha} \approx C \quad \text{and} \quad B(n, 0.20)_{\beta} \approx C.$$

Values of (n, C) equal to (20,1) were found to be appropriate. The operating characteristic of this control design is contained in Figure 5. The probabilities of drawing erroneous conclusions α and β are 26% and 7%, respectively, which is quite close to the same figures for the *E. coli* sample size of 10, where α and β are 33% and 9%, respectively.

RESULTS FROM 1988

Twenty samples of bathing water were taken at each of four control stations. Two of the stations had one detection of *S. aureus* (contents 10/100 mL), while the other two had no detections. At all four stations the water quality was accepted ($Z \leq 1$).

THE ASSOCIATION BETWEEN *E. coli* AND *S. aureus*

The monitoring programme was also used to investigate the correlation between occurrence of *E. coli* and *S. aureus.*

An earlier investigation suggests that a quite close correlation exists between the *E. coli* content and the content of the pathogenic bacteria Salmonella (Grunnet, 1978). As a result, it was concluded that monitoring of *E. coli* content would be sufficient for monitoring Salmonella.

This does not seem to be case for *E. coli* and *S. aureus* as no correlation emerges from the illustration in Figure 6.

Summary

Statistical principles have been employed since 1978 in Denmark to evaluate and control bathing water quality. The control programme is based on measurements of the indicator bacteria *E. coli*. Today, there are control stations at 1100 sites where 5 to 20 samples are taken each season.

The sampling plan and control strategy are based on assumptions of stationarity and log-normality of the *E. coli* data and are designed, based on a fixed AQL point and predefined, fixed sample sizes. Also, rules concerning revision of sample sizes between seasons are based on statistical principles.

During 1988 a special control programme for direct control of content of the pathogenic bacteria *S. aureus* in the bathing water was designed and employed in a Danish bay.

This control programme was based on the general concepts of the bathing water quality programme. Due to analytical limitations, contents of *S. aureus* had to be assumed to be binomially distributed and hence a larger sample size was required to ensure a satisfactory reliability level of the control programme.

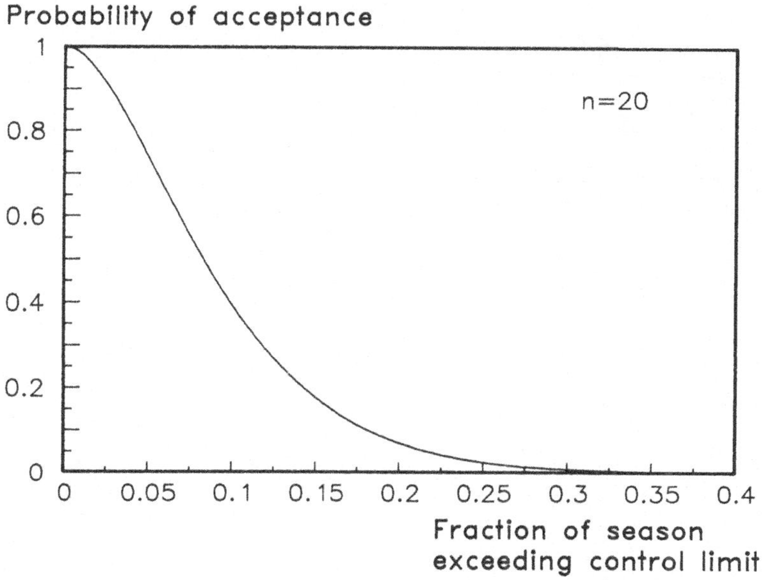

Fig. 5. Operating characteristic curve for control based on 20 samples per season and an acceptance number equal to 1.

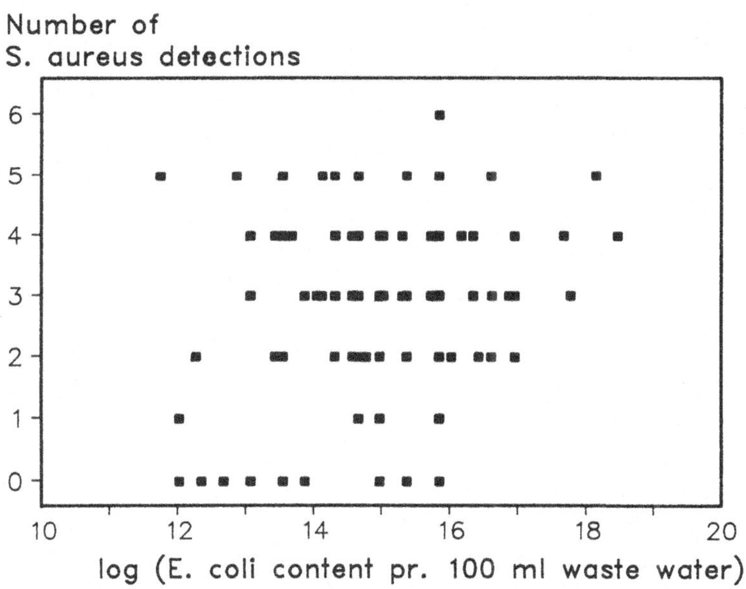

Fig. 6. Illustration of the relationship between *E. coli* and *S. aureus*.

References

Grunnet, K.: 1978, *Selected Microorganisms for Coastal Pollution Studies,* in Coastal Pollution Control, Volume III, WHO training course, pp. 759-775.

Schilling, E. G.: 1982, *Acceptance Sampling in Quality Control,* Marcel Dekker Inc.

Miljøstyrelsen: 1985, *Monitoring of Bathing Water Quality*, Guideline from the National Agency of Environmental Protection, Denmark.

VKI: 1988, *A Monitoring Programme for Staphylococcus in Bathing Water*, Report from the Water Quality Institute by Jansen, Jes la Cour, Sørensen, Per Settergren.

RELATIONSHIPS BETWEEN WATER MASS CHARACTERISTICS AND ESTIMATES OF FISH POPULATION ABUNDANCE FROM TRAWL SURVEYS

STEPHEN J. SMITH,* R. IAN PERRY,** and
L. PAUL FANNING[1]

Department of Fisheries and Oceans, Marine Fish Division

(Received May 1990)

Abstract. The Canadian Department of Fisheries and Oceans conducts annual bottom trawl surveys to monitor changes in the abundance of the major commercially important groundfish populations. Some of these surveys have been in operation for almost 20 yr. The estimates from these surveys often indicate rapid changes in abundance over time beyond that expected from the population dynamics of the fish. Much of this interannual change has been interpreted as variation, the magnitude of which has often made it difficult to measure anything but the most severe effects of fishing, pollution or any other intervention on the population. Recent studies have shown that some of this variation may be attributed to changes in catchability of fish due to the effects of environmental variables on fish distribution. Annual changes in abundance as estimated from such field surveys may be confounded by changes in catchability due to annual changes in environmental conditions. In this study, trawl catches of age 4 Atlantic cod (*Gadus morhua*) from surveys conducted during March 1979–1988 were compared with concurrent measurements of bottom salinity, temperature and depth. Large catches of age 4 cod are more likely to occur in water characterized as the intermediate cold layer defined by salinities of 32-33.5 and temperatures <5 °C. This relationship also appears to be modified by depth. We further show that interannual changes in the estimated abundance from the surveys were, in a number of cases, coincident with changes in the proportion of the bottom water composed of the intermediate cold water layer. The implications that these patterns may have on interpreting trends in the estimates of abundance from trawl surveys are discussed.

1. Introduction

As part of its fisheries management mandate, the Canadian Government conducts annual groundfish trawl surveys of the various fishing grounds on the east coast of Canada to monitor changes in abundance over time of commercially exploited fish species. These surveys use a stratified random design with bottom depth as the major stratifying variable. Other considerations such as prior evidence of fish distributions and general logistics are used to delineate finer areal strata. The history, development and experience with these surveys on the east coast of North America are given in Doubleday and Rivard (1981). Smith (1988) gives an account of developments since 1981 for Canadian surveys.

The abundance estimates from these surveys often have large standard errors within years and also exhibit sudden temporal changes in abundance over time (Smith, 1988). Gavaris and Smith (1987) and Francis (1984) have explored the use of alternate sample-to-strata allocation schemes to improve the precision within any one trawl survey. Parsons

* P.O. Box 1006, Dartmouth, Nova Scotia, B2Y 4A2.
** Biological Station, St. Andrews, New Brunswick, E0G 2X0, Canada.

Environmental Monitoring and Assessment **17**: 227–245, 1991.

(1988) reported substantial improvements in the precision of the abundance estimates from a survey of shrimp using Francis' approach. Statistical distributions have also been used to deal with the large variances associated with abundance estimates from trawl surveys. The use and misuse of these models are discussed in Jolly and Smith (1989) and Smith (1990). However, all of these methods confine their attention to decreasing the variance within a survey and not to the apparent variability observed over time. One exception to this is Pennington's (1985, 1986) approach which combines the use of the Δ-distribution for the number of fish caught per tow with a time series model for smoothing temporal trends.

Changes in location and extent of the water masses in which the fish are generally found could affect the availability of the fish to the trawl and thereby introduce pronounced variation in the annual estimates (Pinhorn and Halliday, 1985). A number of studies have shown that water temperature, salinity, bottom type and depth appear to be related to the distribution of many species of commercial fish (Scott, 1982; Mahon *et al.*, 1984; Perry *et al.*, 1988). Also different age groups of the same species may respond differently to environmental conditions. Tremblay and Sinclair (1985) studied the age-specific distributions of Atlantic cod (*Gadus morhua*) in the Gulf of St. Lawrence during autumn, and noted that older fish occurred at increasing depth and salinity, and decreasing temperature. They were unable to identify any single dominant parameter controlling the distribution, in part due to the strong correlation between depth, temperature, and salinity. The correlation between these parameters is a problem when trying to distinguish which environmental factors may affect spatial distributions. Indeed, the use of depth as a stratifying variable in the trawl surveys was an attempt to incorporate gross limits on the distribution of the fish, however, this does not account for the possibility of changes in temperature and salinity. Smith (1990) has described a method of incorporating environmental information into the abundance estimates from trawl surveys. This method provides a *predictive* estimate which uses relationships between the observed catch of fish and concurrently measured covariates (e.g. hydrographic variables) to predict abundance over areas where only the covariates are known. A model was developed which related salinity, temperature and depth to the catch of 4 year old Atlantic cod in trawl surveys on the eastern Scotian Shelf for March 1986-87. Individual measurements of salinity and temperature would be difficult to obtain however, over the whole survey area for all sample units.

In this paper we investigate the spatial and temporal variability of bottom temperature and salinity on the eastern Scotian Shelf measured during March trawl surveys over the period of 1979–1988 (excluding 1985 for which there was no survey). We suggest that water mass, defined by a specific temperature and salinity range, provides a more informative measure of hydrographic conditions than either temperature or salinity alone. Relationships between water mass and the catch of age 4 cod are also investigated. The model in Smith (1990) was reformulated to include water mass information and used to analyze the age 4 cod data for the 1979–1988 time period. The relationships obtained from this analysis are then used to comment on whether trends in the estimates from trawl surveys reflect changes in abundance and/or availability.

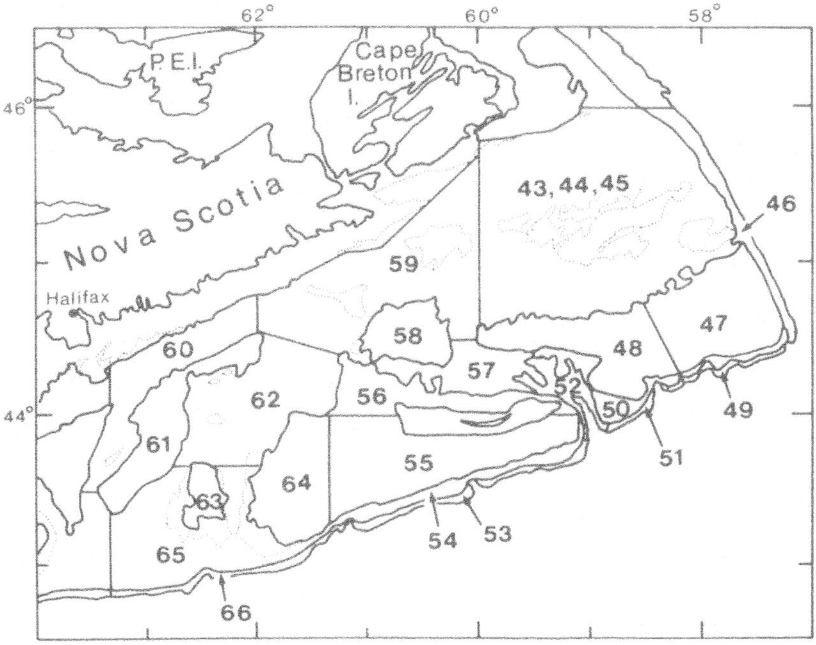

Fig. 1. Stratification Map for the Scotian Shelf (1979–84). Strata primarily based on depth ranges.

2. Survey Methods

Stratification schemes for groundfish trawl surveys on the Scotian Shelf are presented in Figures 1 and 2. The first scheme has been used since 1970 for the July survey series and from 1979–1984 for the March survey series. The strata boundaries are primarily based on depth ranges of 0–50 fm (0–91 m), 50–100 fm (91–183 m) and 100–200 fm (183–366 m). The second scheme (Figure 2) is an experimental design with boundaries based on the spatial distribution of cod from previous surveys. This system has been in place for the March surveys from 1986 to the present. No survey was made of the area in 1985.

The areas covered by the two survey designs are not completely coincident. The major difference between the two is the exclusion of the areas designated as strata 60 and 61 (Emerald Basin) in Figure 1 in the scheme in Figure 2. Cod are rarely found in these strata and they were excluded from the following analysis. The other difference between the two survey designs is in the area designated as strata 43, 44 and 45 in Figure 1. The northern section of these strata extend into the neighbouring management unit known as NAFO (Northwest Atlantic Fishery Organization) Subdivision 4Vn or the Sydney Bight area of Cape Breton. In the experimental design the upper boundaries of strata 401 and 402 are coincident with the upper boundary of the 4VsW cod stock area which is the subject of the analysis in this paper. Very few sets occured in the area of overlap with the 4Vn area during the March surveys of 1979–1984 and therefore this area was retained in the analysis of those years.

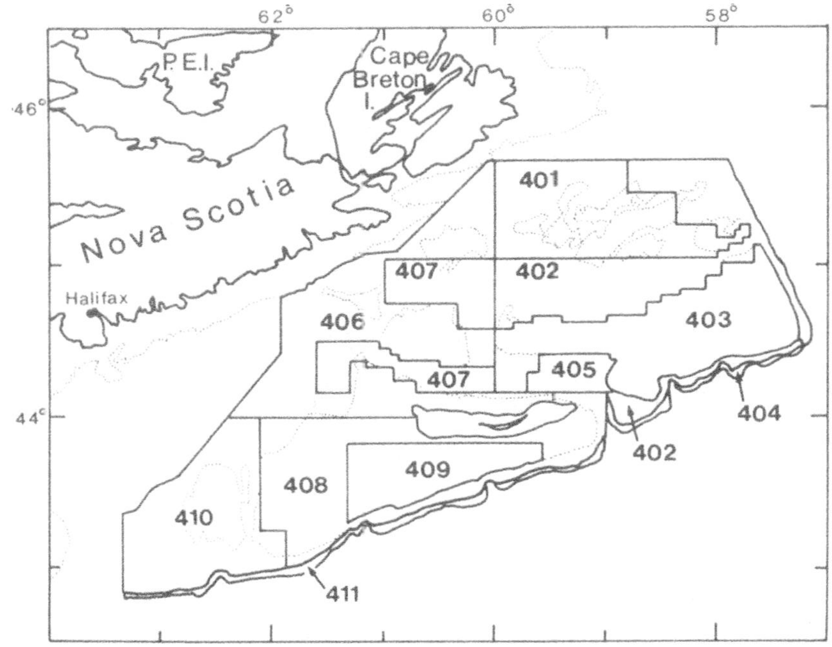

Fig. 2. Stratification Map for the Scotian Shelf (1986–present). Strata primarily based on spatial distribution of cod from previous years.

In both schemes the sample unit is defined as the area over the bottom covered by a trawl of a specific width towed for a distance of 1.75 nautical miles. The sample units or sets are preselected before the cruise and randomly located in each stratum. Following the tow, the net is brought aboard and samples of bottom water are obtained for hydrographic measurements using water sample bottles with paired reversing thermometers. Salinities were determined in the laboratory using a Guildline model 8400 Autosal. Further details on the operations of these surveys are given in Doubleday (1981).

The sample sizes for these cruises ranged from 62 to 92 sets. Spatial coverage was not complete in all years, being limited by sea ice to the northwest which resulted in strata 43, 45, 57 and 58 being excluded in 1980, strata 43, 44, 45 and 46 in 1982 and stratum 401 in 1988.

3. Environment

General description of the oceanographic environment of the Scotian Shelf

The waters of the Scotian Shelf form three distinct vertical layers. Hachey (1942) defined their characteristics as:

	Thickness (fm.)	Salinity	Temperature (°C)
(1) Upper layer	0–63	< 32.0	> 5 in summer < 5 in winter
(2) Intermediate layer	16–47	32.0–33.5	< 5 generally
(3) Bottom Layer	to bottom	> 33.5	> 5 generally

The lower salinity upper layer is derived mainly from the Gulf of St. Lawrence, and is warmed by surface heating in summer (McLellan, 1954a). The cold intermediate layer is also derived predominately from the Gulf of St. Lawrence, but can experience contributions directly from the Labrador Current (McLellan and Trites, 1951). The warm bottom layer is derived from intermediate and deep Slope Water found seaward of the continental shelf. This water flows onto the shelf via the deep channels and gullies between the shallow banks. Its advection shoreward appears to be due in large part to alongshore winds at the shelf break, which generate upwelling and onshore flow at mid-depth (Petrie, 1983). During winter and spring, the temperature in the upper and intermediate layers are similar but the three layer structure remains due to salinity stratification.

The general circulation over the shelf is to the southwest, with maximum currents found on the inner third of the shelf. This, coupled with mixing of shelf water with offshore Slope Water, causes the surface and intermediate layers to become warmer and more saline to the southwest.

The near-bottom temperature and salinity will depend to a large extent on the bottom depth, and whether the location is within the warm bottom or cold intermediate layer. However, the layer thicknesses are variable, both spatially and temporally and depend upon in part the contributions from their respective source regions. Deeper areas usually covered with warm bottom layer waters may experience markedly lower temperatures if the intermediate layer becomes unusually thick (McLellan, 1954b). The temperatures within the layers can also vary with changes in the temperatures of their source regions and with the extent of mixing.

SPATIAL AND TEMPORAL VARIABILITY OF HYDROGRAPHIC PROPERTIES

The stratified mean and standard error of bottom temperature and salinity for each year were calculated for each survey according to the formulae given in Cochran (1977) and are presented in Table I. The proportions of the characteristic water masses near bottom for each survey identified using the definitions of Hachey (1942) are presented in Table II. On average, spring temperatures were warmest in 1984 and coldest in 1987 (Table I). Consistent with this are the proportions of the cold intermediate layer in these two years, with 1984 having the smallest proportion and 1987 the largest (Table II). The lowest mean salinity occurred in 1986, which also had the largest proportion of the low salinity upper water mass (Table II).

The warmest and saltiest years in spring were 1981, 1983, and 1984 (Table I), of which 1981 and 1984 had the highest proportions of the warm, saline bottom water (Table II).

TABLE I

Estimates of stratified mean temperature and mean salinity with associated standard errors for groundfish trawl surveys conducted during March, 1979–1987. Note no survey was conducted during 1985.

	Temperature		Salinity	
Year	Mean	Standard Error	Mean	Standard Error
1979	3.2	0.16	33.1	0.13
1980	3.6	0.18	32.9	0.09
1981	3.9	0.17	33.2	0.07
1982	3.0	0.12	33.1	0.04
1983	4.0	0.21	33.1	0.06
1984	4.6	0.19	33.1	0.11
1986	3.4	0.32	32.8	0.10
1987	2.8	0.30	32.9	0.09
1988	3.3	0.28	33.0	0.10

TABLE II

Estimate of stratified proportion of each water mass type

Year	Upper	Intermediate	Bottom
1979	0.026	0.693	0.281
1980	0.091	0.677	0.232
1981	0.104	0.547	0.349
1982	0.016	0.676	0.308
1983	0.026	0.678	0.296
1984	0.205	0.420	0.375
1986	0.267	0.423	0.310
1987	0.011	0.796	0.193
1988	0.113	0.622	0.265

TABLE III

Estimates of stratified mean depth for March Survey with standard error

Year	Mean (fms.)	Standard error
1979	65.9	0.91
1980	66.0	2.63
1981	68.5	2.27
1982	58.8	1.32
1983	62.1	1.97
1984	67.3	1.52
1986	58.9	2.24
1987	63.3	2.63
1988	62.3	3.00

The extent of the characteristic water masses can therefore influence the mean oceanographic properties for each survey. However, differences of temperature and salinity within water masses can also modify the overall conditions. Both 1982 and 1983 had similar proportions of all three water masses (Table II), yet on average 1982 was relatively cold and 1983 relatively warm (Table I). No systematic differences appeared in the mean depths sampled on each survey (Table III). Instead, mean temperatures within the intermediate and upper layers were warmer in 1983 (1.7 °C and 2.8 °C, respectively) than in 1982 (–0.3 °C and 1.2 °C, respectively). The temperature salinity plots of Figure 3

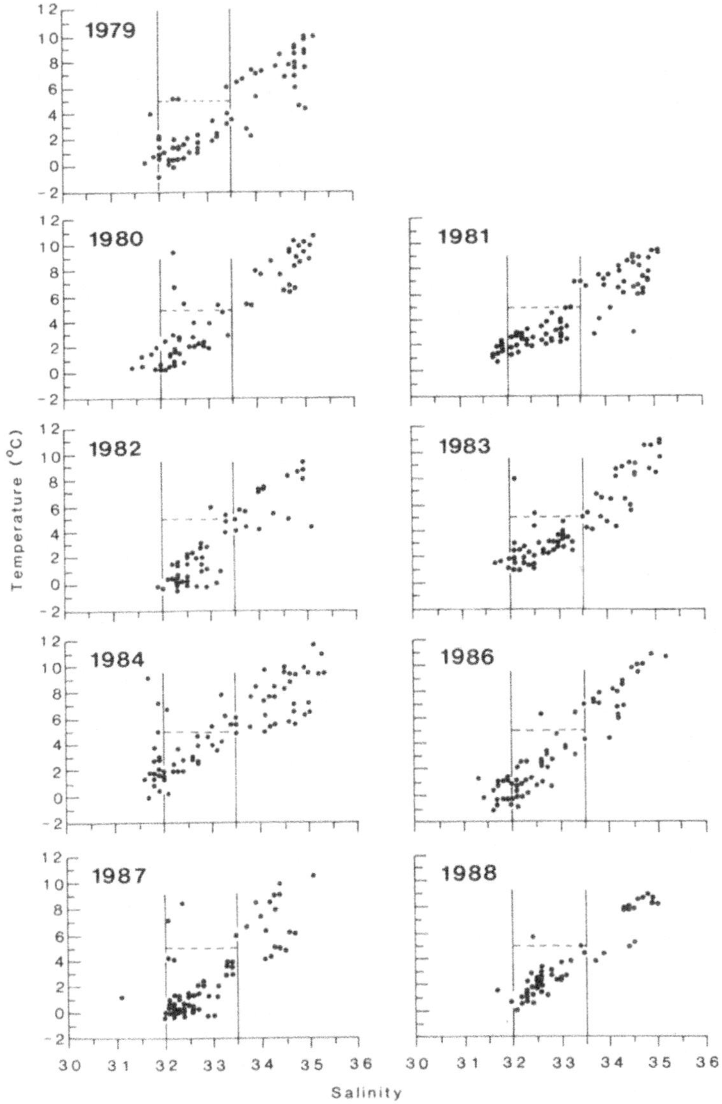

Fig. 3. Temperature vs. Salinity plots for surveys conducted during March, 1979–1988. Salinity and Temperature limits shown for Intermediate cold layer.

indicate the interannual variability in the extent of these properties within specific water masses on the eastern Scotian Shelf.

4. Hydrography and Fish Distribution

Spatial Relationships

The annual number of sets occuring in each water mass and those in which age 4 cod were caught are given in Table IV. Although cod can occur in all three water masses they are relatively rare in the upper layer. A much stronger pattern was evident in the proportion of the total number of age 4 cod caught in each water mass (Table V). The observed proportions in the intermediate cold layer, with the exception of 1980, exceeded that expected based upon the number of sets which contained age 4 cod. Also the largest catches from each survey were always associated with the intermediate water mass.

TABLE IV

Frequency of occurrence of water mass types during March surveys. The number of sets with age 4 cod for each water mass type is shown in parentheses.

	Water mass		
Year	Upper	Intermediate	Bottom
1979	3 (0)	40 (22)	25 (8)
1980	6 (3)	33 (21)	19 (7)
1981	10 (5)	38 (26)	30 (14)
1982	1 (1)	41 (26)	16 (8)
1983	2 (0)	45 (27)	27 (8)
1984	16 (8)	29 (18)	30 (11)
1986	19 (7)	36 (22)	20 (10)
1987	1 (0)	69 (38)	17 (8)
1988	1 (0)	47 (20)	20 (10)

TABLE V

Percent of total catch of age 4 cod by water mass type and year. March surveys NAFO 4VsW. Expected percent of total catch based on relative number of sets with cod present is given in parentheses.

	Water mass		
Year	Upper	Intermediate	Bottom
1979	0.0	74.0 (73)	26.0 (27)
1980	5.2 (9)	63.8 (68)	31.0 (23)
1981	5.5 (11)	82.5 (58)	12.0 (31)
1982	3.6 (3)	92.7 (74)	3.7 (23)
1983	0.0	98.2 (77)	1.8 (23)
1984	5.4 (21)	70.5 (48)	24.1 (30)
1986	2.0 (18)	91.6 (56)	6.4 (26)
1987	0.0	97.1 (83)	2.9 (17)
1988	0.0	98.1 (67)	1.9 (33)

TABLE VI

Observed depth (fms.) ranges for each water mass type. March surveys NAFO 4VsW.

Year	Water mass		
	Upper	Intermediate	Bottom
1979	18–31	11–125	45–190
1980	17–38	32–145	49–192
1981	20–36	25–168	48–197
1982	32	20–128	56–146
1983	19–22	18–133	41–190
1984	16–148	23–176	42–181
1986	19–99	22–130	50–111
1987	56	23–179	41–154
1988	68	18–138	57–143

The minimum and maximum depths for the three layers observed during the surveys are given in Table VI. It may be possible that the observed association between the occurrence of cod and the cold intermediate layer was an artefact of the cod's association with a specific depth range. Smith (1990) found that depth was a significant factor in a model relating the number of age 4 cod caught and the associated salinity, temperature and depth measured at the trawl site. That model was defined as follows. Let y_{hi} be the number of age 4 cod caught in set i and stratum h, and $\mathbf{x}$ be a $p \times 1$ vector of explanatory covariates or factor levels. Observed relationships between the mean and variance and further evaluation of the residuals suggested a Poisson distribution for y_{hi} such that,

$$E[y_{hi}] = \mu_h(\mathbf{x}),$$
$$\text{Var}\ [y_{hi}] = \varnothing\mu_h(\mathbf{x})$$

where $\varnothing$ is a nuisance parameter denoting extra-Poisson variation.

The exponential form, i.e. $\mu_h(\mathbf{x}) = \exp(\beta\mathbf{x})$, where β represents a vector of coefficients, was used to ensure that predicted values were greater than or equal to zero. We modified this model to include water mass characteristics as a grouping variable or factor and nested depth within each water mass. That is, $\mu_h(\mathbf{x}) = \exp(\beta_{jh0} + \beta_j x_{hi})$, where j indexes water mass and x_{hi} is the depth measured at set i in stratum h. Separate intercepts (β_{jh0}) were fitted for each stratum. The procedure for assessing whether or not water mass and depth were important in explaining cod catches involved testing the significance of the water mass terms and then comparing the estimated coefficients for depth within each level of the water mass. All parameter estimates were obtained for the 1979–1988 data using the GLIM software package (Payne, 1986). The effects of water mass and depth on the number of cod caught were evaluated using the analysis of deviance approach and χ^2 test discussed in McCullagh and Nelder (1983).

The results of fitting the Poisson model are given in Table VII. There were very few cod caught in 1979 and 1980 and this may account for the difficulties in fitting the model to these data. Very little pattern was left in the residuals for these two years once the effect due to water mass and stratum had been accounted for. As a result, the iterative algorithm

TABLE VII

Analysis of Deviance results from the fitting of the Poisson Model. Model A=stratum and water mass effects only. Model B=Model A plus depth effects nested within water mass. The term $\hat{\phi}$ refers to the nuisance parameter for extra-Poisson variation. The P-level refers to the χ^2 statistic.

	Model A		Model B	
Year	P-Level	$\hat{\phi}$	P-Level	$\hat{\phi}$
1979	0.0001	0.73	no convergence	
1980	0.8645	4.06	no convergence	
1981	0.0087	7.17	0.0588	6.32
1982	0.9836	98.96	0.2019	93.79
1983	0.6159	111.30	0.0038	86.32
1984	0.0008	14.74	0.1632	13.50
1986	0.0002	79.62	0.0	45.87
1987	0.1300	16.46	0.0	8.82
1988	0.6770	30.64	0.0057	25.36

in GLIM failed to converge when depth was included. Results show that for the remaining years, with the exception of 1982, either the water mass and/or the depth terms were highly significant. In all cases where the depth terms were significantly different from zero, they were so only for the terms nested within the intermediate layer water mass.

Parameter estimates for the intermediate layer depth terms for the 1981–1988 data are given in Table VIII. The predicted values where the P-levels were less than 0.05 are plotted in Figure 4. These values were all derived assuming a zero stratum effect for comparability. The coefficients for these four years (1983, 1986, 1987 and 1988) are within 2 standard errors of each other and therefore are probably indistinguishable statistically. The major difference between the resultant curves lay in the depth of the predicted maximum catch and the magnitude of this maximum. In Figure 4 the maximum catch occurs somewhere between 90 and 111 fathoms depending upon the year. The predicted

TABLE VIII

Parameter estimates for the depth terms within the intermediate layer water mass for years where these terms were significant in Table VII. In the case of the estimates for 1988: (i) all data analyzed; (ii) remove one large catch only; and (iii) remove three zero sets only (see text).

Year	Depth	Std. Error	$(\text{Depth})^2$	Std. Error
1981	0.0829	0.035	−0.0004	0.0003
1982	−0.1119	0.184	0.0023	0.0024
1983	0.2328	0.079	−0.0012	0.0004
1984	0.1621	0.032	−0.0003	0.0002
1986	0.2508	0.056	−0.0014	0.0003
1987	0.2225	0.043	−0.0010	0.0002
1988 (i)	0.1772	0.091	−0.0009	0.0009
1988 (ii)	0.2325	0.083	−0.0010	0.0009
1988 (iii)	0.2549	0.121	−0.0010	0.0012

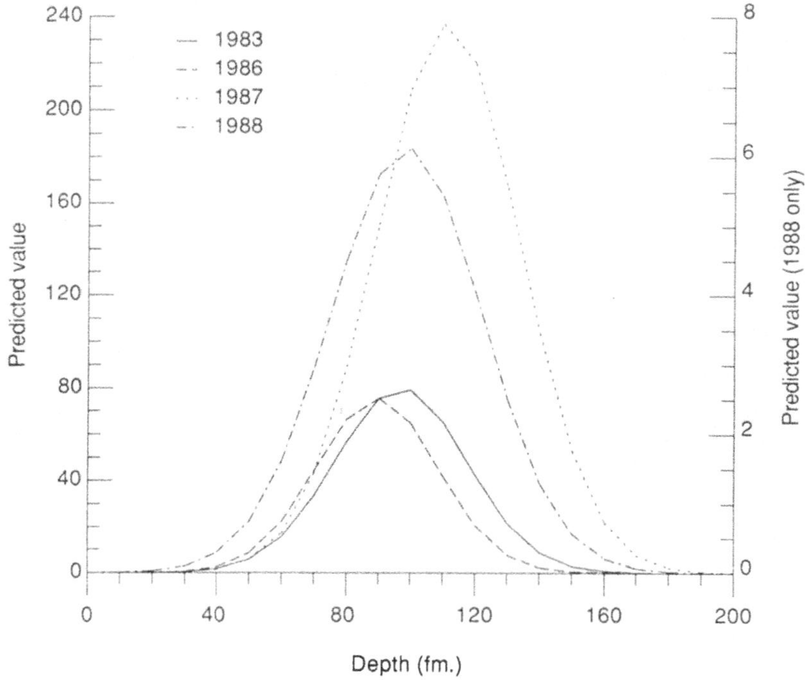

Fig. 4. Predicted values of catch as a function of depth from Poisson model. Scale for catch is arbitrary because a zero stratum effect was assumed for the sake of comparison. Left ordinate axis applicable to 1983, 1986 and 1987. The right ordinate scale refers to 1988.

values for all of the survey years scaled by their respective maximum catch are similarly plotted in Figure 5. Although the coefficients for 1981 (Table VIII) appear to be different from those used to plot the curves in Figure 4, the associated predicted values follow a similar pattern with the maximum catch occurring at 108 fathoms. The very different curves for 1982 and 1984 are due to both surveys having very few observations at the deeper depths. Ice cover in 1982 prevented the survey vessel from sampling much of the deeper water on the shelf while there appeared to be a marked absence of intermediate layer water at deeper depths in 1984. The parameter estimates for the 1988 data exhibited a degree of instability due to limited sampling of the deep water because of ice cover. The general pattern of catches followed that of the other years with a trend towards increasing catches with depth and then a decrease in the number caught for the deeper water. However, the largest catch was made very near to three other locations of similar depth where no cod had been found during the survey a week earlier. The salinities were equal in both areas at the different times although the temperature for the zero catches ranged from 2.1 to 2.9 °C while the temperature was 3.3 °C at the location and time of the large catch. This high catch and the zero catches nearby were highly influential on the parameter estimates. This resulted in large standard errors for the depth coefficients for the 1988 data which was in contradiction to the low P-level given in Table VII for those terms when added as a group. The effect of removing either the large set or the three zero catches nearby are presented in Table VIII in the rows labelled 1988 *ii* and 1988 *iii*,

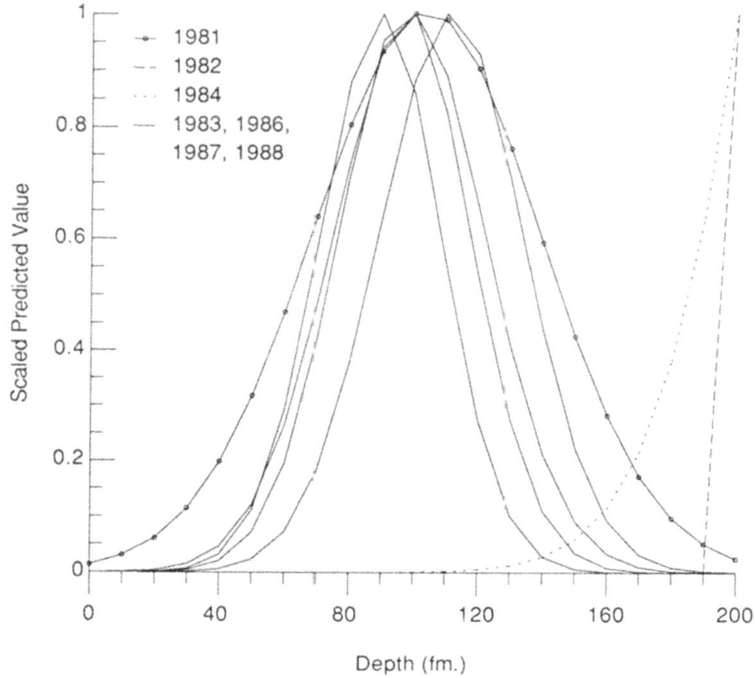

Fig. 5. Predicted values of catch as a function of depth from Poisson model. Catch has been scaled by the maximum predicted catch value for each year.

respectively. The estimates for the depth terms in these cases are similar to those for the 1983, 1986 and 1987 data, however the standard errors for the $(\text{Depth})^2$ coefficients still remain high. The major differences between the predictions from the coefficients from the full data set (labelled as 1988 *i*) and those obtained from the reduced data set (1988 *ii* and *iii*) are the increased magnitude of the predicted maximum catch and the deeper location of this maximum at 130–140 fathoms. For the remaining years there were enough observations of few or no age 4 cod in sets in intermediate layer water greater than 120 fathoms to define the curves predicted by the models. The results for the data from 1982, 1984 and 1988 show that this model is extremely sensitive to the range of depths sampled in the survey. That is, the absence or rare occurrence of intermediate layer water at depths greater than 100 fms. or the failure to adequately sample such depths will result in an unsuccessful fit of the model. The latter can be avoided during the cruise, however the former problem can not be controlled. It may be possible to increase the sampling intensity so that more deep intermediate layer water is observed, given detection to the rare occurrence of intermediate layer on the bottom early in the survey.

Catch per unit effort (CPUE) for cod sets observed on commercial trawlers ($150 < \text{gross tonnage} < 500$) fishing in the survey area during February and March, 1980–88 are plotted against the depth of the respective tows in Figure 6. These data were obtained from the International Observer Program and were confined to sets which designated as cod sets

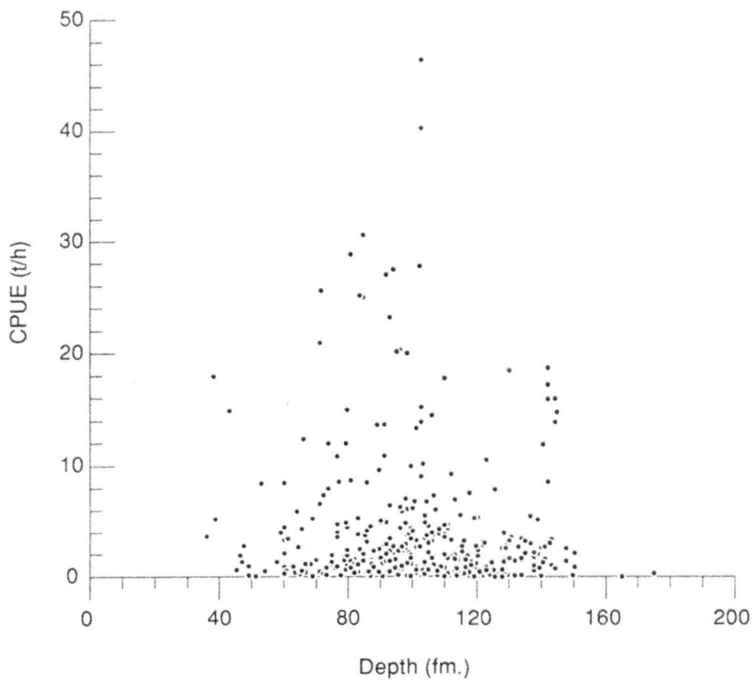

Fig. 6. Commercial catch per unit effort (*t/h*) from commercial trawlers operating in the NAFO area 4VsW during February and March, 1980–1988. Data obtained from the International Observer Program of the Department of Fisheries and Oceans.

prior to the tow being made. All age groups are included in this plot. Unfortunately no observations were made on water mass for these sets. The trend in CPUE with depth is similar to that predicted by the survey-based models with an apparent increase in the catch rate with depth to a maximum around 100 fathoms and then a decrease for water deeper than 120 fathoms. The cluster of high catch rates observed at approximately 140 fathoms are all from the same trip in 1984 and at the same location in the deep basin just north of Middle Bank (stratum 407, Figure 2). The length frequency taken by the observer on board the vessel indicates that the majority of the cod caught were older than 4 yr. These data cannot be used to confirm the predictions from the survey based models because of the lack of data on water mass and incomplete data on age composition. However, they do indicate that the spatial distributions of cod encountered by commercial groundfish trawlers during the same time of the year as the survey also seem to be related to depth.

TEMPORAL RELATIONSHIPS

The estimates of the abundance of age 4 cod from cohort analysis (Fanning and MacEachern, 1989), the July survey and the March survey series are plotted in Figure 7a. The July survey series is often used to derive the cohort estimate and therefore does not always represent an independent estimate of abundance from that of the cohort analysis. All three series give the same general trend for the relative strengths of the year-classes of

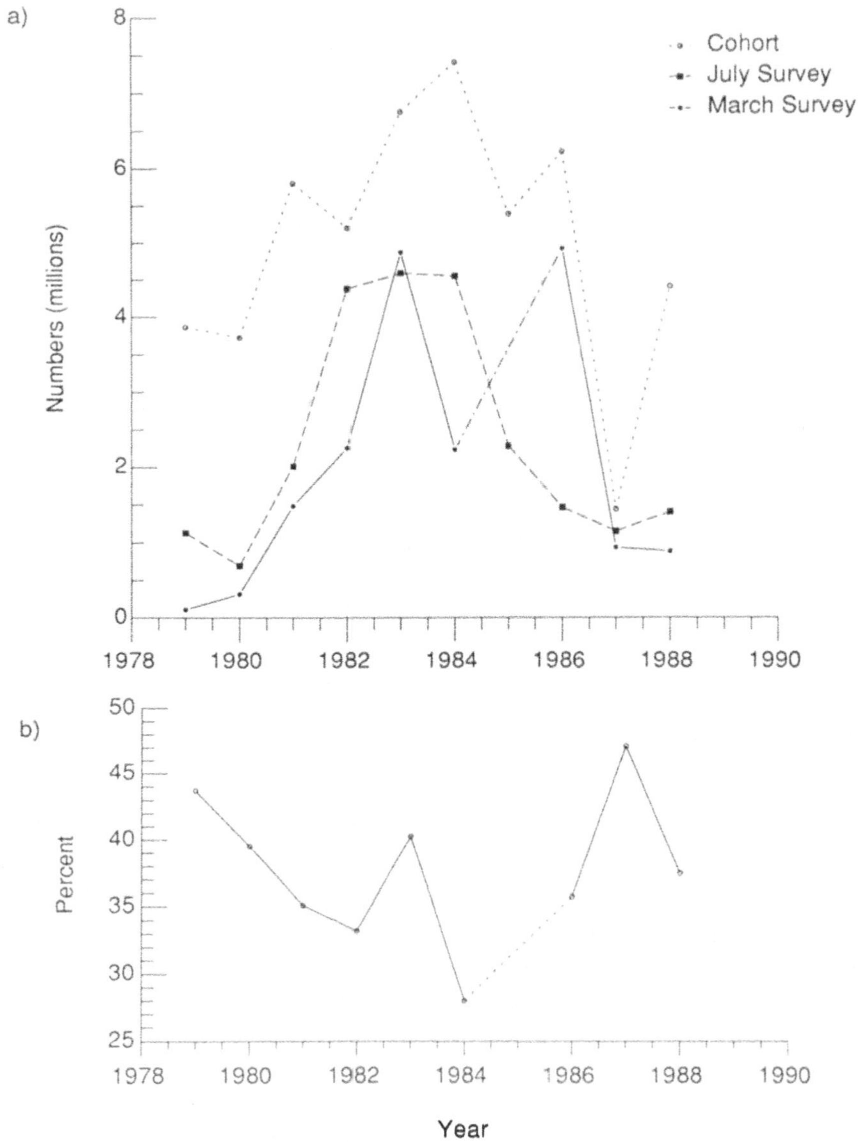

Fig. 7 (a.) Temporal trends in estimates of abundance of age 4 4VsW cod from cohort analysis, the July research survey and the March research survey for 1979–1988. (b.) Temporal trend of the proportion of water on the bottom identified as intermediate cold layer water mass during the March surveys, 1979–1988. Note that there was no survey of the area in 1985.

1975 to 1984 at age 4. That is, a weak year-class in 1979 followed in general, by increasingly stronger year-classes at age 4 until sometime between 1983 and 1986. Thereafter, we observe the extremely weak 1983 and 1984 year-classes in 1987 and 1988, respectively. The major discrepancies between the March series and the other two series are the estimates of relative abundance in 1984, 1986, and 1988. In the latter case the difference between the two survey estimates and the cohort analysis can be ignored because the cohort estimate of age 4 in 1988 was considered to be unreliable in the stock assessment (Fanning and MacEachern, 1989). The cohort estimate of the 1984 year-class at age 4 will become more reliable as more of the year-class is caught at older ages.

The estimates of the proportion of the bottom water at depths greater than 40 fathoms identified as intermediate layer water from each of the March surveys is plotted against time in Figure 7b. The 40 fathoms limit was chosen based on the patterns observed in Figure 4. The peaks in the March survey (Figure 7a) in 1983 and 1986 appear to be associated with relatively high proportions of intermediate layer water, while the lower 1984 March survey estimate is coincident with the lowest proportion of this water mass in the series. It may be that the extent of the bottom water consisting of the intermediate layer water mass can distort the survey's view of actual changes in abundance. That is, the peaks in the March survey abundance trend may reflect increases in availability of the cod to the trawl rather than strong year-classes. The low March survey estimate for 1984 was contrary to the indications from the cohort analysis for the same year. In this case the survey estimate may be indicating a sharp decrease in availability due to the lower than usual quantity of intermediate layer water on the bottom. These interpretations assume that: (a) the availability of the cod to the survey gear is directly related to the presence of this water; (b) the survey estimate of the proportion of intermediate layer water near the bottom is reasonable; and (c) the cohort estimate represents the actual population trend. We also assume that changes in availability are simply due to the fish being either on the bottom or up in the water column depending upon whether or not the intermediate layer water mass is in contact with the bottom.

Although the proportion of intermediate layer water on the bottom decreased from 1979 to 1982, the 1977 and 1978 year-classes may have been strong enough to be detected by the survey despite the decreased availability. Note that these two year-classes were ranked fourth and sixth highest amongst the cohort estimates. The largest proportion of intermediate layer water in 1987 was also coincident with the 1983 year-class which has been estimated in the cohort analysis as the weakest year-class at ages 1–5 in the 1971–1988 period. In fact the cohort analysis estimates the total cod population in 1987 to have been the smallest since 1978. However, the March survey estimate of the 1983 year-class suggests that it was larger at age 4 than either the 1975 and 1976 year classes. This discrepancy with the cohort analysis ranking of the strengths of these year-classes could be attributed to the increased availability of a weak year-class due to the relatively larger amount of intermediate layer water on the bottom.

Recall from Table I, that the average bottom temperature in 1987 was the coldest in the series despite the very low proportion of the usually cold upper layer water (Table II). Bottom temperature measurements taken during shrimp research cruises in the same area indicate that 1987 was the coldest year on record since the beginning of the series in 1982 (Etter and Mohn, 1989). The cod were also found deeper than usual in 1987 with 50% of the catch taken at greater than 109 fathoms. For the remaining years in the March series almost 90% of the 4 yr old cod was caught at depths shallower than 109 fathoms. Indeed the model in Table VIII predicts a deeper depth for the peak catch for 1987 than for the other years in Figure 4. In addition, catches in 1987 were generally restricted to the areas just north of Banquereau Bank (stratum 402, Figure 2) unlike the other years when catches were widespread over the shelf. Finally, the ice cover was, also uncharacteristic in 1987 with the majority of the ice confined to inshore areas of Nova Scotia by the winds. Ice blocked Halifax Harbour for the first time in more than 30 yr and the survey vessel required an icebreaker escort to enter the harbour. Any or all of these events may have been indicative of unusual hydrographic conditions which in turn may have resulted in the intermediate water mass being less suitable for the cod 1987.

The 1984 year class in 1988 was either weaker than the 1983 at the same age or somewhat stronger but less available to the trawl gear because of a decrease in the proportion of intermediate layer water in 1988. The July survey estimate may suggest the latter, however we do not know at this time if this survey has been affected by environmental effects or other factors.

The patterns in the March series are not confined to age 4 cod only. The trends of abundance at age for the 1978–1982 cohorts from the March series are presented in Figure 8 along with the estimate of the proportion of intermediate layer water on the bottom. In Figure 8a the survey estimates have been divided by their respective cohort analysis estimates. Note that, with the possible exception of the 1978 cohort at age 5 (1983), the patterns for each cohort are similar in the same years and not at the same ages. That is, when the survey estimate of total abundance was high (or low) in any year, it was high (or low) for most ages in the population. In the case of the 1978, 1979 and 1980 cohorts there were two peaks in estimated abundance relative to the cohort estimates. For the latter two cohorts these peaks were coincident with increased proportions of intermediate layer water (Figure 8b). The second peak of the 1979 cohort was coincident with the increase in intermediate layer water in 1986 relative to 1984.

5. Conclusions

We have shown that there is a spatial coincidence of survey catches of age 4 cod and the presence of the intermediate layer water mass on the bottom during the March survey. This relationship appears to be modified by depth with the probability of encountering aggregations of cod in this water mass increasing with depth until it peaks somewhere between 90 and 111 fms. This probability decreases for water deeper than 111 fms. The commercial catch rates appear to exhibit a similar pattern with depth but the water mass characteristics at the time of their catches are unknown.

There also appears to be a relationship between the March survey estimates of age 4 cod and the proportion of the bottom water composed of intermediate layer water mass. Discrepancies between the March estimates of abundance and those from the cohort analysis are consistent with changes in availability of cod to the trawl gear due to changes in the amount of intermediate layer water on the bottom. This trend would imply that the research survey estimates of abundance may be confounded by changes in the water mass composition on the bottom. We also noted that this effect appears to be independent of the age of the fish.

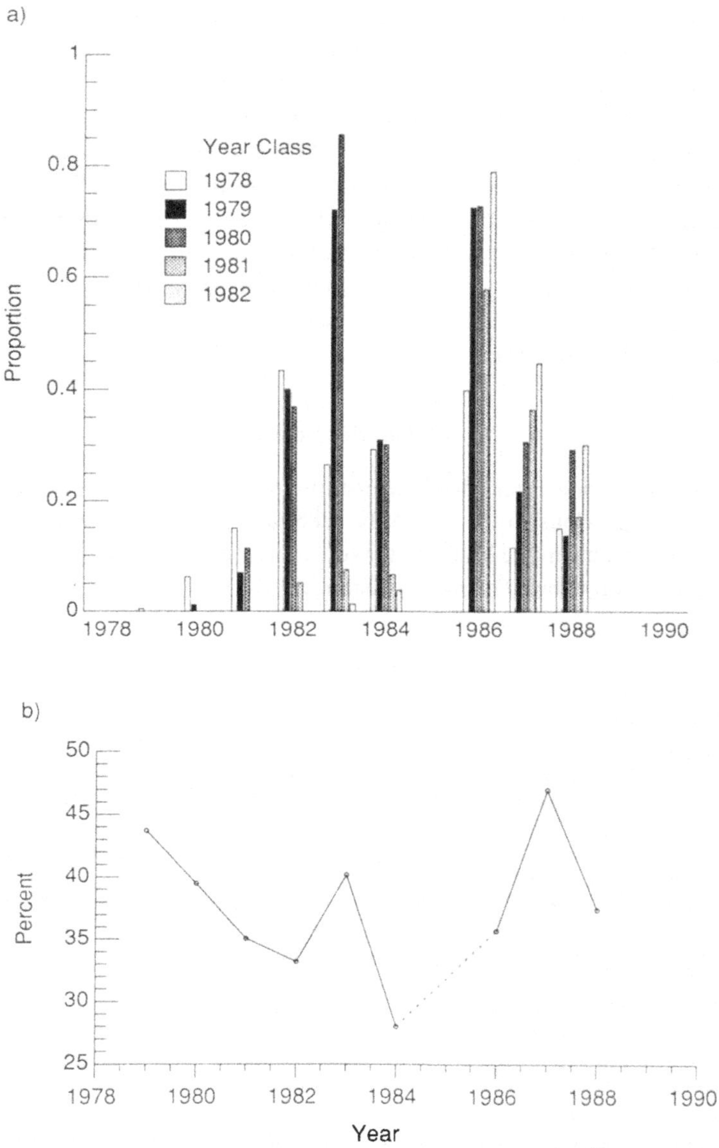

Fig. 8 (a.) Temporal trends in scaled estimates of abundance of the 1978–1982 year-classes of 4VsW cod from the March research survey for 1979–1988. Estimates of abundance have been scaled by their respective cohort analysis estimates. (b.) Temporal trend of the proportion of water on the bottom identified as intermediate cold layer water mass during the March surveys, 1979–1988. Note that there was no survey of the area in 1985.

The possibility that changes in availability of the cod to the survey gear occur due to factors not associated with fish stock dynamics can seriously limit the use of surveys as population monitoring tools. The variability introduced by environmental changes may make all but the most severe changes in abundance undetectable. While it may be possible to use the estimated proportion of intermediate layer water on the bottom as a qualitative diagnostic tool, the relationship of changes in the survey index to changes in the population remain uncertain. Unless we know the relationship between the fish on the bottom and those up in the water column and how their distribution overall can be effected by hydrographic conditions, we cannot properly interpret what annual changes in the survey estimates mean in the context of the total population. In general, it has been assumed that, on average, there is a constant relationship between fish available to the trawl gear and those fish which by virtue of their location in the water column are not available (Smith, 1988). Our investigations have cast serious doubt on that assumption. Further work is required to try and quantify this relationship. Model based methods such as those discussed in Smith (1990) may be useful in providing survey estimates if these relationships can be determined.

Acknowledgements

We thank Dr. K. Drinkwater (Bedford Institute of Oceanography) for his constructive comments on a previous draft. Dr. A. H. El-Shaarawi and two anonymous referees contributed comments on the final draft. We also wish to thank Mr. R. J. Losier and Mr. W. J. MacEachern for their technical assistance.

References

Chapman, D. C. and Beardsley, R. C.: 1989, 'On the Origin of Shelfwater in the Middle Atlantic Bight', *Journal of Physical Oceanography* **19**, 384–391.

Cochran, W. G.: 1977, *Sampling Techniques*, Wiley, New York, 428 p.

Doubleday, W. G. (ed.): 1981, *Manual on Groundfish Surveys in the Northwest Atlantic*, NAFO Scientific Council Studies. No. 2. 55p.

Doubleday, W. G. and Rivard, D. (eds.): 1981, 'Bottom Trawl Surveys', *Canadian Special Publication of Fisheries and Aquatic Science* **58**, 273p.

Etter, M. L. and Mohn, R. K.: 1989, 'Scotia-Fundy Shrimp Stock Status – 1988', CAFSAC Research Document. 89/4. 25p.

Fanning, L. P. and MacEachern, W. J.: 1989, 'Stock Status of 4VsW cod in 1988', CAFSAC Research Document. 89/57. 71p.

Francis, R. I. C. C.: 1984, 'An Adaptive Strategy for Stratified Random Trawl Surveys', *New Zealand Journal of Marine and Freshwater Research* **18**, 59–71.

Gavaris, S. and Smith., S. J.: 1987, 'Effects of Allocation and Stratification Strategies on Precision of Survey Abundance Estimates for Atlantic Cod (*Gadus morhua*) on the Eastern Scotian Shelf', *Journal of the Northwest Atlantic Fishery Science* **7**, 137–144.

Hachey, H. B.: 1942, 'The Waters of the Scotian Shelf', *Journal of the Fisheries Research Board of Canada* **5**, 377–397.

Houghton, R. W., Smith, P. C. and Fournier, R. O.: 1978, 'A Simple Model for Cross-Shelf Mixing on the Scotian Shelf', *Journal of the Fisheries Research Board of Canada*, **35**, 414–421.

Jolly, G. M. and Smith, S. J.: 1989, 'A Note on the Analysis of Marine Survey Data', *Proceedings of the Institute of Acoustics* **11**, (3), 195–201.

Mahon, R., Smith, R. W., Bernstein, B. B. and Scott, J. S., 1984, 'Spatial and Temporal Patterns of Groundfish Distributions on the Scotian Shelf and in the Bay of Fundy, 1970–1981'. Canadian Technical Report Series of Fisheries and Aquatic Sciences. 1300: *ix*+164 p.

McCullagh, P. and Nelder, J. A.: 1983, *Generalized Linear Models*, Chapman and Hall. New York. 580 pp.

McLellan, H. J.: 1954a, 'Temperature-Salinity Relations and Mixing on the Scotian Shelf', *Journal of the Fisheries Research Board of Canada* **11**, 419–430.

McLellan, H. J.: 1954b, 'Bottom Temperatures on the Scotian Shelf', *Journal of the Fisheries Research Board of Canada* **11**, 404–418.

McLellan, H. J. and Trites, R. W.: 1951, 'The Waters on the Scotian Shelf, June 1950 – May 1951', Manuscript Report of the Atlantic Oceanographic Group, St. Andrews, N. B.

Parsons, D. G.: 1988, 'Two-Phased Survey Design for Shrimp off Labrador', CAFSAC Research Document. 88/18:10p.

Payne, C. D. (ed.): 1986, 'The GLIM (Generalized Linear Interactive Modelling) System', Numerical Algorithms Group. Oxford, U.K.

Pennington, M.: 1985, 'Estimating the Relative Abundance of Fish from a Series of Trawl surveys', *Biometrics* **41**, 197–202.

Pennington, M.: 1986, 'Some Statistical Techniques for Estimating Abundance Indices from Trawl Surveys', *Fishery Bulletin* **84**, 519–525.

Perry, R. I., Scott, J. S. and Losier, R. J.: 1988. 'Water Mass Analysis and Groundfish Distributions on the Scotian Shelf, 1979–84', NAFO SCR Doc. 88/81. 14p.

Petrie, B. D.: 1983, 'Current Response at The Shelf Break to Transient wind Forcing', *Journal of Geophysical Research* **88C**: 9567–9578.

Pinhorn, A. T. and Halliday, R. G.: 1985, 'A Framework for Identifying Fisheries Management Problems Associated with the Influence of Environmental FActors on Distribution and Migration of Marine Species', NAFO Scientific Council Studies. 8:83–92.

Scott, J. S.: 1982, 'Depth, Temperature and Salinity Preferences of Common Fishes of the Scotian Shelf', *Journal of Northwest Atlantic Fishery Science* **3**, 29–39.

Smith, S. J.: 1988, 'Abundance Indices from Research Survey Data', in Rivard,D. (ed.), *Collected Papers on Stock Assessment Methods*, CAFSAC Res. Doc. 88/61:16–43.

Smith, S. J.: 1990, 'Use of Statistical Models for the Estimation of Abundance from Groundfish Trawl Survey', *Canadian Journal of Fisheries and Aquatic Science* **47**, 894-903.

Tremblay, M. J. and Sinclair, M.: 1985, 'Gulf of St. Lawrence Cod: Age-specific Geographic Distributions and Environmental Occurrences from 1971 to 1981', Canadian Technical Report of Fisheries and Aquatic Science. 1387: iv + 43p.

SAMPLING INFERENCE, AN ALTERNATE STATISTICAL MODEL

WALTER W. ZWIRNER

The University of Calgary, Dept. of Educational Psychology, 2500 University Drive N.W., Calgary, Alberta TN IN4, Canada

(Received June 1989)

Abstract. Sampling inference is proposed as an alternate paradigm. This method is discussed with respect to classical, Bayesian and a proposed probabilistic real world system. A probabilistic real world system is suggested as more valid and useful in practical as well as research situations. Sampling inference applied to a probabilistic model allows for valid inferences for volunteer and representative as well as random samples. A concept of time and distance based samples is introduced.

Statistics is, in a colloquial sense, the study of how information should be employed to reflect on possible interpretational models, and also to give guidance for action in a practical situation involving uncertainty. Although this describes the aim of this paper, a more technical definition, dealing with statistical inference to be used in this paper is from Fraser (1979):

> Statistical inference is the process by which conclusions about unknown characteristics and properties of a real world system are reached from background information and current data from an investigation of the system.
>
> The starting point for this process is the formal mathematical presentation of the background information together with the data – the model with the data, called the inference base.
>
> The common statistical model is really just a class of density functions, or even just a class of probability measures.

The proposed technique for statistical inference stays within these definitions. The first statement describes a very general aim and could be considered to include statistical decision theory; the second set delineates in more detail what might be used to describe statistical inference.

Observing the usage of statistical results one notices that inferences and decisions are often made for samples which a researcher or end user of statistics plans to use. Of course, statistical inferences from any of the proposed inference models, be that the classical, Bayesian, empirical Bayesian, fiducial, likelihood, plausibility, structural, conditional, or pivotal inference model may be used to predict characteristics about samples. The optimization of estimators is for the parameters used in the population model. For this reason these methods can be called models for population inference. To achieve completion of such population models might include the acceptance of an hypothesis selected from a set of proposed hypotheses. Such selection depends, in addition to the optimization criteria used, on the necessity of random selection to allow for the determination of involved risk factors for the method employed.

The new proposed inference is from sample to sample. For this reason this type of

Environmental Monitoring and Assessment **17**: 247–252, 1991.

inference is called sampling inference. This means that optimization for newly arising parallel concepts to hypotheses testing and estimation have to be examined. First a word about the sample to sample concept. This involves a number of different types of inference dealing with sampling over time, distance, or time-distance. The proposed models might validly use independence or dependence assumptions between sampling even if only the same sampling units are observed. Time sampling inferences are considered for the same composition of sample units at different times t, where t is a continuous variable which can take any value, $-\infty < t < \infty$. The distance dependency concept is used to discuss observations from different sampling units. Distance could be measured in physical distance or of the psychometric type, such as socio-economic or attitudinal differences. We need to consider a distance measure d, $0 \leq d < \infty$. Since population concepts such as parameters are not used in the inference statement, we do need different types of information for decision making and inferences describing or estimating world system characteristics. The parallel concept to Type I and Type II risk levels and hypotheses testing is the probability level of being able to identify if a sample is the result of a particular treatment or experiment. Such a probability will depend on the frequency distributions used for estimating the characteristic of the function of interest based on the two observed samples. For example, if we have samples $\{x_{i1,1}\}$ and $\{x_{i1,2}\}$ for two experimental conditions these would lead to distribution functions for $g(x_{k2,1})$ and $g(x_{h2,2})$ for a function g say $f_1(g(x_{k2,1})|\{x_{i1,1}\})$ and $f_2(g(x_{h2,2})|\{x_{j1,2}\})$. Here the first subscript identifies the sampling unit, the second identifies the sampling occasion, and the third identifies the experiment or treatment. The overlap of f_1 and f_2 corresponds to the probability that a new sample will not be uniquelly assignable to be the result of experiment one or two. What we need is a risk factor for f_1 and f_2 depending on the method employed to estimate these functions.

Next consider the widely used measurement models for classical, Bayesian and probabilistic real world systems. Following Fraser (1968) we consider for classical inference the model:

$$f(e)de,$$
$$x = \theta + e.$$

For n measurements $(x_1,...,x_n) = x'$ we have the simple measurement model describing the operation of the instrument for the n instances of measurement as:

simple measurement model, classical inference

$$\Pi f(e_i) \Pi de_i$$
$$x_i = \theta + e_i.$$

x' is known, θ is an unknown (fixed) parameter, and e has an assumed-known distribution. This model assumes random selection of x' and since θ is a fixed parameter, the possibility of repeated measurements. Based on a number of well-known optimization criteria we can find an estimate for the parameter thus completing the information needed for the real world system and allowing us to draw conclusions about the population and also to make decisions for actions if so required. This model assumes that we can act as if a stable population existed, or at least assume such stability for the time needed for action or inferences. Pearson (1907) expressed the necessity for such an assumption with respect to a ratio:

> ...One and all, we act on the principle that the statistical ratio determined from our past experience will hold, at any rate approximately, for the near future. This category of the stability of statistical ratios is all important not only in statistical theory but in practical conduct, as is from a second standpoint in physical theory and also in practical life the principle that the same causes will reproduce the same effects. Neither principle admits of an ultimate logical demonstration; both rest on the foundation of commonsense and the experience of what follows their disregard. Both need considerable care in their application, but what is quite clear that practical life cannot progress without them.

Using an estimate for the parameter θ allows us to predict the characteristics of future samples or functions of samples. Such an approach is discussed in Cox and Hinkley (1974). For example if we are interested in the mean of $\{x_{k1,2}\}$, we can use the estimate for μ from $\{x_{i1,1}\}$ and Student's *t*-distribution, if we are willing to make the needed assumptions to find estimates for such a mean. Of course such an approach would have an obvious weakness; the fact that we do not know how 'good' the estimate is. We need a risk factor for such a statement. The confidence interval concept supplies us with such a risk factor. Using estimates for μ from within the confidence interval based on a confidence coefficient of 1-α establishes the risk as of magnitude α for not having used a 'true' estimate of μ. The idea of a needed maximum size of confidence intervals indicates an approach we could take to establish 'usefulness' as a concept of statistical inference. Simulation studies using this approach are planned.

Bayesian inference uses the prior distribution function of θ, $\pi(\theta)$ and we need to examine a measurement model of the following type:

simple measurement model, Bayesian inference

$$\Pi f(e_i)\ \Pi\ de_i$$
$$x_i = \theta + e_i$$
$$\pi(\theta).$$

Using this model we arrive at a predictive distribution of $\{x_{k2}\}$:

$$p(\{x_{k2}\}|\{x_{i1}\}) = [\int p\theta(\{x_{k2}\})p_\theta(\{x_{i1}\})\pi(\theta)]/[\int p_\theta(\{x_{i1}\})\pi(\theta)].$$

The integration is over the space Θ for all $\theta \in \Theta$. Since we do not know the value of θ precisely, we again need a risk factor for $p(\{x_{k2}\}|\{x_{i1}\})$ the conditional probability distribution of $\{x_{k2}\}$ for a given $\{x_{i1}\}$. This is of course sample dependent. To use sampling inference one needs further information on the possible impact of the observed sample. Combining confidence interval results with the Bayesian prediction for simulation studies is planned.

A new measurement model is proposed next. This model extends the previous ones in introducing a set of random variables θ_i which make each measurement 'unique' in the sense suggested by Vollenweiler (1989) and also describes realistically a wide variety of measurements from the social sciences:

simple measurement model, 'probabilistic' inference

$$\Pi f(e_i) \, \Pi \, de_i$$
$$x_i = \theta_i + e_i$$
$$\pi(\theta_i).$$

For this model only sampling inference is justified since the parameters θ_i are only measured 'instantaneously' and do not allow for repeated measurements. It is assumed that this model would prove of use if one measures a quantity which is influenced by a large number of random variables and thus makes each θ_i a random observation and thus unique or non-replicable. The model represents observations which could be called unique in the sense described by Vollenweiler (1989) or not 'strictly deterministic' according to Menges and Skala (1974) which describe measurements where one cause can have more than one effect. This problem of probabilistic measurements is confounded for social science observations. Because of the specific concept formation for such issues as attitudes and impressions, we have results which are more vague than those in the physicial sciences. Environmetrics has to deal with both types of problems, the uniqueness and the vagueness of measurements.

A number of consequences arise from a probabilistic measurement model. We have to consider two random sources of variation, the traditional one, e_i, is due to measurement unreliability and can be decreased by improving the measurement instrument; it is thus controllable although it can never be completely removed. The second source is the unique-vague-probabilistic character of the measurement. Because there are a number of possible random causes, we cannot replicate a measurement. Since θ_i can only be considered with respect to $p(\theta_i)$ we need to 'generalize' to samples and not to populations.

Three types of sampling inferences need to be considered: one, inferences for the same sampling units at a different time t, where, as previously mentioned $-\infty < t < \infty$. When $t=0$ we consider the estimate for the sample based on the randomness of θ_i. In other words, what we might have reasonably expected for the measurements given the random character of θ_i. Every sample for this model has to be considered a random sample. When $t \neq 0$ we have a time shift and an estimate for the sample or for a function $g(\{x_{i1}\})$.

Two, inferences for sampling units $\{x_{k2}\}$ different from $\{x_{i1}\}$ are to be considered. This time the estimate is for some other group. A measure d is considered which indicates how far away, either in a physical or conceptual sense, the other sampling units are. This corresponds to the 'spatial' difference Vollenweiler (1989) called for, or in the social sciences indicates a different background factor for $\{x_{k2}\}$.

Three, a combination of time and distance, typically a future sample in a different location.

Inferences and decisions for these types of sampling units are made on a regular basis in the social sciences. To illustrate the different interpretations which have to be made let us consider the case of $t=0$ and $d=0$. Namely estimation for the same sample considering the fact that the parameters θ_i are random variables. Let us assume that a researcher is considering the results of an experiment where the sampling units were volunteers, not an uncommon situation. Since the sample is obviously not a random sample from any population, estimation and generalizations should not be made using a classical statistical inference model. Often the argument is made that one should proceed on an 'as if' basis. Statements of this type on attitudes, aptitudes, or intelligence measures have often been made on this 'as if' basis and conclusions advanced for population descriptions. If we assume a probabilistic model we arrive at a different interpretation of for example the confidence intervals for IQ measures, or the comparison between samples resulting from different ethnic groups. The confidence interval statement indicates the probable spread of $g(\{x_{i1}\})$ which could be expected for a volunteer sample. Of course the assumption, if Student's t-distribution has been used, would be that the θ_i are normally distributed and that the variance between sampling units is similar to the variance within sampling units. For comparing $g(\{x_{i1.1}\})$ with $g(\{x_{j2.1}\})$ we are now discussing the difference or identifiability of volunteer samples, and not a comparison between populations.

We need to examine the restriction placed on sampling inference by the different statistical models in use and the proposed probabilistic model. For this we need a density function $f(g(x_{k2,1}),t,d|\{x_{i1,1}\})$ with a corresponding risk factor α. The probabilistic model allows for a valid interpretation for volunteer samples (t=0, d=0), representative samples (t≠0, d=0) and random samples (t≠0, d≠0).

Simulation studies are planned to illustrate the concepts outlined in this short report.

References

Bartlett, M. S.: 1975, *Probability, Statistics and Time.* Chapman and Hall.
Bartlett, M. S.: 1962, *Essays on Probability and Statistics.* Methuen.
Baumol, W. J. and Oates, W. E.: 1988, *The Theory of Environmental Policy.* Cambridge.
Beekman, J. A.: 1974, *Two Stochastic Processes.* Halsted Press.
Box, G. E. P. and Tiao, G. C.: 1972, *Bayesian Inference in Statistical Analysis.* Addision-Wesley.
Bratley, P., Fox, B. L., and Schrage, L. E.: 1987, *A Guide to Simulation.* Springer Verlag.
Cox, D. R. and Hinkley, D. V.: 1974, *Theoretical Statistics.* Chapman and Hall.
de Finetti, B.: 1972, *Probability, Induction and Statistics.* John Wiley.
DeGroot, M. H.: 1970, *Optimal Statistical Decisions.* McGraw-Hill.
Dubins, L. E. and Savage, L. J.: 1965, *How to gamble if you must.* McGraw-Hill.
Fraser, D. A. S.: 1968, *The Structure of Inference.* John Wiley.
Fraser, D. A. S.: 1979, *Inference and Linear Models.* McGraw-Hill.
Goel, P. K. and Zellner, A. (eds.): 1986, *Bayesian Inference and Decision Techniques.* North-Holland.
Gottfried, B. S.: 1985, *Elements of Stochastic Process Simulation.* Prentice Hall.
Gupta, S. S. and Huang, D. Y.: 1980, *Multiple Statistical Decision Theory: Recent Developments.* Springer Verlag.
Hacking, I.: 1975, *The Emergence of Probability.* Cambridge U. Press.
Iverson, G. R.: 1984, *Bayesian Statistical Inference.* Sage.
Kiefer, J. C.: 1987, *Introduction to Statistical Inference.* Springer.
Kleinen, J. P. C.: 1987, *Statistical Tools for Simulation Practitioners.* Marcel Dekker.
Kuhn, R. S.: 1977, *The Essential Tension.* U. of Chicago Press.
Kuhn, T. S.: 1970, *The Structure of Scientific Revolutions.* U. of Chicago Press.
Lazarsfeld, P. F. and Henry, N. W.: 1968, *Latent Structure Analysis.* Houghton Mifflin.
Lehmann, E. L.: 1959, *Testing Statistical Hypotheses.* John Wiley.
Lindgren, B. W.: 1971, *Elements of Decision Theory.* Macmillan Co.
Lindley, D. V.: 1071, *Bayesian Statistics. A Review.* Arrowsmith.
Maritz, J. S.: 1970, *Empirical Bayes Methods.* Methuen.
Matloff, N. S.: 1988, *Probability Modelling and Computer Simulation.* PW-Kent.

Menges, G. and Skala, H. J.: 1975, 'Vagueness in the Social Sciences', in Menges (ed.), *Information, Inference and Decision.* D. Reidel Publ. Co.

Namboodiri, N. K. (ed.): 1978, *Survey Sampling and Measurement.* Academic Press.

Pearson, K.: 1907, 'On the influence of Past Experience on Future Expectations'. *Phil Mag. S. 6.* **13** (75).

Phillips, L. D.: 1973, *Bayesian Statistics for Social Scientists.* Nelson.

Rao, C. R.: 1965, *Linear Statistical Inference and its Applications.* John Wiley.

Scharnberg, M.: 1984, 'The Myth of Paradigm-shift, or How to Lie with Methodology' *ACTA Universitatis Upsaliensis.*

Vollenweider, R. A.: 1989, 'Environmetrics: Objectives and Strategies', *Conference on Environmetrics*, Cairo.

STATISTICAL NEEDS IN NATIONAL WATER QUALITY MONITORING PROGRAMS

ROY E. KWIATKOWSKI

Office of Environmental Affairs, Department of Energy, Mines and Resources, 580 Booth Street, Ottawa, Ontario, Canada

(Received April 1990)

Abstract. The concept that a few well chosen, strategically placed, water quality stations can provide valuable scientific information to water managers is common to many countries. Historically within Canada, water quality regional networks (Great Lakes network, Prairie Provinces Water Board network, Long Range Transport of Airborne Pollutants network, etc.) have been successfully operating for many years. This paper will describe the difficulties associated with developing a national water quality network for a country the size of Canada. In particular, it will describe some of the statistical tools presently being used in regional networks which are suitable for a national network, and discuss the need to develop new statistical tools for environmental monitoring in the 1990's.

1. Introduction

Canada possesses abundant aquatic resources covering 7.6% of its surface (9% of the world's freshwater supply). However, despite the apparent abundance of water in Canada, several authors (Harvey, 1976; Johnson, 1980; Foster and Sewell, 1981) have repeatedly warned of the critical situation with respect to, not only the quantity, but also the quality of freshwater resources in Canada. There are several reasons for these concerns:

- 60% of Canada's freshwater drains north (Figure 1) while 90% of Canada's population can be found within 300 km of the Canada-United States border.
- Canadians use more than 2000 L of water per person per day, for domestic, commercial, agricultural and industrial purposes. This represents the second highest consumption rate in the world.
- At an average cost of \$ 0.47 m^{-3}, Canadians have one of the lowest costs for water in the world, approximately one half of that of the United States and one fifth that of European countries. As a result there is no financial incentive to conserve water.
- Many users of water (domestic, commercial, agricultural and industrial) return this water to the environment in a deteriorated state. As a result conflict between water users is increasing.

The Canadian government has accepted the concept of water quality conservation (e.g. maintaining the present aquatic ecosystem or an improved condition, so as not to eliminate future options for use). To achieve this concept, water managers have recognized the need for a scientifically sound measure of water demand (defined as the amount of water consumed plus the degree to which wastewater is degraded, Brooks and Peters, 1988). Only after an accurate measure of water demand is made will alternative

Environmental Monitoring and Assessment **17**: 253–271, 1991.

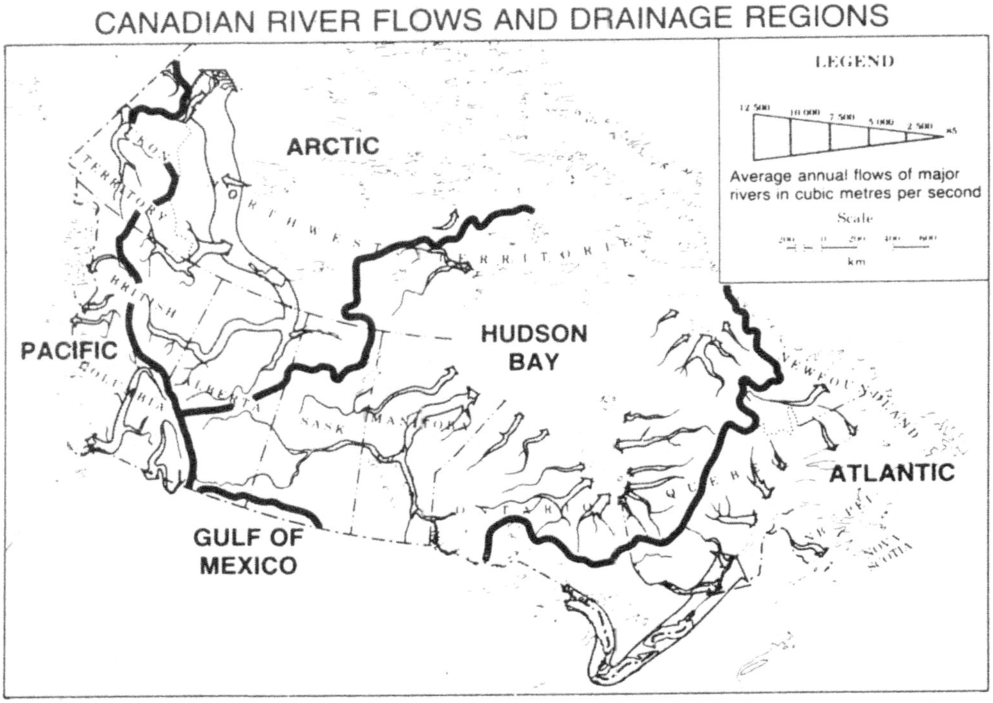

Fig. 1. Average annual flows of major rivers within drainage regions of Canada.

approaches to water demand management be developed, verified and implemented.

Within Canada, management of water resources is a provincial responsibility. However, the federal government has a mandate to show leadership on national issues. National assessment of water quality falls within the federal mandate. With the ever expanding list of man-made chemicals being introduced into the environment and the increased costs of monitoring these anthropogenically produced chemicals, a mandatory need for any national water quality network in Canada will be the close cooperation among the various agencies responsible for water. Areas requiring harmonization include network design, chemical analyses and data interpretation. Statistics plays an important role in all these aspects. The view that statistics are solely an end application, (e.g. for interpretation) will result in poor data interpretation and unresolved environmental issues.

The application of statistics to environmental assessments has dramatically increased in the last decade. The objective of this paper is to review presently used techniques for characterizing the quality of waters in Canada. This overview will provide the unitiated water manager with the salient areas within environmental assessment where statistical application plays a paramount role.

Before doing so, a brief review of three 'Areas of Concern' when dealing with large scale networks will be given. These 'Areas of Concern' are universal to all large networks and are of critical importance to the statistical conversion of data to information. Each has been discussed extensively in the literature and therefore are only briefly discussed here.

Statistical 'Areas of Concern'

(1) Sample representativeness: In all natural aquatic systems, a complex interaction of physical and biochemical cycles exists. The annual and the long-term hydrographs of a river basin are a result of the basin's hydrological regime. Superimposed on these are the biochemical cycles such as the diurnal cycle which is measured in terms of hours, and the seasonal cycle which is measured in terms of months. Water temperature affects saturation values for dissolved gases, alters metabolic rates of aquatic organisms and affects the specific gravity of water, producing substantially altered mixing characteristics. Light supplies the driving force for primary production and therefore influences the uptake and depuration rates of toxics. Sediments act as a transport mechanism for adsorbed substances, alters light regimes and directly affects aquatic organisms. Typical values of natural variations range from 100 to 400% of observed mean values for physical, chemical and biotic variables (Mar *et al.*, 1986). As a result, all aquatic systems are undergoing change, spatially and temporally. Unfortunately, many monitoring programs are continuously describing only this variation, without attempting to understand it. Due to the stochastic nature of natural processes and the short duration of many monitoring programs, or the tendency to carry out synoptic (snapshot) sampling at fixed points in space and time, the interactions of these natural cycles with the physical, chemical and biological components of the aquatic ecosystem are often missed or misinterpreted. Proper placement of sampling both spatially and temporally is of paramount importance to the generation of scientifically sound information. Most of the scientific and technical problems associated with the assessment of environmental impact can ultimately be traced back to the fact that the natural variability inherent in the aquatic ecosystem was not adequately characterized. Differences in natural variability between parameters of concern influences not only the timing, location and method of collection, but also will determine the accuracy of predicting parameter responses to the impact. Many papers have discussed the importance of network design with respect to macro versus micro scale monitoring, as well as cross stream and interbasin variability (O'Sullivan, 1979; Lotspeich, 1980; Sanders *et al.*, 1983; Désilets, 1988).

Contrary to what many water managers think, 'more' is not necessarily better. Serial positive correlation may be thought of as redundancy of information and results from the fact that samples taken close together in time are often correlated thereby prohibiting the use of standard parametric statistics, because of the violation of the sample independency. This redundancy in successive observations means less information is obtained from the data than can be expected from independent samples. As a result, the confidence limit associated with the mean value is indeed larger than indicated by the statistics. This effect is compounded in small sample sizes ($n<30$) where serial correlation reduces the number of independent samples even further. With water quality variables, the degree of correlation generally increases as samples are taken closer together in time (Loftis and

War, 1980; Lettenmaier, 1978; Sanders *et al.*, 1983). Sanders *et al.* (1983) indicated that a large degree of redundancy in observations exists for water quality data collected two days apart, while observation spaced 30 days apart are nearly independent.

(2) Quality Assurance/Quality Control (QA/QC): As environmental concerns moved from eutrophication in the 60's to toxic organics and pesticidesin the 80's, detection limits were lowered from the milligrams per litre (mg l^{-1}or parts per million) range to picograms per litre (pg L^{-1} or parts per quadrillion) range. At these ultra-trace levels the potential for analytical error, in either precision or accuracy, is high. All too often, concentrations of various parameters are reported as absolute values, with no information available on the laboratory's ability to report at ultra-trace levels. Data files store information, not only on the variable's concentration, but also on station location, time of sampling, and field and laboratory methods used. Few data archives also store the associated QA/QC information necessary to validate the accuracy of the numbers. QA/QC is also needed in the field: during sample collection, preservation, storage and transportation; as well as in the office at the data storage and data retrieval stage. Contamination of the sample in either the field or laboratory, or a transcription error at the data storage or retrieval stage, can produce outliers which can lead to erroneous management decisions. Often in large networks either contract or other agency laboratories are involved cooperatively in the analyses of the samples. When this occurs the water manager should take note of the fact that the standard deviation of the systematic error on samples between laboratories is often two to three times greater than the within-laboratory variability. The Government Accounting Office (1981) emphasized the need for quality assurance and quality control protocols in all aspects of data collection and analyses. Manuals providing statistical techniques for collaborative tests for QA/QC are readily available (Youden and Steiner, 1975; Agemian, 1985; American Society for Testing Materials, 1986), and should be used to develop laboratory protocols prior to sample analysis.

(3) Detection Limits: Two separate and distinct concerns are associated with detection limits and long-term national networks. The first is the constant improvement in analytical capabilities through improved analytical techniques or equipment, resulting in lower detection limits. As detection limits lower, the frequency of non-detectable values in the data sets is reduced. In the future, issues will develop not because the contaminant has been recently released into the environment, but rather because our ability to identify the compound has increased so greatly. Indeed measurements in femtograms per litre (fg L^{-1} or parts per quint-trillion) are now possible. Parameters previously not detected, are now found at these ultra-trace levels. As a result of these analytical improvements, the statistician can be faced with long-term trend analysis for a series with a non-stationary censor.

Historically the use of the detection limit, one-half of the detection limit, or absolute zero, for the non-detected value in the calculation of population statistics has been used. The appropriateness of nonparametric tests for censored data is discussed by Prentice and Marek (1979). The most robust estimation method for minimizing errors in estimates of the mean, standard deviation, median and interquartile range of censored data was determined by Gilliom and Helsel (1986), Helsel (1986), and Helsel and Gilliom (1986) to be the log-probability regression method (LR). However, El-Shaarawi (1989) has indicated that the LR method is technically inadmissible for estimating the unobserved water quality censored data which are of type I censoring, and has modified the LR method.

A second difficulty associated with detection limits is their definition. The Method Detection Limit (MDL) is the minimum concentration that can be measured with a 95% confidence that the true concentration is above zero.

$$\mathrm{MDL} = \mathrm{Sb} + t(n-1)S \qquad [1]$$

where Sb is the average blank signal, t is the appropriate quartile of t distribution with n-1 degrees of freedom, S is the standard deviation of the blanks, and $n > 7$.

The Instrument Detection Limit (IDL) is the lowest concentration of analyte that an analytical instrument can detect and which is statistically different ($P<0.05$) from the response obtained from the background instrumental noise.

$$\mathrm{IDL} = \mathrm{t}(n-1)S. \qquad [2]$$

This implies that these limites are determined using a statistical test and hence the results are subject to a false positive rate of 5%, for each measurement.

Often no mention is made of accuracy or precision of the MDL or IDL. It can be safely assumed that as concentrations approach the MDL, analytical precision decreases significantly (C.V. of 50-100% are not uncommon). Any statistical analyses of data which assumes constant variance over the full range of the analytical procedure is likely to be incorrect.

The limit of quantitation (LOQ) is defined as the level above which quantitation is reliable and is expressed as:

$$\mathrm{LOQ} = \mathrm{Sb} + 10S. \qquad [3]$$

A Practical Detection Limit (PDL) also exists. It is the lowest concentration of analyte in a real sample-matrix that a method can reliably detect and is statistically ($P<0.05$) different from the response obtained from a blank. The PDL is calculated in the same manner as the MDL.

The relationship between these terms (Agemian 1985) can be graphically summarized as:

absolute	analyte	IDL	Sb	MDL	PDL	LOQ
zero	not detected					Sb+10S

Before any statistical comparisons of data sets can occur, consensus between laboratories on which detection limit to use is needed, as well as recognition of the fact that accuracy and precision are not constant between IDL, MDL, PDL, and LOQ.

Water Quality Network Objectives

As has often been pointed out, well defined, stated objectives are of paramount importance to the design and successful implementation of any monitoring network for water quality assessment. The concept of hypothesis testing is often mentioned, and with it, the need for the correct statistical tools to test the hypothesis. Unfortunately this concept is often forgotten once network objectives are developed. Indeed, often an inverse relationship exists between the size of the network and the clarity of the stated objectives. As a result, for national networks the establishment of null and alternate hypotheses, and the associated statistical tools needed to test the hypotheses, are only superficially discussed at the network design stage. The concept being that the objective of the network is to collect analytically sound data, and, independently it is the statistician's objective to transform this data into valuable information. However, without clearly defined objectives, or without a definitive null hypothesis to test, the statistician introduced to the program after the data is collected is often at a loss to provide the statistical interpretation required or expected by the water manager.

Though by no means definitive, the following objectives are common to many large networks:

(1) to provide information on the location, severity, areal or volumetric extent, frequency and duration of non-compliance of variables of concern;

(2) to provide information for measuring site specific or whole network responses to increased anthropogenic inputs or control measures, using trend analyses or cause and effect relationships; and to determine the presence of new or hitherto undetected problems, leading to proactive, rather than reactive pollution control measures;

(3) to provide information for development and application of predictive models for assessing the impact of new pollution sources and assessing various enforcement and management strategies; and,

(4) to determine ecosystem health and identify significant changes from normal succession or expected sequential changes which occur naturally in aquatic ecosystems.

Though the above objectives seem reasonable, they are entirely qualitative. No clear performance indicators are given, and as a result, evaluation of the network is often impossible, leading to the data rich, information poor syndrome described by Ward *et al.* (1986). Additionally at the national level, it is assumed that all of the above stated objectives, and that all issues of concern associated with each objective, will be addressed with the same degree of statistical effort, irrespective of the present knowledge base. A sound statistical approach provides the only mechanism whereby water managers can hope to meet their informational needs within realistic monetary constraints. Unfortunately once objectives are set, budgets are often frozen. As the number of issues increase, the network is left with more and more issues to address over time, with a budget deteriorated by inflation. As a result, the number of samples (number of sampling locations, frequency of sampling, or number of samples devoted to QA/QC) are often reduced. Unfortunately, the network objectives remain the same. It is necessary for water managers to adopt an iterative process. As the network changes, so must the network objectives and management expectations.

The need for acceptable statistical procedures in network design for environmental impact assessments has been described by Eberhardt (1976), Lucas (1976) and Thomas *et al.* (1978). An ideal network design would make use of replication and controls in both the space and time dimensions (Green, 1979). Lucas (1976) stressed the need to have paired sampling stations (equal numbers in both the control and impacted areas) in order to adequately estimate the sampling error.

A major concern in any large network is reducing complexity through data compression, either spatially or temporally, and then developing some estimate of central tendency. Assuming that the data, or a transformation of that data, approximately satisfies the assumptions required for an analysis of variance model (Conner and Myers, 1973; El-Shaarawi and Shah, 1978; Lachance *et al.*, 1979; Millard and Lettenmaier, 1986) large data sets can be conveniently reduced to statistically homogeneous zones or strata at any required significance level (Figure 2, Table I).

Once the lake or river basin is divided into homogeneous zones or strata, either areal, volumetric or flow weighted mean values can be calculated. If the costs to obtain any given sample are equal, then the strategy is to allocate more samples to strata in which the estimated natural variation and measurement errors are the largest (Mar *et al.*, 1986). However, care must be taken to ensure that changes in network design over time to improve measurement error do not lead to erroneous trend-in-time results. The network average for water quality variables which are non-homogeneously distributed are biased by the relationship which exists between sampling location and the underlying spatial distribution of that variable (Kwiatkowski, 1986). A number of different methods to calculate the optimum sampling required within each strata to achieve an accuracy of Y% (expressed as a fraction of the true concentration u_0) within a predetermined confidence are available (Mandel, 1964; Wallin and Schaeffer, 1979; Ward *et al.*, 1979; Dunnette, 1980; Loftis and Ward, 1980; El-Shaarawi, 1987; Lesht, 1988).

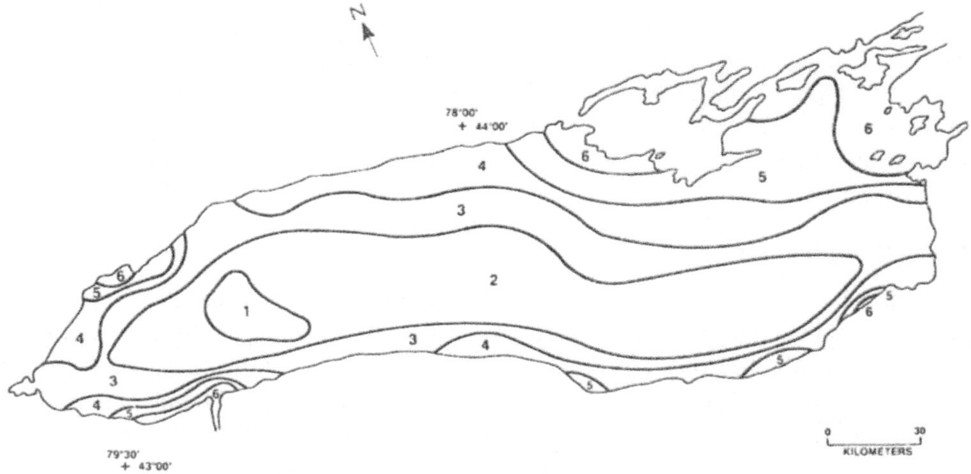

Fig. 2. Statistical zonation of chlorophyll *a* in Lake Ontario, 1974. Redrawn from El-Shaarawi and Kwiatkowski (1977).

TABLE I

Zone annual means, standard deviations and number of observations for chlorophyll *a* zonation model, Lake Ontario, 1974. Data taken from El-Shaarawi and Kwiatkowski (1977

Zone	Mean (μg L_{-1})	Standard deviation	Number of observations
1	3.29	1.66	52
2	3.94	2.04	347
3	4.84	2.38	274
4	5.91	3.16	200
5	6.85	2.76	135
6	8.00	3.93	125
7	11.60	7.91	27

Objective 1: Non-compliance

Non-compliance is the comparison of the mean of a given variable X, in zone i, at time t, against a given tolerance limit. If μ and σ are known, the tolerance limit for normal (or transformed normal) populations takes the form of $\mu + Z\sigma$. In environmental work the one sided tolerance limit is usually used.

The Lake Ontario Toxics Categorization Workgroup utilized one-sided tolerance limits to evaluate and develop a comprehensive toxics categorization scheme. Lake Ontario was divided into statistical zones following a classicifation scheme of Neilson and Stevens (1986). Water chemistry data, by statistical zone, was tested for normality, transformed if necessary and subjected to a one-sided tolerance limit test using the 95% confidence limit, and compared to federal, provincial or state water quality standards, criteria and guidelines. Whole lake fish data were divided into age classes, with the 95% confidence limit of each age class subjected to the one-sided tolerance limit test. Age classes were used rather than lake zonation because large game fish are highly mobile (10–100 km day^{-1}) and the contration of many contaminants often increase with age class. Thus for fish, age class becomes the strata of concern rather than aerial location. Either whole fish (Figure 3) or edible portions (fillets) can be used depending on the requirements of the standards, criteria or guidelines. The advantage of dividing the data into strata by zones (water chemistry) or age classes (fish) is that detection of an environmental problem is enhanced.

Objective 2: Trends

Detection of the effect of anthropogenic inputs over time has become the major activity of large scale water quality monitoring networks. Water quality managers are acutely aware of the need: to establish long term trends in water quality; to establish an estimate of when various water uses will be impaired; to determine the effect on historical trends of remedial action; and, to determine abrupt (step) changes in historical trends, so as to establish a proactive rather than reactive stance to water quality management.

The Student's t-test and linear regression are probably the two most commonly used parametric tests for detecting change in water quality over time. The t-test was evaluated

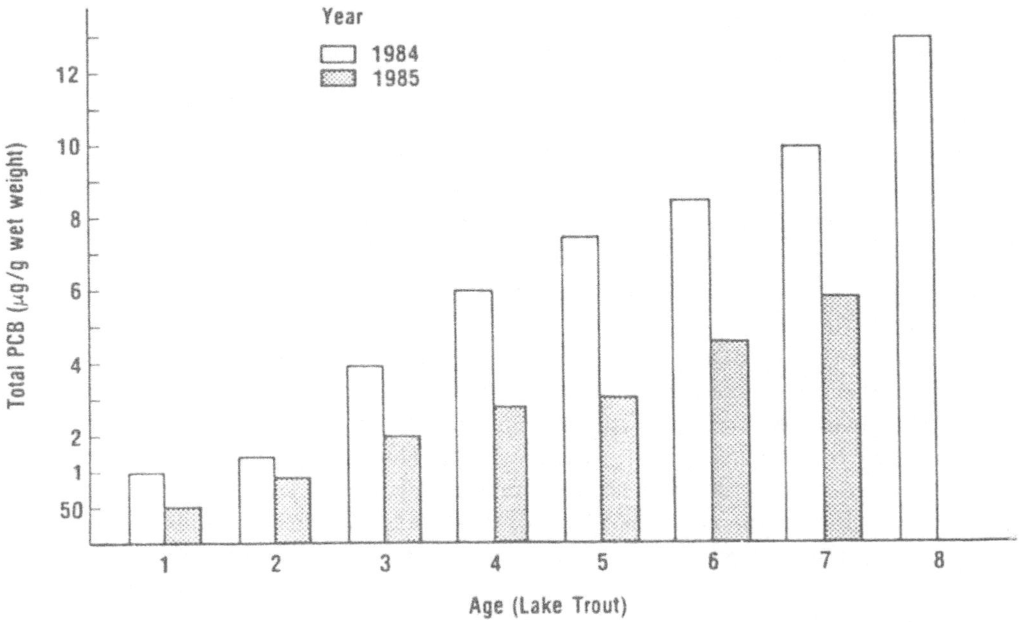

Fig. 3. Concentrations of total PCBs in various age classes of whole lake trout, Lake Ontario, 1984 and 1985. Figure graciously supplied by Baumann, from Baumann and Whittle (1988).

by Montgomery and Loftis (1987) as a water quality statistical tool. They confirmed what has been stated in the statistical literature for decades, that the t-test was not robust when samples had different distributions, unequal variances or lengths, when serial dependence was present, or if seasonal changes were not removed.

Linear regression is another common approach to trend analysis. The regression analysis can be further used to detrend, deseasonalize a time series. The seasonal Kendall test, the seasonal Kendall slope estimator and estimates of change over time in the relationship between constituent concentration and flow (Hirsch *et al.*, 1982) have gained rapid acceptance in the U.S. National Stream Quality Accounting Network (Smith *et al.*, 1982).

It should be noted that detection of a change in any variable from baseline conditions, implies that the baseline condition for that variable is known. The aquatic environment, as already pointed out, is a dynamic bio-geochemical system dominated by physical factors resulting in a high 'noise' to signal ratio with respect to trend-in-time analyses. Standard parametric statistical tools are generally inappropriate for the following reasons:

- non-normality of data
- non-homogeneity of variances
- auto correlation
- pronounced seasonal variation
- numerous outliers or censored data

- often missing data points.
- strong flow dependence.

Box and Jenkins (1970) were one of the first to develop time series univariate or multi-variate modelling with seasonality, suitable for water quality trend analyses. The Box-Jenkins (ARIMA) model is made up of three parts: an autoregressive (AR), and integrated (I) and a moving average (MA) component.

Tukey (1977) proposed a two stage approach – Exploratory Data Analysis (graphical or numerical techniques to discover important patterns and statistical characteristics) and Confirmatory Data Analysis (rigorous statistical confirmation of characteristics of interest). Confirmatory data analysis can be further divided into parametric and nonparametric testing.

Hipel *et al.* (1978) described in detail an approach to predicting and measuring the variables that are expected to follow long-term trends. Reviews on the applicability of several nonparametric tests to the detection of trends in water quality data sets have been carried out by Berryman (1984), Van Belle and Hughes (1984) and Helsel (1987). Berryman *et al.* (1988) established the asymptotic relative efficiency for nine nonparametric tests for monotonic trends, seven for step trends and three for multi-step trends. The choice of which test to use depends not only on the series characteristics but also on the type of trend to be detected. A general approach in identifying the appropriate test for a given monotonous time series was summarized by Berryman *et al.* (1988) and is presented in Figure 4.

OBJECTIVE 3: MODELLING

Water quality modelling efforts have been in place since the classical studies of Thienemann on oxygen conditions of oligotrophic and eutrophic lakes (Thienemann 1928). Environmental modelling efforts can be divided into two major categories, conceptual and quantitative. A conceptual model is simply a flow diagram of the system, describing the important components of the ecosystem and indicating which components interact. Because this model type only provides a qualitative description of the system's structure, statistics plays no role in its development or use. However, its utility for network planning can not be understated. Obvious information gaps can occur during the data collection stage unless all the interacting components are identified at the planning stage.

A quantitative model, on the other hand, is where mathematical representations of the ecological interactions is described. A quantitative modelling effort gaining rapid acceptance and use in Canada is the Quantitative Water-Air-Sediment Interaction (QWASI) model (Figure 5). The QWASI model describes the fate of a chemical in a lake or river system and it not only is used to establish the ultimate fate of contaminants but also the possible recovery times of an air-water-sediment system (Paterson and MacKay, 1985).

The advantage of using fugacity models (MacKay and Paterson 1986) is the holistic approach which considers the entire environmental system, allowing for simple (equilibrium, steady-state) to highly complicated (non-equilibrium, non steady-state) applications. The values for the system-specific variables for the model distinguishes a

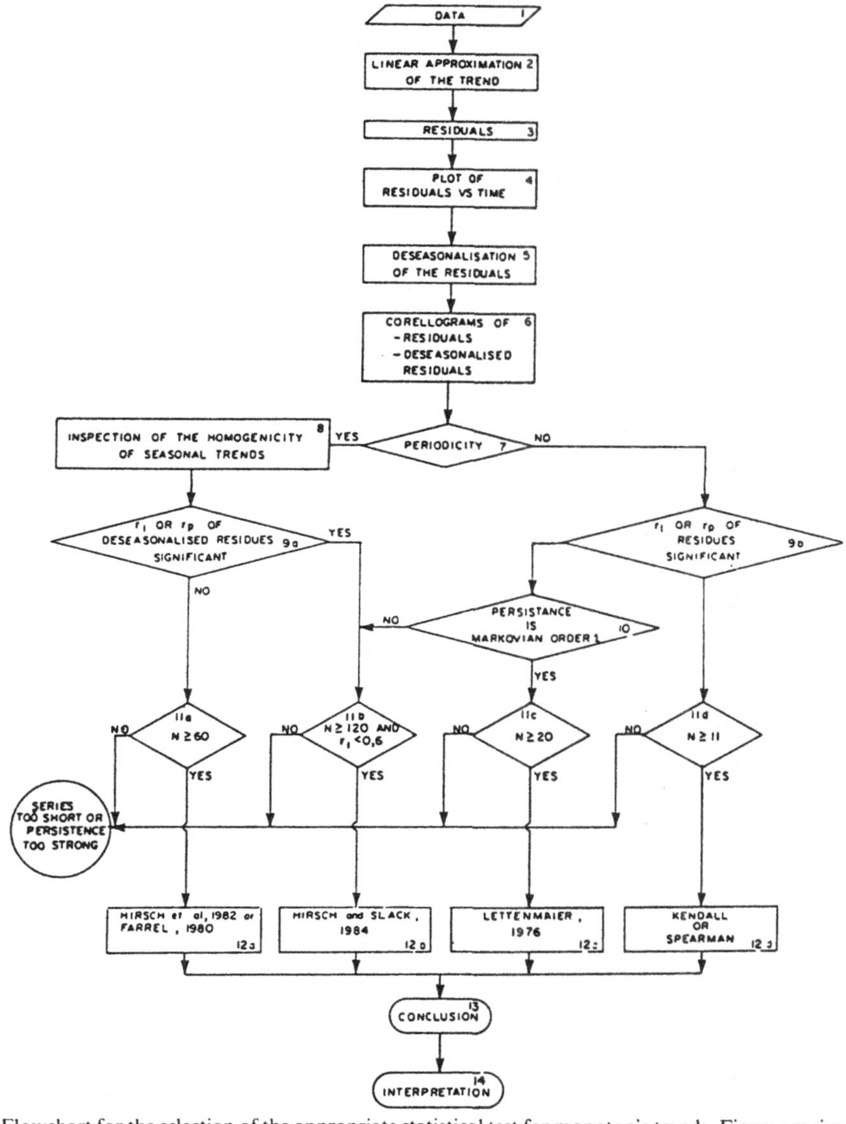

Fig. 4. Flowchart for the selection of the appropriate statistical test for monotonic trends. Figure graciously supplied by Berryman, from Berryman *et al.* (1988).

river from a lake. For a river, water flow processes control system dynamics, while within-river processes are overshadowed. Conversely, within-lake processes dominate over water flow processes in a lake due to the greater water residence time in lakes.

The need for fugacity type models in water quality monitoring is financial. In the 1950's when eutrophication was the environmental issue of most concern, inexpensive analytical costs allowed the water manager the opportunity to obtain nutrient measurements within all media (water, sediment and biota). Even analyses of heavy metals, though substantially more expensive than nutrients, pose no major threat to the budget of water quality

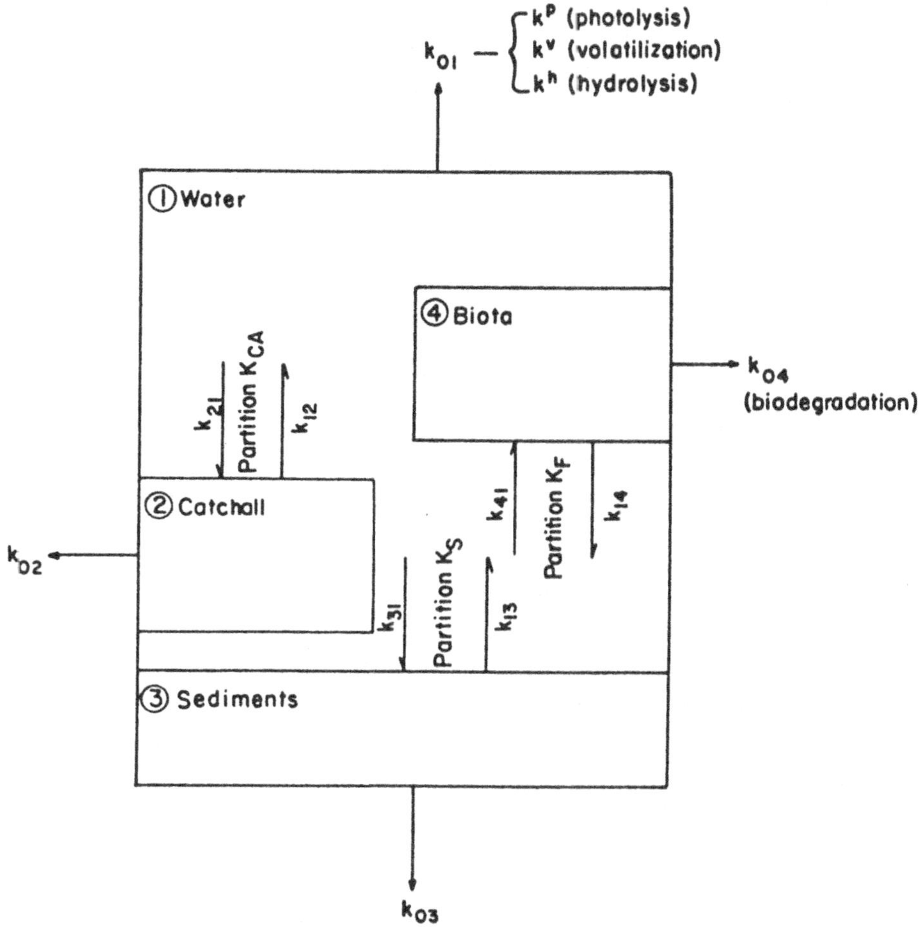

Fig. 5. Schematic of the four compartment persistent model. k_{12}, k_{21}, k_{13}, k_{31}, k_{14}, and k_{41} are first-order transfer constants. K_{ca}, K_s and K_f are partition coefficients between each compartment and water. k_{01}, k_{02}, k_{03}, and k_{04} are first-order removal rate constants. Figure redrawn from National Research Council of Canada (1981).

managers. However, analytical costs for toxic organic compounds or organic pesticides have now reached the \$ 1000 to \$ 2000 per sample range, greatly restricting the number of samples (to stay within a monitoring budget) which can be taken. In order to get the maximum 'bang for the buck', water managers must divert their monitoring efforts to locations or media within the aquatic system most likely to produce positive analytical results. Clark *et al.* (1988) indicated that it may be possible to determine the fugacities at which individual animal populations are adversely affected by any given chemical or even by a combination of chemicals. The environmental strategy would then be to ensure that the chemical's fugacity in the environment is kept below the tolerance limit at which adverse environmental effects are manifested.

Statistics plays an important role in modelling efforts in field (model) confirmation, surveillance of predicted tolerance limits, and in providing realistic estimates of the rate constants and their associated error terms. All too often rate constants are reported in the literature as absolute numbers (e.g. no error terms given). Sensitivity analysis (defined as the response of the output of a model to changes in input variables, Majkowski *et al.* (1981) and Silvert (1981)) is mandatory to any modelling efforts to ensure that inappropriate management decisions are not made. Sensitivity analyses also clearly identifies which input variables require further refinement at the research level. A discussion plus example (chlorobenzenes) of the importance of model type, conditions (steady vs. non-steady state), the environment (river *vs* lake) and the compound type to the output of a QWASI model is given in Asher *et al.* (1985).

OBJECTIVE 4: ECOSYSTEM HEALTH

Water quality guidelines and objectives are used by Canadian agencies in their efforts to assess water quality issues and to manage competing uses of water resources. Though the water quality guidelines and objectives approach is noteworthy, it by itself, is insufficient to ensure the attainment of a biological non-degradation (healthy ecosystem) objective because:

- water quality guidelines, objectives and standards are prepared on a chemical by chemical basis, ignoring their synergistic effects once released into the environment;
- no true measure (instream measurement) of biological integrity is obtained through the use of guidelines, objectives or standards to ensure that biological integrity is not affected by the interaction of the pollutant of concern with the other environmental factors which influence biological integrity;
- ecotoxicological effects are a result of exposure (concentration over time). Both components of the exposure equation must be evaluated to determine effects. Guidelines, objectives and standards often only contain the concentration component of the equation.

Indigenous biological organisms are integrators of the prevailing and past chemical, physical and biological condition of the water body in which they reside. Biological organisms reflect the dynamic interactions of stream flow, pollution loadings, toxicity, habitat and chemical quality. As a result, bioassays and biological indices are two approaches used in environmental impact assessments to obtain a measure of biological non-degradation. Both have advantages and disadvantages, and both require new and innovative statistical tools to improve their interpretability.

Bioassays are tests in which the toxicity of a material is determined by the reaction of a living organism to its presence. The concept being that individuals are likely to display effects before they appear in population characteristics. It should be noted that an individual organism's response varies according to both the concentration and duration of the toxicant. Performing bioassays on a key (valued) species, or at a number of trophic levels, provides a valuable insight into that chemical's effect on the ecosystem. However, the use of bioassays has also come under criticism. Extrapolation of laboratory bioassay results with little or no variability, and performed on standard organism(s), through the dynamic hierarchies of biological organization found within the natural aquatic ecosystem (in which the test organism may not be a significant component, or may be naturally absent) is precarious. Also, laboratory based bioassays often fail to identify the effects of sporadic spills or unintentional non-compliance effluent discharges, nor do they incorporate the concept that chemicals from several discharges that are harmless independently, may act synergistically.

A major concern not adressed is the issue of concentration versus loadings. High concentrations of many effluents have an immediate toxic effect which can be identified through bioassays. However, a second situation exists, where concentrations of the effluents are very low, well below toxic levels, but the volume is high, resulting in a high loading of effluent to the environment. Under this scenario, no local effects on biota are found (since the concentration of the effluent is low). However, the high loading results in the degradation of the receiving waters with long term deleterious effects on the associated biota (Thomas, 1988). The effects of loadings are complicated by bioaccumulation and biomagnification through the food chain. An example given by Thomas (1988) to highlight the importance of loadings to the aquatic environment was the estimation that an annual loading from all sources of 5 gr yr^{-1} of TCDD (Tetra chlorinated dibenzo dioxin) to Lake Ontario (which has a volume of 1720 km^3) would result in concentrations in predator fish to be above the edible guideline of 25 ppt.

From the headwaters to the mouth of a river, the physical variables within the river basin present a continuous gradient of physical conditions. This gradient elicits a series of responses within the constituent populations resulting in a continuum of biotic adjustments along the length of the river (Vannote *et al.*, 1980). The first attempt to interpret the results of biological surveys carried out within a river basin was made by Kolkwitz and Maisson (1908), who devised their Saprobien system. Since then, a variety of biological indices have followed, as well as the concept of indicator species. Karr *et al.* (1986) established a five compartment conceptual model (chemical variables, flow regime, energy source, biotic interactions and habitat structure) of the major classes of environmental factors that affect aquatic biota. Anthropogenically induced changes to any compartment manifest themselves into alterations in habitat quality and thereby affect the biological integrity of the system. Indicators of environmental stress have been developed for individual organisms, populations, communities and ecosystems. A review of the evolution in complexity and refinement of indicators for the purpose of assessing water quality is provided by Averett (1981).

The first step to the incorporation of biotic indices into the water quality network is the delineation of the study area into its principle eco-regions. Eco-regions that are very different from one another will support very different biological communities. However, once characterization of the eco-region is done, particularly if fish and macroinvertebrates indices are used, a true measure of the attainment/non-attainment of ecosystem health can be obtained at impacted sites.

The Sediment Quality Triad (SQT) approach (Chapman, 1986; Chapman *et al.*, 1987) has gained rapid acceptance in Canada. SQT consists of three components: sediment chemistry, sediment bioassays, and biological community structure analysis. The use of

SQT offers water managers a holistic approach to water quality impact assessment. Its major advantages over the standard water chemistry approach are that: the entire aquatic ecosystem is evaluated; deterioration in any component will result in mitigation; collection of data to simply describe changes in water chemistry without interpreting their effects will cease; and finally, sampling for SQT need not be conducted during the worst case (spill) situation, because the effects of the spill will be detected well after the spill, in the sediment, in the toxicity of the sediment and in the response of the biotic community.

The major area requiring statistical development with respect to bioassays, biological indices and SQT approaches is a method of determining significant differences between areas or between samples. Esterby (1988) pointed out that the fundamental statistical concept is the separation of the variation associated with natural fluctuations of the aquatic ecosystem from that associated with the toxic effluents (either in the short or long term). The strength of the cause/effect relationships deteriorate as one moves from the controlled laboratory situation to the uncontrolled field observations. As a result there is a need for a corresponding change in statistical methods and the nature of the statistical inferences that can be drawn, between the two activities (Esterby 1988). The importance of taking replicate samples, before the discharge begins and after the discharge has begun, at both the reference and impacted sites is discussed by Green (1979), Hurlbert(1984) and Stewart-Oaten *et al.* (1986). Pseudoreplication (defined as the use of inferential statistics to test for treatment effects with data from experiments where either treatments are not replicated, or replicates are not statistically independent) is probably the single most common fault in the design and analysis of ecological field studies (Hurlbert, 1984). It is quite common for individuals to be clustered in biological populations. Line transect methods distinguish the density of clusters from the density of the individual organisms. Various methods on estimation of density from line transect sampling of biological populations is given in Burnham *et al.* (1980).

As measures of ecosystems health are incorporated into monitoring networks, water managers will need to be able to compare results from various locations, or at one location over a period of time. Ecosystem health depends to a certain degree on the interactions which occur between species that have been associated over evolutionary time (Schaeffer *et al.*, 1988). Therefore, simply providing an estimate of the percentages of producers, consumers and decomposers, does not provide sufficient information to make a judgement on ecosystem health. It is the differences in these percentages from expected that is important. Once aware of the variability which can occur within any given population response to environmental stress, a number of questions arise. How will statistical differences in response at the individual or at the community level, be assessed? What levels of significance will become important? Do the traditional 1 and 5% significance levels, so commonly used with water chemistry data, still have ecological significance with respect to these new biological approaches, where within sample variability can be so great? Once the stress at the community or ecosystem level can be statistically ascertained at the 1% level, it may be too late to mitigate the cause. These are the challenges of the future for environmental statisticians.

Conclusions

All economic activity depends on a healthy environment. Over 40% of Canada's Gross Domestic Product is directly related to economic activities which are directly dependent on the environment. Concerns over degradation in the environment has lead to strengthened environmental legislation. Ecological monitoring plays a crucial role in environmental impact assessment and statistics plays an important role in ecological monitoring. Statistical tools are applied to the characterization of baseline conditions at reference sites or before discharge has begun; to the establishment and testing of hypotheses; to the detection of standards' violations; to model impact predictions; and to the determination of changes in ecosystem health as a result of anthropogenic impacts.

Many of the problems associated with environmental monitoring (i.e. the collection of large amounts of data, but little information, Ward *et al.*, 1986) can be attributed to the fact that managers continue to think of statisticians as simple number crunchers and statistics as crude mathematical tools. Non-statisticians within the organization are simply asked to apply textbook statistics or to use the statistical packages supplied on their personal computers to environmental data. As a result, often either the wrong statistics are used, or the statistical assumptions required for the proper use of the statistical tool are violated. One way to improve this situation is through improved technology-transfer between statisticians, those who use statistics and the water managers. Until this improvement occurs it is safe to assume that environmental monitoring will continue as it has for the past two decades, with data collection being the major objective of the network. Monitoring consists of two major elements, observation, or the collection of certain facts (data), and interpretation, or the process by which the meaning of the collected facts, their interrelationships and their relation to an existing body of knowledge, are converted into information. As environmental problems grow, the need to integrate various disciplines and fields of specialization in environmental assessments will also grow. The common thread which transcends the various disciplines is statistics.

Specific recommendations which will promote the proper use of statistics in aquatic studies include:

- state data requirements of the network in a hypothesis testing format, rather than in bureaucratise;
- pay special attention to sample representativeness in network design;
- store associated field and laboratory quality assurance/quality control information with the data on the computer file, as well as the detection limits of the analytical method;
- do not simply state the level of significance, but rather give both the probability statistics and the degrees of freedom when stating statistical results;
- have independent reviews (by qualified personnel) of the various statistical analyses proposed and establish protocols for their proper use;
- give error terms associated with rate constants in all modelling efforts so that sensitivity analysis can be independently carried out; and,
- devote greater effort to the development and application of statistical analyses of

bioassay and biological community structure data, including the development of a working definition for ecologically (*vs* statistically) significant difference.

References

Agemian, H.: 1985, *Quality Assurance in the National Water Quality Laboratory*, Water Quality Branch, Environment Canada.

American Society for Testing and Materials: 1986, *Manual on Presentation of Data and Control Chart Analysis*, 6th ed. ASTM Special Technical Publication 15D, Part 3, Philadelphia, Pa.

Asher, S. C., Lloyd, K. M., MacKay, D., Paterson, S. and Roberts, J. R.: 1985, *A Critical Examination of Environmental Modelling – Modelling the Environmental Fate of Chlorobenzenes Using the Persistence and Fugacity Models*, National Research Council of Canada. Publication No. NRCC 23990. Ottawa, Canada.

Averett, R. C.: 1981, 'Species Diversity and its Measurement', in Greeson, P. E. (ed.), *Biota and Biological Parameters as Environmental Indicators*, Geological Survey Circ. 848-B, U.S. Dept. of Interior, Washington, U.S., pp. B3–B6.

Baumann, P. C. and Whittle, D. M., 1988, 'The Status of Selected Organics in the Laurentian Great Lakes: An Overview of DDT, PCBs, Dioxins, Furans, and Aromatic Hydrocarbons', *Aqu. Toxicol.* **11**, 241–257.

Berryman, D.: 1984, 'La detection des tendances dans les series temporells de parametres de la qualité de l'eau a l'aide des tests non-parametriques', M. Sc. Thesis, Institut national de la recherche scientifique (INRS-Eau), Sainte-Foy, Quebec.

Berryman, D., Bobee, B., Cluis, D., and Haemmerli, J.: 1988, 'Non-Parametric Tests for Trend Detection in Water Quality Time Series', *Water Resources Bulletin* **24**, 545–555.

Box, G. E. P. and Jenkins, G. M.: 1970, *Time Series Analysis Forecasting and Control*, Holden-Day, San Francisco.

Brooks, D. B. and Peters, R.: 1988, *Water: The Potential for Demand Management in Canada*, Science Council of Canada. Ottawa, Canada.

Burnham, K. P., Anderson, D. R., and Laake, J. L.: 1980, 'Estimation of Density from Line Transect Sampling of Biological Populations', Wildl. Monogr. No. 72.

Chapman, P. M.: 1986, 'Sediment Quality Criteria from the Sediment Quality Triad – an Example', *Environ. Toxicol. Chem.* **5**, 957–964.

Chapman, P. M., Dexter, R. N., and Long, E. R., 1987, 'Synoptic Measures of Sediment Contamination, Toxicity and Infaunal Community Composition (The Sediment Quality Triad) in San Francisco Bay', *Mar. Ecol. Progr. Ser.* **37**, 75–96.

Clark, T., Clark, K., Paterson, S., MacKay, D., and Norstrom, R. J.: 1988, 'Wildlife Monitoring, Modelling and Fugacity', *Environ. Sci. Technol.* **22**, 120–127.

Conner, J. J. and Myers, A. T.: 1973, 'How to Sample a Mountain', in *Sampling Standards and Homogeneity*. ASTM STP 540, American Society for Testing and Materials. Philadelphia. pp. 30–36.

Désilets, L.: 1988, *Criteria for Basis Selection and Sampling Station Macrolocation*, Environment Canada, Scientific Series No. 164. Ottawa, Canada.

Dunnette, D. A.: 1980. 'Observed Frequency Optimization Using a Water Quality Index', *Journal WPCF* **52**, 2807–2811.

Eberhardt, L. L.: 1976. 'Quantitative Ecology and Impact Assessment', *J. Environ. Mgmt.* **4**, 27–70.

El-Shaarawi, A. H.: 1989, 'Inferences About the Mean from Censored Water Quality Data', *Water Resources Research* **25**, 685–690.

El-Shaarawi, A. H.: 1987, 'Frequency of Sampling Required for Monitoring the Niagara River', *Can. J. Fish. Aquat. Sci.* **44**, 1315–1319.

El-Shaarawi, A. H. and Kwiatkowski, R. E.: 1977, 'A Model to Describe the Inherent Spatial and Temporal Variability of Parameters in Lake Ontario, 1974', *J. Great Lakes Res.* **3**, 177–183.

El-Shaarawi, A. H. and Shah, K. R.: 1978, 'Statistical Procedures for Classification of a Lake, Inland Waters Directorate, Environment Canada, Scientific Series No. 86.

Esterby, S. R.: 1988, 'Toxic Contaminates and Ecosystem Health: Great Lakes Focus', in M. S. Evans (ed.), *Advances in Environmental Science and Technology*, John Wiley and Sons, pp. 447–475.

Farrel, R.: 1980, *Methods for Classifying Changes in Environmental Conditions*, Tech. Rep. VRF-EPA7.4-FR80-1, Vector Res. Inc., Ann Arbor, Michigan.

Foster, H. D. and Derrick-Sewell, W. R.: 1981, *Water – The Emerging Crisis in Canada*, James Lorimer and Company, Publishers, Toronto, Ontario, Canada.

Gilliom, R. J. and Helsel, D. R.: 1986, 'Estimation of Distributional Parameters for Censored Trace Level Water Quality Data 1: Estimation Techniques', *Water Resources Research* **22**, 135–146.

Green, R. H.: 1979, *Sampling Design and Statistical Methods for Environmental Biologists*, John Wiley and Sons, Inc. Toronto, Canada.

Government Accounting Office: 1981, *Better Monitoring Techniques Are Needed to Assess the Quality of Rivers and Streams*, Vol. 1. U.S. General Accounting Office, Washington. D. C. Report, CED-81-30.

Harvey, H. H.: 1976, 'Aquatic Environmental Quality: Problems and Proposals', *J. Fish. Res. Board Can.* **33**, 2634–2670.

Helsel, D. R.: 1986, 'Estimation of Distributional Parameters for Censored Water Quality Data', in El Shaarawi and Kwiatkowski (eds.), *Statistical Aspects of Wate Quality Monitoring*, Elsevier Vol 27: Developments in Water Science, pp. 137–157.

Helsel, D. R.: 1987, 'Advantages of Non-Parametric Procedures for Analysis of Water Quality Data', *Journal of Hydrological Sciences* **32**, 179–190.

Helsel, D. R. and Gilliom, R. J.: 1986, 'Estimation of Distributional Parameters for Censored Trace Level Water Quality Data 2: Verification and Applications', *Water Resources Research* **22**, 147–155.

Hipel, K. W., Lettenmaier, D. P., and McLeod, A. I.: 1978, 'Assessment of Environmental Impacts. Part 1: Intervention Analysis', *Environ. Mgmt.* **2**, 529–535.

Hirsch, R. M., Slack, J. R., and Smith, R. A.: 1982, 'Techniques of Trend Analysis for Monthly Water Quality Data', *Water Resources Research* **18**, 107–121.

Hirsch, R. M. and Slack, J. R.: 1984, 'A Nonparametric Trend Test for Seasonal Data with Serial Dependence', *Water Resources Research* **20**, 727–732.

Hurlbert, S. H.: 1984, 'Pseudoreplication and the Design of Ecological Field Experiments', *Ecol. Monogr.* **54**, 187–211.

Johnson, M. G.: 1980, 'Great Lakes Environmental Protection Policies from a Fisheries Perspective', *Can. J. Fish. Aquat. Sci.* **37**, 1196–1204.

Karr, J. R., Fausch, K. D., Angermier, P. L., Yant, P. R., and Schlosser, I. J.: 1986, 'Assessing Biological Integrity in Running Waters: A Method and its Rationale', Ill. Nat. Hist. Surv. Spec. Publ. 5. Champaign, Illinois, U.S.A.

Kolkwitz, R. and Maisson, M.: 1908, 'Okologie der pflanzlichen Saprobien', *Ber. dt. Bot. Ges.* **26**, 505–519.

Kwiatkowski, R. E.: 1986, 'The Importance of Design Quality Control to a National Monitoring Program', in El-Shaarawi, A. H. and R. E. Kwiatkowski (eds.), *Statistical Aspects of Water Quality Monitoring*, Elsevier Scientific. Developments in Water Science **27**, 79–98.

Lachance, M., Bobee, B., and Gouin, D., 1979, 'Characterization of the Water Quality in the Saint Lawrence River: Determination of Homogeneous Zones by Correspondence Analysis', *Water Resources Research* **15**, 1451–1462.

Lesht, B. M.: 1988, 'Nonparametric Evaluation of the Size of Limnological Sampling Networks: Application to the Design of a Survey of Green Bay', *J. Great Lakes Res.* **14**, 325–337.

Lettenmaier, D. P.: 1976, 'Detection of Trends in Water Quality Data from Records with Dependent Observations', *Water Resources Research* **12**, 1037–1046.

Lettenmaier, D. P.: 1978, 'Design Considerations for Ambient Stream Quality Monitoring', *Water Resources Bulletin* **14**, 884–901.

Loftis, J. C. and Ward, R. C.: 1980, 'Water Quality Monitoring – Some practical Sampling Frequency Considerations', *Environmental Management* **4**, 521–526.

Lotspeich, F. B.: 1980, 'Wastershed as the Basic Ecosystem: This Conceptual Framework Provides a Basis for a Natural Classification System', *Water Resour. Bull.* **16**, 581–586.

Lucs, H. L.: 1976, 'Some Statistical Aspects of Assessing Environmental Impact', in Sharma, R. K., J. D. Buffington, and J. T. McFadden (eds.), *Proc., Workshop on the Biological Significance of Environmental Impacts*, NR-CONF-002, U.S. Nuclear Regulatory Commission, Washington, D. C., pp. 295–306.

MacKay, D. and Paterson, S.: 1986, 'Calculating Fugacity', *Environ. Sci. Technol.* **15**, 1006–1014.

Mandel, J.: 1964, *The Statistical Analysis of Experimental Data*, John Wiley and Sons, Inc.

Majkowski, J., Ridgeway, J. M., and Miller, D. R., 1981, 'Multiplicative Sensitivity Analysis and its Role in Development of Simulation Models', *Ecol. Model.* **12**, 191–208.

Mar, B. W., Horner, R. R., Richey, J. S., Palmer, R. N., and Lettenmaier, D. P.: 1986, 'Data Acquisition:

Cost-Effective Methods for Obtaining Data on Water Quality', *Environ. Sci. Technol.* **20,** 545–551.

Millard, S. P. and Lettenmaier, D. P.: 1986, 'Optimal Design of Biological Sampling Programs Using the Analysis of Variance', *Estuarine, Coastal and Shelf Science* **22**, 637–656.

Montgomery, R. H. and Loftis, J. C.: 1987, 'Applicability of the t-Test for Detecting Trends in Water Quality Variables', *Water Resources Bulletin* **23**, 653–662.

National Research Council of Canada: 1981, 'A Screen for the Relative Persistence of Lipophilic Organic Chemicals in Aquatic Ecosystems – An Analysis of the Role of a Simple Computer Model in Screening. I. A Simple Computer model as a Screen for Peristence. II. An Introduction to Process Analyses and Their Use in Preliminary Screening of Chemical Persistence', NRCC No. 18570. Ottawa, Canada.

Neilson, M. A. and Stevens, R. J. J.: 1986, 'Determination of Water Quality Zonation in Lake Ontario Using Multivariate Techniques', in El-Shaarawi A. H. and R. E. Kwiatkowski (eds.) *Statistical Aspects of Water Quality Monitoring*, Elsevier Scientific, Developments in Water Science, Vol. 27. Pp. 99–116.

O'Sullivan, P. E.: 1979, 'The Ecosystem – Watershed Concept in the Environmental Sciences – A Review', *Int. J. Environ. Stud.* **13,** 273–281.

Paterson, S. and MacKay, D. 1985, 'The Fugacity Concept in Environmental Modelling', in O. Hutzinger (ed.), *The Handbook of Environmental Chemistry.* Vol. 2. Part C. Springer-Verlag, New York.

Prentice, R. L. and Marek, P.: 1979, 'A Qualitative Discrepancy Between Censored Data Rank Tests', *Biometrics* **35**, 861–867.

Sanders, T. G., Ward, R. C., Loftis, J. C., Steele, T. D., Adrian, D. D., And Yevjevich, V. J.: 1983, *Design of Networks for Monitoring Water Quality*, Water Resources Publications, Littleton, Colorado.

Schaeffer, D. J., Herricks, E. E., and Kerster, H. W., 1988, 'Ecosystem Health: 1. Measuring Ecosystem Health', *Environmental Management* **12**, 445–455.

Silvert, W.: 1981, 'Principles in Ecosystem Modelling', in *Multiplicative Sensitivity Analysis and its Role in Development of Simulation Models. Ecol. Model.* **12,** 191–208.

Smith, R. A., Hirsch, R. M., and Slack, J. R.: 1982. 'A Study of Trends in Total Phosphorus Measurements at NASQUAN Stations', United States Geological Survey Water-Supply Paper 2190. U.S. Government Printing Office.

Stewart-Oaten, A., Murdoch, W. W., and Parker, K. R.: 1986, 'Environmental Impact Assessment: 'Pseudo-replication' in Time?' *Ecology* **67**, 929–940.

Thienemann, A.: 1928, 'Der Sauerstoff in eutrophen und oligotrophen Seen', *Die Binnengewasser* Vol. 4. Schweizerbart'sche Verlagsbuchhandlung, Stuttgart, Germany.

Thomas, J. M., Mahaffey, J. A., Gore, K. L., and Watson, D. G., 1978, 'Statistical Methods used to Assess Biological Impact at Nuclear Power Plants', *J. Environ. Mgmt.* **7**, 269–290.

Thomas, R. L.: 1988, 'Concluding Remarks', in *Biology in the New Regulatory Framework for Aquatic Protection*, Proceedings of the AllistonWorkshop, April 26–28, 1988. Environment Canada, Ottawa. ISBN No. 0-662-16818-6.

Tukey, J. W.: 1977. *Exploratory Data Analysis*, Addison-Wesley, Reading, Massachusetts.

Van Belle, G. and Hughes, J. P., 1984, 'Non-Parametric Tests for Trends in Water Quality', *Water Resources Research* **20**, 127–136.

Vannote, R. L., Minshall, G. W., Cummins, K. W., Sedell, J. R., and Cushing, C. E.: 1980, 'The River Continuum Concept', *Can. J. Fish. Aquat. Sci.* **37**, 130–137.

Wallin, T. R. and Schaeffer, D. J.: 1979, 'Illinois Redesigns its Ambient Water Quality Monitoring Network', *Environmental Management* **3**, 313–319.

Ward, R. C., Loftis, J. C., and McBride, G. B.: 1986, 'The "Data-Rich but Information-Poor" Syndrome in Water Quality Monitoring', *Environmental Management* **10**, 291–297.

Ward, R. C., Loftis, J. C., Nielson, K. S., and Anderson, R. D.: 1979, 'Statistical Evaluation of Sampling Frequencies in Monitoring Networks', *Journal WPCF* **51**, 2292–2300.

Youden, W. J. and Steiner, E. H.: 1975. *Statistical Manual of the Association of Official Analytical Chemists*, Washington, D.C. 20044.

AN APPLICATION OF MULTIVARIATE ANALYSIS TO ACID RAIN DATA IN NORTHERN ITALY TO DISCRIMINATE NATURAL AND MAN-MADE COMPOUNDS

GIOVANNA FINZI

Centro Teoria dei Sistemi CNR, Dipartimento di Elettronica, Politecnico di Milano, Italy

ALBERTO NOVO

ENEL Centro Ricerca Termica e Nucleare, Milano, Italy

and

SILVIO VIARENGO

METIS Srl, Torino, Italy

(Received April 1990)

Abstract. This paper presents the preliminary results of a study, the aim of which was to analyse the pluviometric and chemical rain data, recorded by a wet only network located in Northern Italy.

More in detail, precipitations were collected on a weekly basis and chemical analysis was performed on pH, electric conductivity and Ca, Mg, Na, K, NH_4, NO_3, SO_4, Cl concentrations.

The Principal Components Analysis pointed out that the first three components are enough to explain more than 90% of the variability of the parameters. Moreover each component may have a different physical interpretation, that is the first one is mainly related to the precipitation amount, while the second to the man made and natural sources and the last one to the sea/soil contribution.

Introduction

Acid precipitation is at present one of the most complex problems that environmental researchers have been dealing with during the last ten years. In fact, in addition to the difficulty in identifying the chemical reactions leading to the formation of acid compounds in the atmosphere, it is even more difficult to identify the pollution sources which caused the phenomenon.

Pollutants may be transported for thousands of miles over many days before their deposition takes place. Sulphur and nitrogen oxides, in particular, may accumulate in the atmosphere and give rise to acid compounds affecting the terrestrial ecosystem in many different ways, either directly or indirectly through, for instance, the corrosion of materials, the deterioration of monuments and the decline of forests.

According to the different sources, the chemical composition of acid precipitations changes in terms of the ion species which can be detected through the analysis of the rain samples. Generally speaking, three main different origins can be distinguished as follows:

- An *anthropic* source (of urban, industrial or agricultural type).

  In this case, sulphates and nitrates together with high acidity can be ascribed to an urban or industrial prevailing contribution, while high values of ammonium point to an agricultural source.

Environmental Monitoring and Assessment **17**: 273–280, 1991.

- A *terrestrial* source, which gives a prevailing contribution in terms of calcium, magnesium and potassium, due to the transport of particulate matter lifted from soils, unpaved roads or desert lands.
- A *sea* source, mainly connected to the presence in sea water of sodium and chloride ions together with a minor amount of magnesium and sulphate.

In this paper, the results of a multivariate statistical analysis performed on precipitation data recorded in Northern Italy will be shown. More specifically, two measurement stations were chosen, one in the city of Milan, representative of an urban precipitation quality, and one in Alpe Gera, located in the Alps at an elevation of 2100 m.

The aim of this study is to quantify the existing relationships among some of the ionic species in acid precipitation by means of a cross correlation analysis and to try to discriminate between the different sources of pollution by examining the results of a principal components analysis on the data.

The Data Set Collection and Handling

During 1983–84 a precipitation quality measurement network was built and run in Northern Italy through the cooperation of various institutions (Gruppo di Studio, 1987). In subsequent years, ENEL (the Italian National Electricity Board) extended its own network to the entire Italian area in collaboration with the Ministry of Agriculture and Forestry (Figure 1).

Moreover, the samplers have been changed from the preceding bulk type to the present wet-dry instruments (Novo *et al.*, 1988) which allow for the discrimination between wet and dry depositions. Unfortunately, the two different sampling methods produce measures which are hardly comparable; this is the main reason why the following analysis was performed solely on data recorded by the second kind of instruments, starting from 1986. Furthermore, the samples are collected on a weekly basis, so that they are representative of all the precipitation events over a seven days period.

The two stations examined in this study are shown in Figure 1. The first station is located in the centre of Milan, a highly populated city (~ 2000000 inhabitants), where there is often severe atmospheric pollution mainly due to domestic heating and traffic. The second station is located at the base of the Bernina glacier, at Alpe Gera (2100 m elevation), in an uncontaminated mountain site, far from any significant urban or industrial area.

Each weekly sample was analyzed for its main chemical constituents, and thus the data base reports the following ionic contents: H^+, Ca^{++}, Mg^{++}, K^+, Na^+, NH_4^+, $SO^-{}_4^-$, NO_3^-, Cl^-, HCO_3^-, which all have the units mg/l.

The last compound (HCO_3) was not considered in the statistical analysis, because the values were always negligible, due to the prevailing acidity (pH $<$ 5.6) of the water samples. The potassium data were also not taken into account, because of low values due to the scarce recycling of potassium from soils and wood combustion.

Statistical techniques used in the analysis include cross correlation and principal components.

Fig. 1. The ENEL precipitation chemistry network.

Results

(1) Cross-correlations analysis

Tables Ia and b show the two correlation matrices (Morrison, 1978) computed on the Alpe Gera and Milano data respectively. It is noticed that their structure is quite similar. In fact both show significant values in correspondence with couples of chemical compounds connected to a common origin. The examination of the highest values of the coefficient may lead to the following considerations:

(i) chloride ion is linked both to sodium and magnesium due to their common origin from

TABLE I

Cross-correlations matrices

	H	Ca	Mg	Na	NH_4	NO_3	SO_4	Cl
(a) Alpe Gera								
H	1	−0.23	−0.19	−0.19	0.52	0.43	0.45	−0.11
Ca	−0.23	1	−0.75	0.36	0.04	0.22	0.47	0.34
Mg	−0.19	0.75	1	0.60	0.15	0.48	0.62	0.74
Na	−0.19	0.36	0.60	1	−0.12	0.25	0.28	0.84
NH_4	0.52	0.04	0.15	−0.12	1	0.68	0.73	−0.01
NO_3	0.43	0.22	0.48	0.25	0.68	1	0.67	0.39
SO_4	0.45	0.47	0.62	0.28	0.73	0.67	1	0.37
Cl	−0.11	0.34	0.74	0.84	−0.01	0.39	0.37	1
(b) Milan								
H	1	−0.22	−0.17	−0.11	0.25	0.55	0.65	−0.08
Ca	−0.22	1	0.90	0.38	0.43	0.38	0.49	0.38
Mg	−0.17	0.90	1	0.57	0.39	0.37	0.47	0.58
Na	−0.11	0.38	0.57	1	0.31	0.19	0.27	0.97
NH_4	0.25	0.43	0.39	0.31	1	0.84	0.77	0.34
NO_3	0.55	0.38	0.37	0.19	0.84	1	0.81	0.26
SO_4	0.65	0.49	0.47	0.27	0.77	0.81	1	0.29
Cl	−0.08	0.37	0.58	0.97	0.34	0.26	0.29	1

sea aerosol;

(ii) the couple calcium-magnesium is typical of a crustal source;

(iii) ammonium, nitrate and sulphate can all be ascribed to the anthropic pollution of the atmosphere.

Furthermore, in the Milan case, a significant correlation value is noticed between hydrogen ions and sulphates (i.e. sulphuric acid), characteristic of the typical urban pollution from domestic heating emissions. This component of the precipitation seems to be a consequence of the local atmospheric wash out.

(2) Principal Components Analysis

Principal Components (PC) analysis (Harman, 1976; Morrison, 1978) allows a greater discrimination among the possible sources of the different ionic compounds of the rain samples. Prior to the application of PC, the two time series of data have been standardized by subtracting the overall mean and dividing by the standard deviation. For each of the examined sites, the following computational and graphic outputs have been considered:

- the percentage of variance explained by each component (Table II);
- the values of the weights computed for the first three principal components (Table III);
- the graphic representation of the weights in the plane component 1 versus component 2 (Figure 2a and b); and,
- the scatter graphic of the original data points in the plane component 1 versus component 2 (Figure 3a and b).

The examination of Table II points out that, in both cases, more than 75% of the

TABLE II

Percentage of variance explained by the principal components

Component number	Alpe Gera: Percent of variance	Alpe Gera: Cumulative percentage	Milan: Percent of variance	Milan: Cumulative percentage
1	45.78706	45.78706	50.34979	50.34979
2	29.23518	75.02224	26.81523	77.16502
3	11.65508	86.67731	13.18922	90.35424
4	5.56965	92.24696	6.50625	96.86049
5	3.63913	95.88609	2.04670	98.90719
6	2.50695	98.39304	0.71240	99.61959
7	1.20262	99.59566	0.24499	99.86458
8	0.40434	100.00000	0.13542	100.00000

variance is explained by the first three components; the first component accounting for approximately 50% of the whole variability of the phenomenon.

It is then possible to deduce the physical meaning of each of the main components by looking at the numerical values of their respective weights (Table III).

As for the first one, all the coefficients are negative and the value of the component is proportional to the concentration of the ionic contents, that is, indirectly, to the total volume of recorded precipitation. As a matter of fact, the relationship between volume and every other compound, except hydrogen ion, is inverse: that is, low volumes of rain are generally associated with high ion concentrations and vice versa (Mosello *et al.*, 1988). Furthermore, the relatively low weight of ammonium may be ascribed to the strong seasonality of this compound and sometimes to its partial consumption due to the activity of bacteria, which often are present in low rain volumes.

The coefficients of the second principal component (second column in Table III) clearly point out the difference between the natural and the anthropogenic sources. In fact,

TABLE III

Weights of the first three principal components

	Alpe gera			Milan		
	1	2	3	1	2	3
H	−0.084	0.529	0.288	−0.122	0.561	0.317
Ca	−0.331	−0.225	−0.642	−0.364	−0.226	−0.542
Mg	−0.460	−0.223	−0.203	−0.392	−0.290	−0.318
Na	−0.338	−0.346	0.449	−0.327	−0.371	0.496
NH_4	−0.250	0.507	−0.110	−0.396	0.226	−0.062
NO_3	−0.389	0.299	0.141	−0.387	0.360	−0.028
SO_4	−0.437	0.270	−0.168	−0.411	0.336	−0.034
Cl	−0.390	−0.284	0.449	−0.340	−0.345	0.502

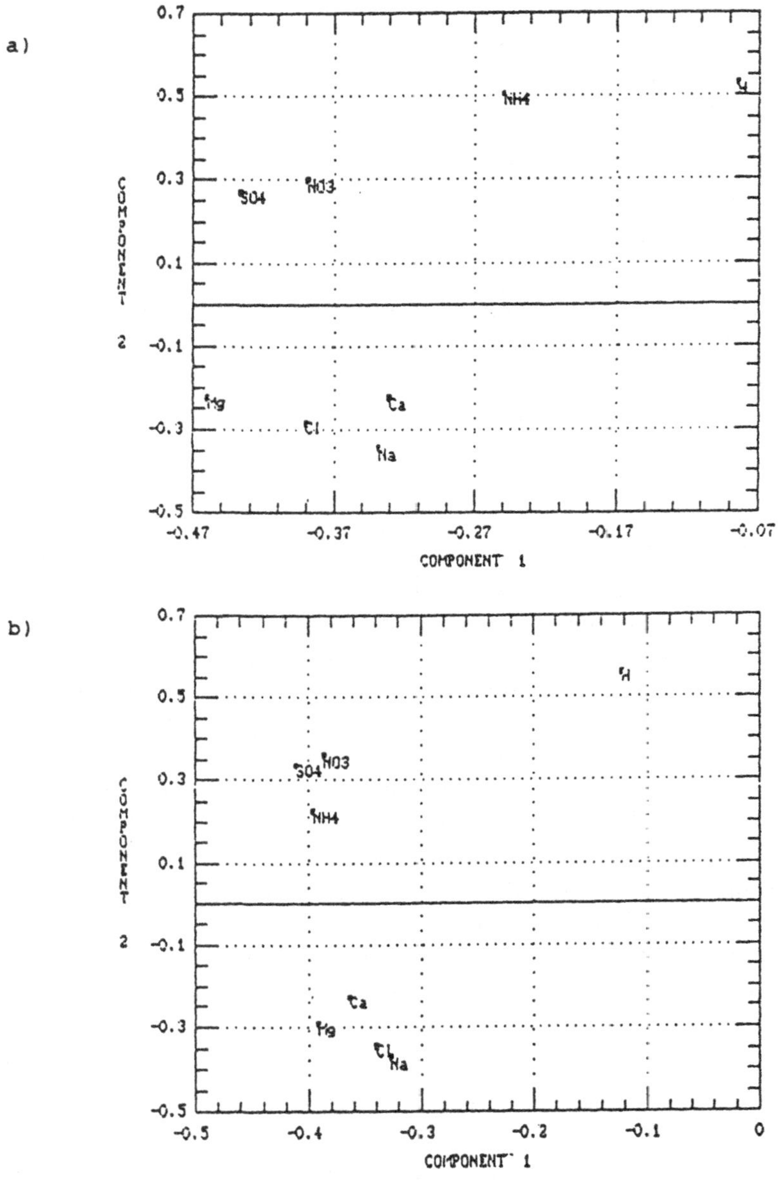

Fig. 2. Representation of the ionic components in terms of the weights of the first two principal components. (a) Alpe Gera; (b) Milan.

positive values are assigned to H, NH_4, NO_3 and SO_4, all derived from agricultural, urban and industrial activities, while negative values come out for Ca, Mg, Na and Cl, which are derived from the soil and sea. High positive values of this component indicate anthropogenic pollution, while negative values are characteristic of the natural chemical composition of rain. The examination of Figure 2 may allow for an easier understanding of these deductions.

Furthermore, the third principal component helps to distinguish between sea and soil contributions, leading to significantly positive values in the first case (Na and Cl from marine source), while negative in the second (Ca from terrigenous sources).

Finally, it may be interesting to examine the scattering of the two samples (53 events for Alpe Gera and 45 for Milan) in the plane of the first two principal components (Figure 3a

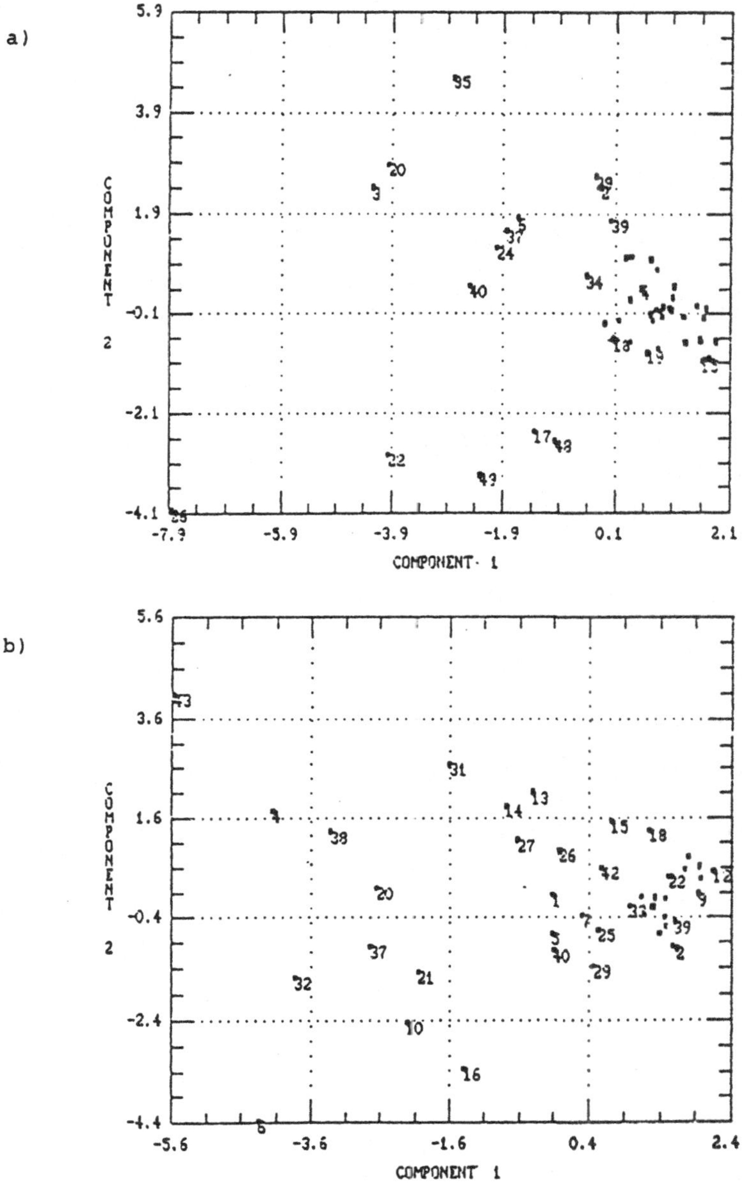

Fig. 3. Scatter diagram of the original data points in the plane component 1 *vs* component 2. (a) Alpe Gera; (b) Milan.

and b). On the basis of the preceding comments, it is possible to identify the highly concentrated events (on the left of the graph) and the significantly polluted ones (on the upper part of the graph).

Conclusions

The results of this preliminary statistical analysis performed on the precipitation chemistry data from two stations in Northern Italy are promising and encourage further investigation in the following directions.

Firstly the analysis will be extended to other stations in the network, increasing the frequency of sampling to produce more reliable data.

Secondly, by means of a long range transport model, it will be possible to compute backward trajectories of the air masses leading to the precipitation. An attempt to go back to the actual pollution source, such as in the case of anthropogenic prevailing contribution to the ionic compounds of the rain, will be less difficult than it is at present.

Acknowledgment

The authors wish to thank dr. P. Bacci of ENEL-CRTN (Milan) for his useful criticism and the students G. De Leo, L. Del Furia and R. Paolucci for the collaboration during their thesis work at the Politecnico of Milan.

References

Gruppo di Studio delle Caratteristiche Chimiche delle Precipitazioni dell'Italia Settentrionale: 1987, 'Deposizioni atmosferiche nel Nord Italia. Rapporto finale anni 1983–1984', *Quaderni di Ingegneria Ambientale* **6**, 1–63.

Harman, Harry H.: 1976, *Modern Factor Analysis*, Chicago: The University of Chicago Press.

Morrison, Donald F.: 1978, *Multivariate Statistical Methods*, McGraw-Hill Book Company.

Mosello, R., Marchetto, A., and Tartari, G. A.: 1988, 'Bulk and Wet Atmospheric Deposition Chemistry at Pallanza (N. Italy)', *Water, Air and Soil Pollution* **42**, 137–151.

Novo, A., Viarengo, S., and Bacci, P.: 1988, Alcuni aspetti delle precipitazioni acide nel nord Italia: Confronto tra due sistemi di campionamento e valutazione delle deposizioni. Proceedings of SEP Pollution: Città e Ambiente, 10–14 aprile 1988 at Padova (Italy).

CHARACTERIZATION OF HYDROCARBON CONTAMINATED AREAS BY MULTIVARIATE STATISTICAL ANALYSIS: CASE STUDIES

GUADALUPE SAENZ[1,2] and NICHOLAS E. PINGITORE[2]

[1] *Department of Geological Sciences, The University of Texas at El Paso, El Paso, Texas*
[2] *Entry-Envirosphere Geochemistry – El Paso, Texas*

(Received April 1990)

Abstract. Analysis of soil gases is a relatively rapid and inexpensive method to delineate and measure hydrocarbon contamination in the subsurface caused by diesel or gasoline. Techniques originally developed for petroleum exploration have been adapted to tracking hydrocarbons which have leaked or spilled at or below the earth's surface.

Discriminant analysis (a multivariate statistical technique) is used to classify soil gas samples of C_1 to C_7 hydrocarbons as biogenic (natural soil gases) or thermogenic (contaminant hydrocarbons). Map plots of C_1 to C_7 total interstitial hydrocarbons, C_2 to C_7 interstitial hydrocarbons, and $C_1/\Sigma C_n$ rations are used to further delineate and document the extent and migration of contamination.

Three case studies of the technique are presented: each involves leakage of hydrocarbons from underground storage tanks. Soil gas analysis clearly defines the spread of contamination and can serve as the basis for the correct placement of monitoring wells. The method proved to be accurate, rapid, and cost-effective; it therefore has potential for widespread application to the identification of soil and groundwater contaminated by hydrocarbons.

Introduction

One of the main sources of groundwater contamination is the spillage and leakage of crude oil and refined petroleum products, which seep into and disperse spatially through the unsaturated zone. The petroleum hydrocarbons move in the subsurface by fluid flow, diffusion and dispersive transport. During this process the volatile hydrocarbons migrate vertically through the soil interstices to the surface (Walter *et al.*, 1987). Borings and monitoring wells normally are employed to detect and define the areal extent of organic contaminants in the subsurface.

An alternative tool for the mapping of groundwater contaminant plumes is the sampling and analysis of volatile organic compounds in the overlying soil (Albertsen and Matthess, 1978; Marrin and Thompson, 1984; Marrin and Thompson, 1987). Characterization and monitoring of subsurface contamination, such as that caused by gasoline leaks from underground storage containers and subsequent reclamation activities, has made use of this technique. A similar technique was developed many years ago for exploration in the oil industry (Laubmeyer, 1933; Sokolov, 1933; Horvitz, 1939), and it has served as a model in the development of soil-gas surveying techniques.

This specific study involves the observation of hydrocarbon leakage from underground storage tanks at three separate localities. Our purpose is to demonstrate the value of multivariate statistical analysis in the differentiation of contaminant hydrocarbon bearing soils from biogenic or background non-polluted areas. The statistical analysis is applied to

Environmental Monitoring and Assessment **17**: 281–302, 1991.

a data base of hydrocarbon concentrations obtained by bulk sampling of soils and subsequent C_1–C_7 analysis of interstitial gases.

Exploration for Hydrocarbons by Soil Gas Analysis in the Petroleum Industry

Hydrocarbons migrate to the earth's surface from sources as deep as 5 to 6 km. According to several investigators, subsurface accumulations of hydrocarbons can be discovered by surface geochemical prospecting methods. In the petroleum and natural gas exploration, Laubmeyer (1933) in Germany, and Sokolov (1933) in the U.S.S.R. initiated the analysis of soil gas. According to the studies of these scientists, the soil atmosphere which accumulates above gas and oil deposits is rich in hydrocarbons, and their detection forms the basis of the exploration technique. Horvitz and Rosaire initiated a related study in the United States in the 1930s, but stressed the analysis of gas adsorbed onto the soil particles and their technique was based on bulk soil samples rather than on the gases of the soil voids. Summaries of their contributions may be found in Horvitz (1939) and Rosaire (1940).

The theoretical basis of these techniques rests on the concept that component gases of hydrocarbon accumulations may escape and move vertically from a deposit by diffusion and effusion in response to pressure gradients and the lower densities of the gases compared to those of water, overlying rocks, and petroleum. The transport rate to the surface depends on the pressure gradient, the viscosity and molecular size of the gas, and the permeability of the rocks along the migration path (Horvitz, 1985).

Detection of Contamination by Hydrocarbons by Soil Gas Analysis

At present, the most widely used method for pinpointing and localizing the areal extent of volatile organics in the subsurface is sampling by borings and the installation of monitoring wells. Although monitoring wells provide the best basis for the quantitative measurement of pollutants in the subsurface, many wells and considerable expenditures are required to define the extent of the pollution plume.

The evaluation of the extent of subsurface contamination can be made by sampling the shallow soil gas, an alternative approach discussed by a number of authors (Silka, 1988; Marrin, 1985; Spittler *et al.*, 1985; Marrin and Thompson, 1987; Lappala and Thompson, 1984). For example, Marrin and Thompson (1987) have shown a linear correlation between contaminant trichloroethylene (TCE) concentrations in groundwater and in soil gas. Lappala and Thompson (1984) compared concentration contours constructed from water samples from nine monitoring wells and from soil gas concentrations. The agreement between the plume patterns mapped by both methods was excellent.

To succesfully apply the technique of soil gas mapping it is important to consider the size distribution of the soil particles and its correlation with the soil gas concentrations. Thus, when comparing gas contents in soils, one must realize that different soil characteristics will affect the quantities of gases found, e.g., clays have a greater adsorption capacity than silts and sands.

Distinguishing Biogenic Hydrocarbons from Contaminants

To correctly interpret soil gas analyses, it is necessary to distinguish biogenic from thermogenic (petrolelum related) gases. Bacterial action on recent sedimentary organic matter forms gases; these gases are rich in methane but generally devoid of C_2 and higher hydrocarbons (Schoell, 1983). Note however, that Hunt *et al.* (1980) documented the biogenic generation of C_4 to C_7 alkanes from terpenoids. Gases of biogenic origin are characterized by a high methane content. $C_1/\Sigma C_n \geq 0.99$, where C_n is the sum of hydrocarbons C_1 through C_7, and by the δ ^{13}C enrichment of methane (Stahl, 1973). Hydrocarbon gases derived from petroleum and natural gas are rich in saturated hydrocarbons (alkanes or paraffins) but do not contain significant quantities of such unsaturated hydrocarbons as alkenes and olefins (Siegel, 1974). Bacterial processes in recent sedimentary organic matter produce minor amounts of paraffins (Horvitz, 1972). Paraffins do not exceed the amount of olefin generated from the sediments, so another criterion for differentiation between biogenic and contaminant hydrocarbons can be the ethane/ethylene ratio. If this ratio exceeds one, it is suggested that at least some of the gas has a petroleum origin (Saenz, 1984).

Statistical Treatment of Soil Gas Data: Discriminant Analysis

Discriminant analysis is a specific tool for the statistical examination of differences between populations. Samples are divided into two or more groups on some *a priori* basis, e.g., location, age, etc. A Wilk's Lambda test establishes whether the groups are distinguishable in n-dimensional space, where n is the number of variables that were tested. The Wilk's Lambda test is thus the *n*-dimensional analogue of a one-dimensional Student's test. If the groups prove to be different, the analysis then searches for the direction in n-space which best discriminates between the groups; this direction is discriminant function one. In applications involving more than two sample groups, a second discriminant function is calculated, representing the next best direction of discrimination. Even with multiple groups, the first two discriminant functions normally account for most of the discrimination between groups. Samples may be plotted in discriminant space, the reduced dimension of a plot of discriminant function one versus discriminant function two. The discriminant plot subsequently may be used to classify additional samples whose group associations are unknown. Descriptions of discriminant analysis are found in Dillon and Goldstein (1984) and Nie *et al.* (1975).

Multivariate statistical analysis has considerable potential in the organic geochemical method of prospecting for subsurface hydrocarbons as demonstrated by Saenz (1987), and Saenz and Pingitore (1989) in retrospective studies. By performing discriminant analysis on soil gas data they distinguished areas underlain by commercial accumulations (at depth) of petroleum from barren areas. Figure 1, adapted from Saenz and Pingitore (1989) shows the Varimax-rotated discriminant scores of the samples used in their oil field study. Note how the hydrocarbon bearing areas plot along the horizontal axis whereas samples from the barren areas plot along the vertical axis. Discriminant function one

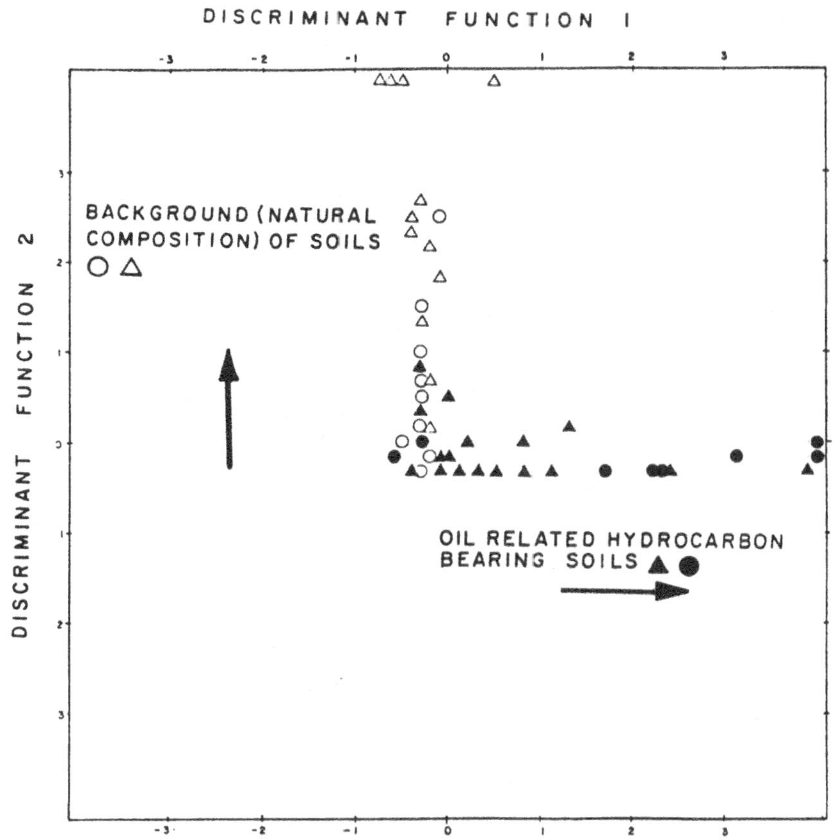

Fig. 1. Soil samples taken over hydrocarbon bearing and background (natural composition) areas plotted in rotated discriminant space.

(horizontal axis in Figure 1) had its highest loading from C_3, i.e., it was defined mainly by C_3, and discriminant function two (vertical axis) had its highest loading from C_1.

The presence of C_3 in soils is the result of thermogenic gases from hydrocarbon accumulations at depth. The presence of C_1 in soils typifies the biogenic production of gases near or at the surface. Thus the discriminant analysis is consistent with our knowledge of the origin of these gases; since the discriminant functions are defined by gases other than C_3 and C_1, they can be very sensitive in distinguishing biogenic and thermogenic gases.

Graphic Analysis of Soil Sample Data

Normally, environmentalists analyze the volatile hydrocarbons as a single group without attention to the values of the individual C_1 to C_7 hydrocarbons. In this study, we utilize up to three separate map plots of hydrocarbons. Total C_1 to C_7 interstitial hydrocarbon plots map the areal distribution (from sample locations) of all the volatile hydrocarbons. Plots of $C_1/\Sigma C_n$ indicate, in general, the proportion of the total hydrocarbons that are biogenic, i.e., that is likely due to natural background concentrations in the soil. Plots of C_2 to C_7 indicate the probable absolute quantity of contaminant gases present at each site. Examples of each type of plot are seen in Figures 3, 4, and 5. Judicious use of these different types of plots permits a more precise delineation of the soil spread of contaminants than using a single plot of total volatile hydrocarbons.

Examples of Soil Gas Analysis Technique

In the present study we present three applications of the exploration technique of Saenz (1987) and Saenz and Pingitore (1989) to delineate and map the near-surface contamination by hydrocarbon leaks. First, the raw data are presented as tables. Then, soil gas analyses are plotted in the discriminant space generated in the exploration study and the position of each sample on that plot indicates whether the sample's composition is consistent with a biogenic (i.e. natural) or thermogenic (i.e. contaminant) origin. Next, a graphical analysis of the data is presented in plot maps to delineate the areal extent of the contamination. Finally, an evaluation and interpretation of these data is presented. Full details about the location and ownership of the studied sites cannot be presented due to the sensitive and proprietary nature of such information; a general description, however, suffices to reveal the details needed by the environmentalist. While some features of Case I were discussed in Saenz *et al.* (1989) Cases II and III have not been previously reported.

Results and Discussion: Case I

Objective

The purpose of the investigation was to define the horizontal extent of the contamination caused by the leakage of hydrocarbons from an underground gasoline storage tank.

Site Location and Sampling

The site is located at El Paso, Texas, in the floodplain of the Rio Grande. An underground storage tank was leaking at the site of a former shallow lake, marginal to the Rio Grande, which had been intentionally filled with sediment. Core samples 1 through 9 were taken at a depth of 2.74 m; depth to the water table is 3.66 m. The subsurface soils encountered at the site are predominantly clays and silty sands. They are moist to wet and have a firm to stiff consistency. Bulk soil samples were collected, and C_1–C_7 insterstitial hydrocarbons were analyzed by gas chromatography.

Statistical Analysis

Soil gas measurements are presented in Table I; in Figure 2 these samples are plotted in the discriminant space of Saenz (1987) and Saenz and Pingitore (1989). Note how the samples are arranged in the discriminant space both along discriminant function 1 (thermogenic gases) and discriminant function 2 (biogenic gases), indicating the presence of gases from

TABLE I

Interstitial concentrations of individual hydrocarbons[a]

Sample Number	Methane	Ethane	Propane	i-Butane	n-Butane	i-Pentane	n-Pentane	Hexane	Heptane
1	78.921	0.510	0.370	0.000	0.098	0.122	0.244	0.000	0.000
2	1.161	0.204	0.222	0.000	0.098	0.000	0.000	0.000	0.000
3	238.761[b]	0.000	0.592	0.098	0.196	0.122	0.122	0.000	0.000
4	19.089	0.612	0.370	0.000	0.098	0.122	0.122	0.000	0.000
5	60.372	2.754	0.592	0.000	0.098	3.294	0.244	272.300	364.200
6	4.374	0.561	0.370	0.098	0.098	0.244	0.244	27.500	13.600
7	2.403	0.204	0.222	0.098	0.098	0.244	0.244	1.450	7.390
8	38.502	1.122	0.444	0.098	0.196	3.172	1.708	94.250	360.000
9	2412.099[b]	0.000	8.140	2.744	6.860	23.668	13.298	59.600	199.250

[a] Values expressed in PPB by weight.

[b] The high methane reading made it difficult to get a reading for ethane.

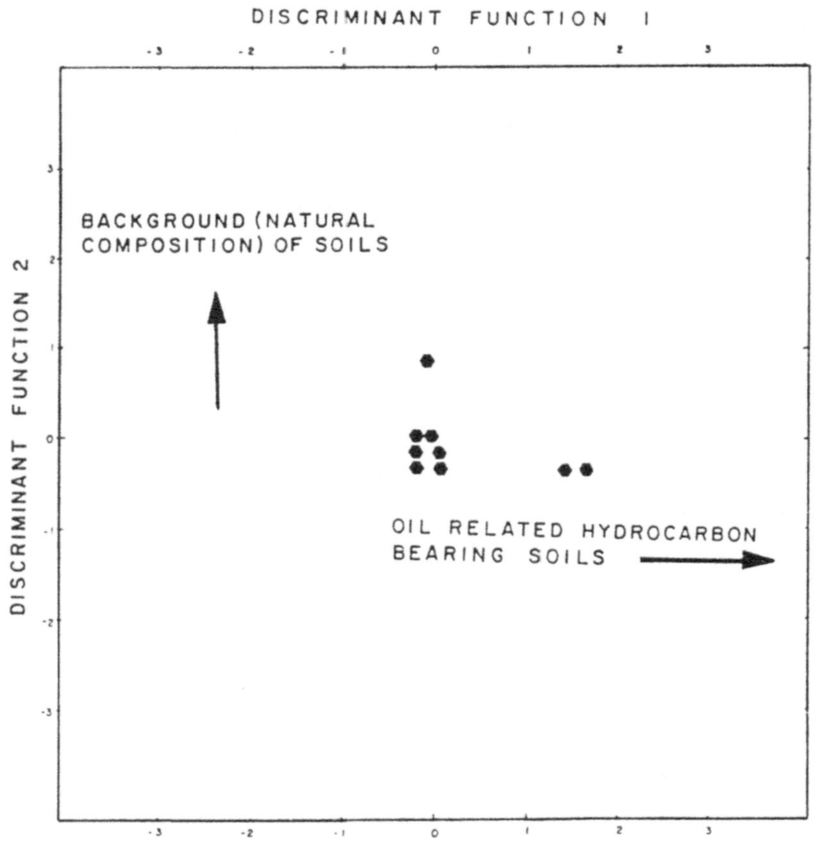

Fig. 2. Mathematical Classification of Study Case I.

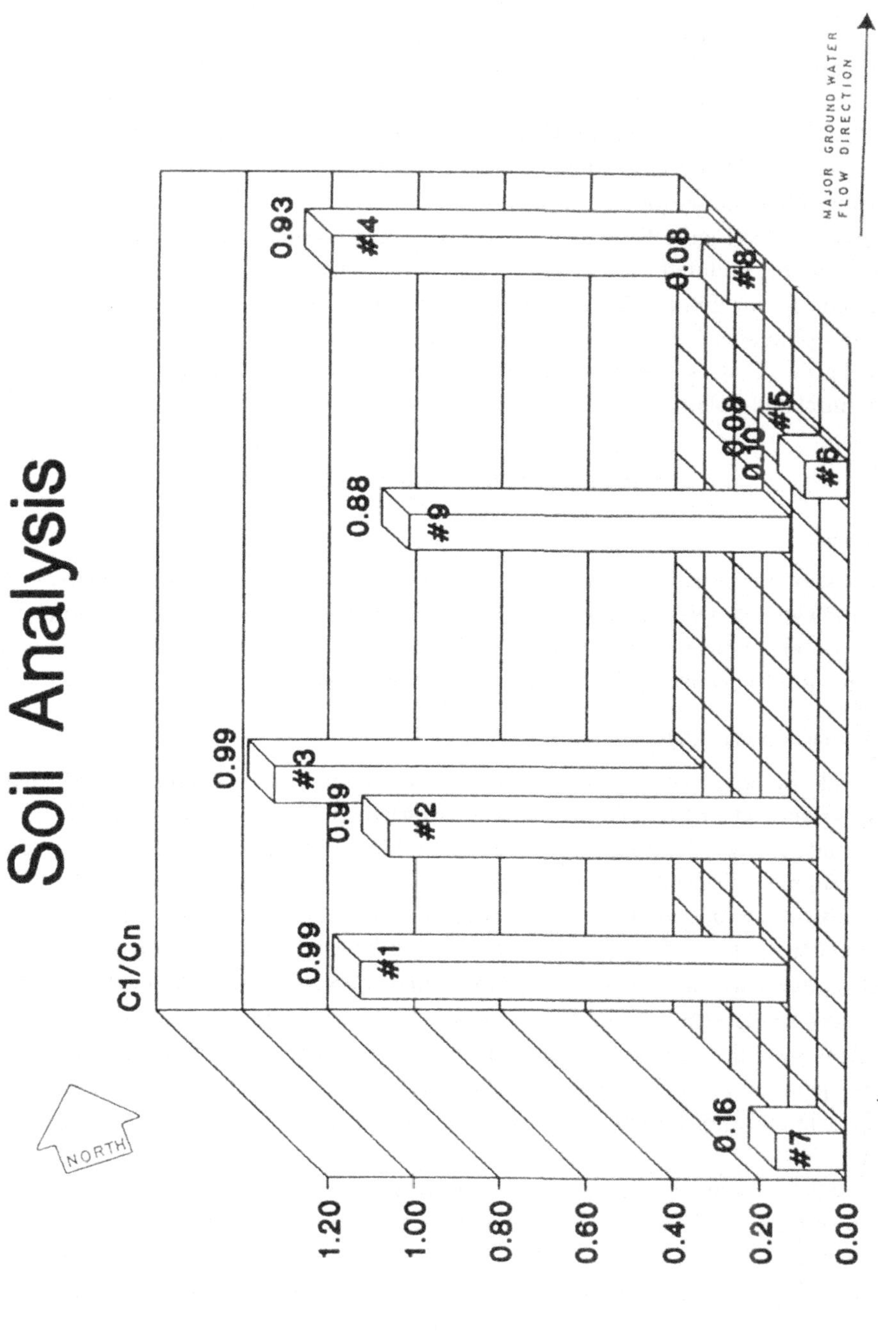

Fig. 3. Proportion of the total hydrocarbons that are biogenic in origin (i.e. natural background in the soil) for Study Case I.

both sources (bacteria and contaminant). Although the regional climate is desert, the site was close to shallow lake presumably containing a considerable amount of recent organic matter, which generated the biogenic gases by bacterial activity. The volatiles produced by the hydrocarbons which leaked from the underground storage tank mixed with these biogenic cases. Thus, the statistical classification of the area proves that only some of the gases are due to the leaking underground storage tank. To further validate and explain this, $C_1/\Sigma C_n$ ratios were calculated (Figure 3). As mentioned in a previous section of this report, ratios of 0.99 or greater indicate biogenic gases. The classification coincides with the geoenvironmental setting of this area.

Isocontent plots (Figure 4 and 5) reveal a natural background area marked by biogenic gases and an area polluted with gasoline. Note in particular the differences in samples 1 and 3 in comparing Figures 3 and 4. Figure 4 suggests the pollution plume may have reached sampling sites 1 and 3 whereas Figure 5 indicates these sites are not contaminated. Slightly elevated C_2–C_7 values at site 7 (Figure 5) may reflect a separate plume emanating from an unrelated pollution source to the left of the diagram.

Results and Discussion: Case II

OBJECTIVES

The purpose of this investigation was to: (1) define the degree of contamination in the soils and groundwater caused by a leaking underground diesel fuel storage tank, (2) define the direction of groundwater movement, (3) define the main direction of contaminant migration, and (4) delineate the horizontal extent of the pollution.

SITE LOCATION AND SAMPLING

Our second case involved an underground diesel storage tank in Chihuahua, Mexico, which was removed long before the area was investigated. It is believed that the fuel leak was intermittent due to fluctuations of the water table. For that reason, the development of contaminant slugs instead of a plume is possible (Figure 6). The site is generally level and the subsurface soils encountered are predominantly clays. These soils have a low to medium plasticity index. They are moist to wet and have a firm to stiff consistency.

Groundwater was encountered at an approximate depth of 1.52 m below the surface. The groundwater table is higher towards the center of the studied area. Knowledge of the general geology of the area indicates that the shallow groundwater found in the study area is possibly perched water. The nearest body of surface water is an open sewer drain that runs approximately northeast to southwest. It is located about 6.10 m to the southeast of the area.

This site is surrounded predominantly by residential buildings. A mechanical shop is located to the north of the site and an inspection of this property indicated it to be a possible source of contamination due to surface fuel spills and washing of mechanical parts. There is no evidence of any other nearby soil excavations, basements, subgrade structures or potential sources of contamination.

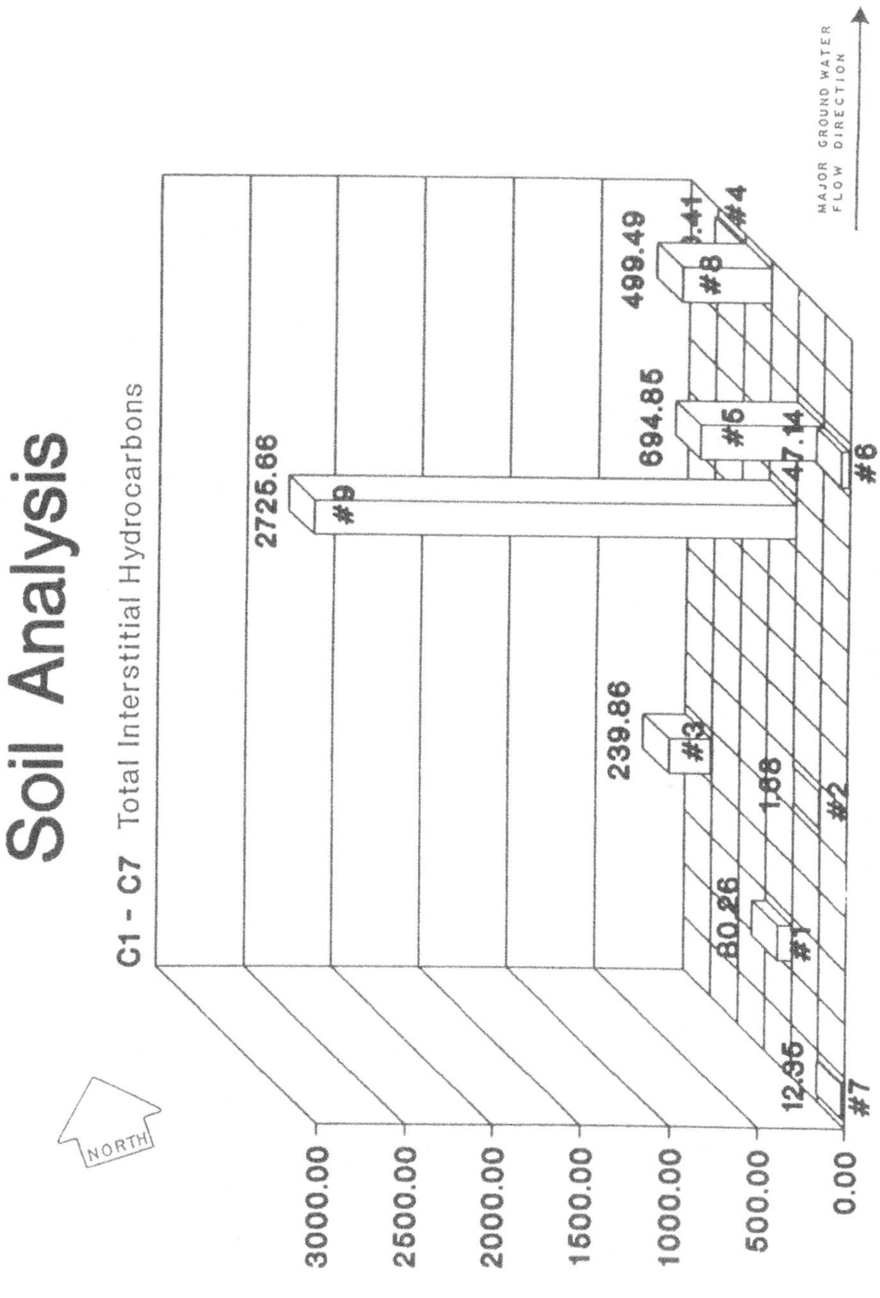

Fig. 4. Distribution of both biogenic and contaminant hydrocarbons in soils for Study Case I.

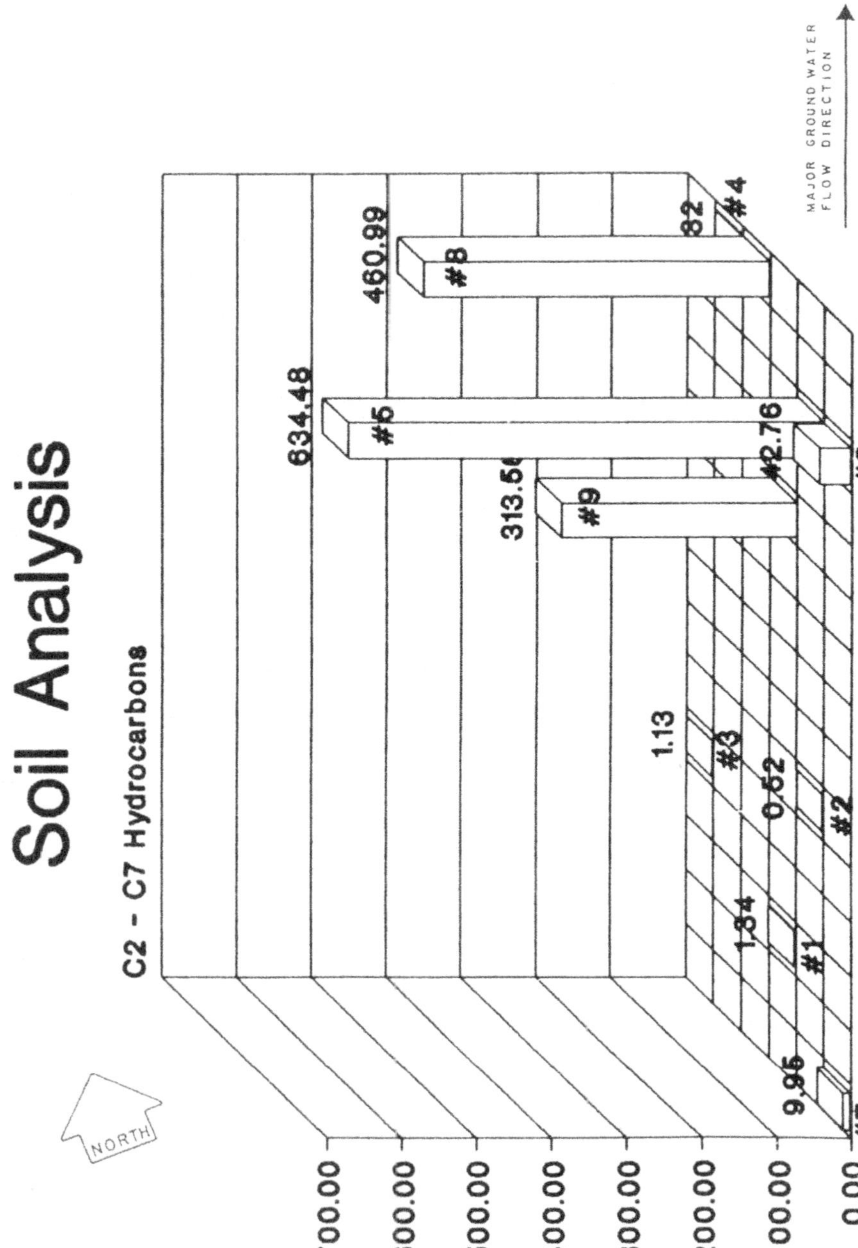

Fig. 5. Distribution of contaminant gases in soils for Study Case I.

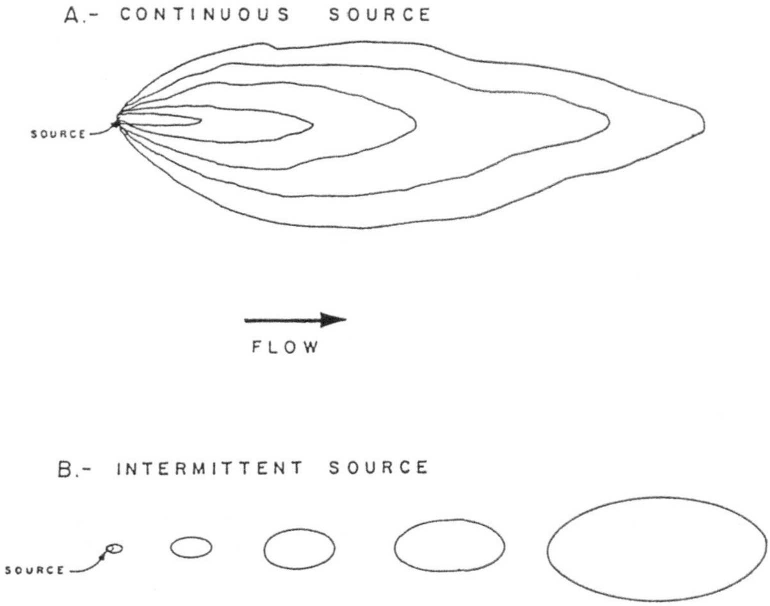

Fig. 6. Continuous contaminant source vs. intermittent source.

Five test borings were performed at the site in order to establish permanent groundwater monitoring wells. Seven subsurface soil samples were recovered at a depth of 0.91 m. Coring was performed by means of a standard 5.08 cm O.D. split spoon sampler. Groundwater samples were recovered from the monitoring wells in accordance with the ASTM D4448-85 method. In addition, two water samples were taken from the sewer drain.

STATISTICAL ANALYSIS

Results of the soil gas analysis (interstitial gases) for this study are presented in Table II.

TABLE II

Interstitial concentrations of individual hydrocarbons[a]

Sample Number	Methane	Ethane	Propane	Isobutane	Normal Butane	Iso-pentane	Normal Pentane	Hexane	Heptane
B # 1	282.0	4.2	22.7	1.0	4.6	1.7	0.6	37.0	197.2
B # 2	46.4	1.1	18.3	1.0	5.4	0.2	0.2	0.0	0.0
B # 3	61.6	1.6	20.7	1.0	4.0	0.1	0.2	0.0	0.0
B # 4	0.8	1.0	19.5	0.9	4.8	0.1	0.2	0.0	0.0
B # 5	0.4	0.4	9.0	0.5	2.3	0.1	0.1	0.0	0.0
B # 6[b]	1.8	1.4	11.0	0.5	3.4	0.1	0.2	0.2	0.0
B # 7[b]	2.0	0.6	6.0	0.5	1.1	0.2	0.2	0.5	0.0

[a] Values expressed in parts per billion by weight.

[b] Corrected values.

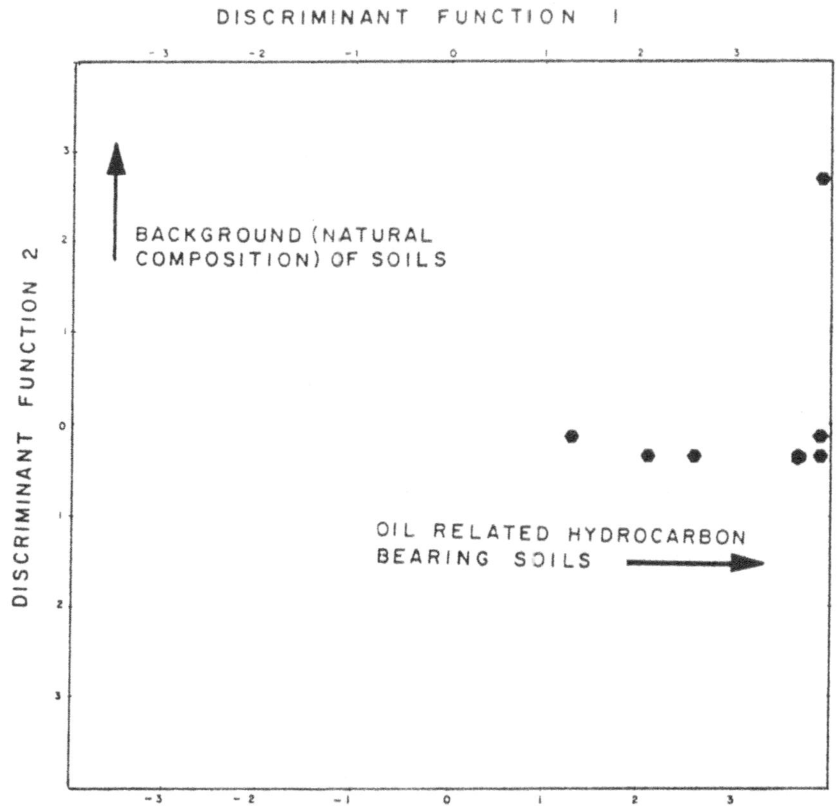

Fig. 7. Mathematical classification of Study Case II.

These data were entered as unknowns to the data base used by Saenz and Pingitore (1989). Figure 7 shows the mathematical classification of each sample site. Note that the samples plot in discriminant space along discriminant function 1 (petroleum related gases), indicating they are contaminants in a soil devoid of recognizable quantities of biogenic gases. This is consistent with the desert climate of the site. Figure 8 presents the $C_1/\Sigma C_n$ values which substantiate the interpretation that all gases found in the area are petroleum related.

Soil analyses were used to construct Figure 9. In this particular case, the need to eliminate C_1 to differentiate biogenic from man-made contamination does not exist. After examination of Figure 9, it can be concluded that the majority of the contamination is migrating to the south and southwest of the area.

The analysis of total petroleum hydrocarbons (TPH) in groundwater is presented in Table III. Sewer drain analyses are shown in Table IV. Groundwater analyses of benzene (B), toluene (T), xylene (X) and ethyl benzene (E) were also performed. The B, T, E, X, test yielded negative results in all cases (<1.0 ppb). The highest concentrations of TPH in groundwater are found in borings 1 and 2 (Table IV and Figure 10). These values possibly correspond to the nearest contaminant slug in the area. Note that the distribution of TPH

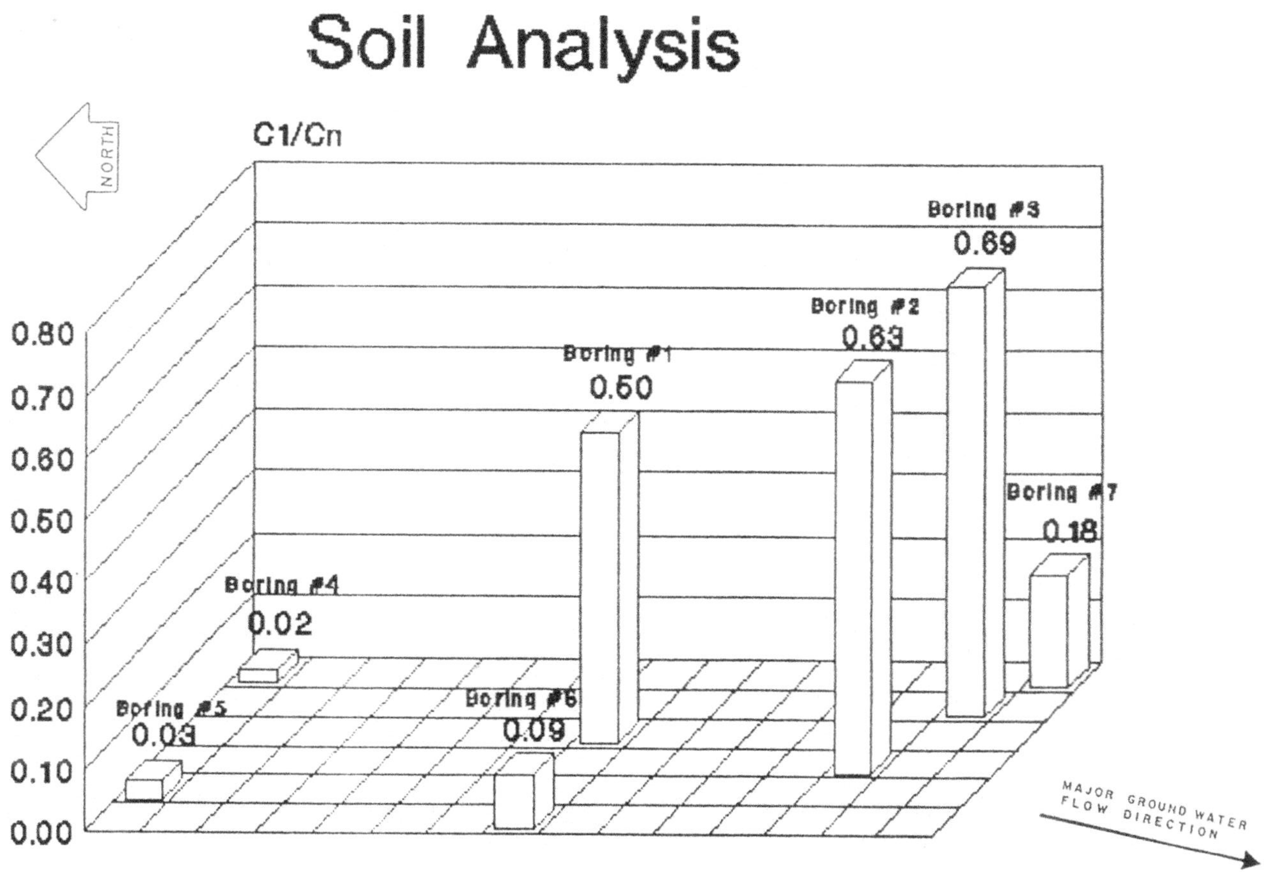

Fig. 8. Proportion of the total hydrocarbons that are biogenic in origin (i.e., natural background in the soil) for Study Case II.

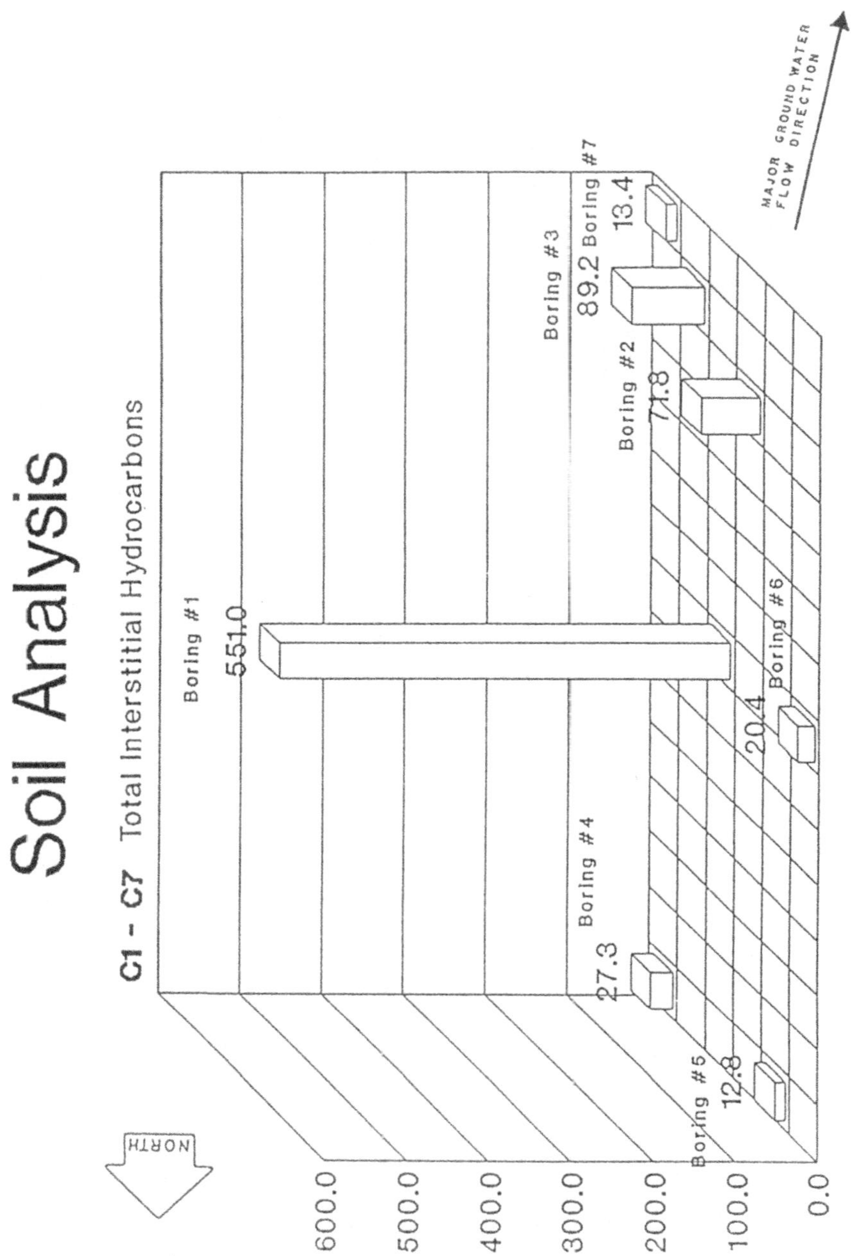

Fig. 9. Distribution of contaminant gases in soils for Study Case II.

TABLE III

Groundwater analysis[abcdefg]

Monitoring well	B	T	X	E	TPH
1	<1.0	<1.0	<1.0	<1.0	1500
2	<1.0	<1.0	<1.0	<1.0	4200
3	<1.0	<1.0	<1.0	<1.0	52
4	<1.0	<1.0	<1.0	<1.0	17
5	<1.0	<1.0	<1.0	<1.0	24

[a] Values in parts per billion.
[b] A less than (<) indicates the compound is not detected at the level indicated.
[c] B indicates benzene.
[d] T indicates toluene.
[e] X indicates xylene.
[f] E indicates ethyl benzene.
[g] TPH indicates total petroleum hydrocarbons.

in groundwater (Figure 10) indicates migration of the contamination towards the south and southwest of the area. It should be noted that neither the timing between leaking events, their duration, nor the groundwater velocity are known. For that reason the proper assessment of the actual existence and magnitude of the contaminant slugs cannot be estimated with confidence at this point.

If a comparison between the hydrocarbon analysis of soils (Figure 9) and the groundwater analytical data (Figure 10) is made, a good correlation is found. However, when a more detailed analysis of both sets of data is made only a fair correlation is found in boring number 1 between TPH in the water and the quantities of interstitial hydrocarbons present in the soils. The discrepancy can be explained if we consider the following points:

(1) Boring number 1 was drilled closest to the location of the underground storage tank,

TABLE IV

Sewer drain analysis[abcdefg]

Sample	B	T	X	E	TPH
Downstream	<1.0	<1.0	<1.0	<1.0	2500
Upstream	<5.0	<5.0	<5.0	<5.0	440

[a] Values in parts per billion.
[b] A less than (<) indicates the compound is not detected at the level indicated.
[c] B indicates benzene.
[d] T indicates toluene.
[e] X indicates xylene.
[f] E indicates ethyl benzene.
[g] TPH indicates total petroleum hydrocarbons.

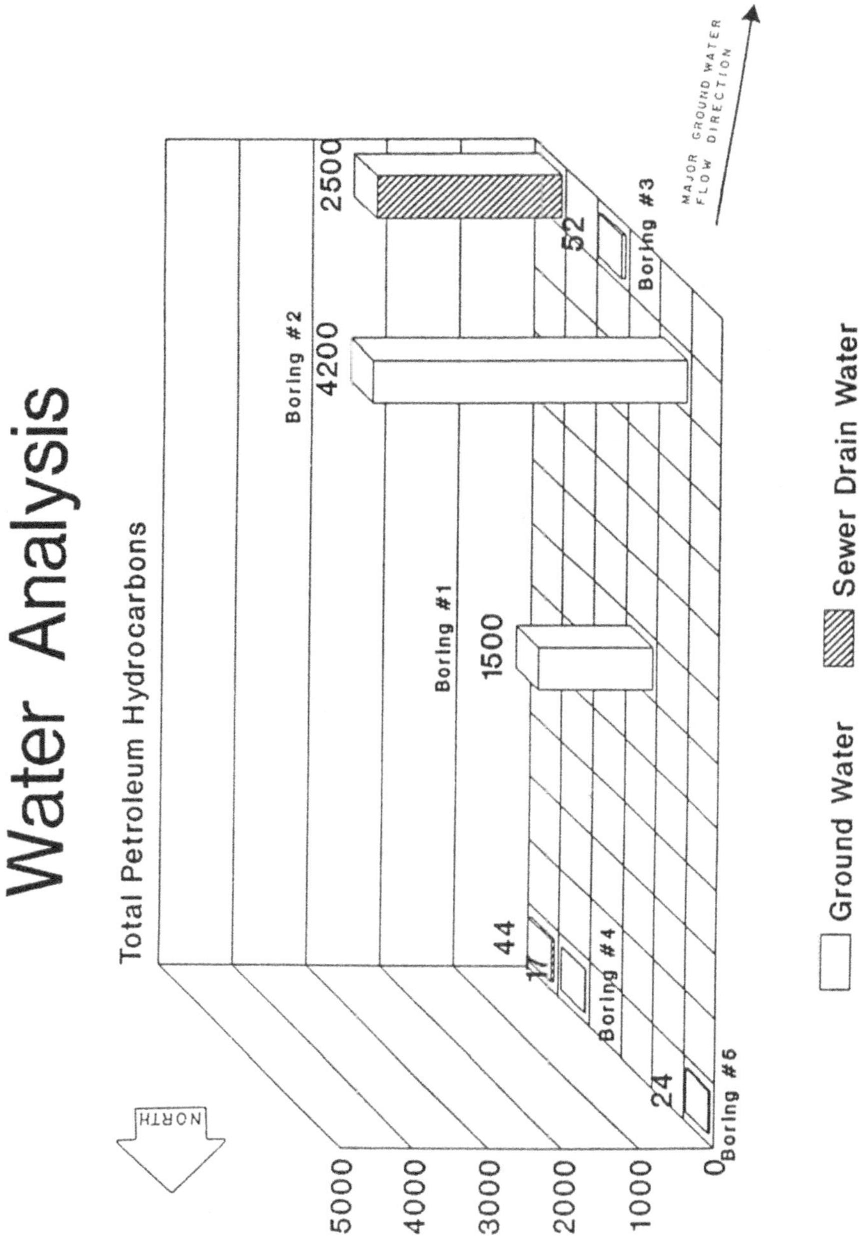

Fig. 10. Distribution of total petroleum hydrocarbons in water.

and thus its soils were exposed to the largest quantities of hydrocarbons leaking from the tank.

(2) Soils found in the area are composed mainly of clays which are excellent adsorbers of organic compounds. Hydrocarbons cannot be desorbed easily from clays; this contributes to the buildup of gases in clay soils close to the leaking source (boring number 1).

Sewer drain water analyses show a higher concentration of TPH downstream (Table IV and Figure 10), indicating a possible contribution of diesel from the underground storage tank.

Considering that the main direction of contamination movement was to the south and southwest of the area, boring number 4 shows slightly higher concentrations of interstitial hydrocarbons than expected (Figure 9). Records indicate that this contamination may be due to the contribution of hydrocarbons from the mechanical shop located near the sampling hole.

Results and Discussion: Case III

Objective

The purpose of this investigation was to define the vertical and horizontal extent of the contamination caused by a leaking underground gasoline storage tank placed at a depth of 3.66 m. The tank was removed before the investigation started.

Site Location and Sampling

The site is located in downtown El Paso, Texas. Subsurface soils encountered in the area at a depth of 3.05 m are mainly fine sands, poorly graded, and gravels. At a depth of 9.14 m an apparently continuous layer of sandy, stiff clay was found. Depth to the water table is 15.24 m.

Core samples 1 through 9 were taken at a depth of 3.05 m. Coring was performed by

TABLE V

Interstitial concentrations of individual hydrocarbons[a]

Sample Number	Methane	Ethane	Propane	i-Butane	n-Butane	i-Pentane	n-Pentane	Hexane	Heptane
1	4.833	0.918	0.814	0.294	0.588	0.366	0.366	1.000	0.000
2	5.184	0.816	0.814	0.196	0.294	0.244	0.244	0.870	0.000
3	0.972	0.153	0.148	0.000	0.098	0.000	0.000	0.000	0.000
4	1.161	0.102	0.148	0.000	0.098	0.000	0.000	0.000	0.000
5	0.972	0.153	0.148	0.000	0.098	0.122	0.122	0.000	0.000
6	3.159	0.612	0.370	0.098	0.196	0.122	0.122	0.435	0.000
7	2.484	0.357	0.370	0.098	0.098	0.122	0.244	1.450	0.000
8	3.996	0.225	0.296	0.098	0.098	0.122	0.122	0.000	0.000
9	5.890	4.540	1.840	3.720	9.630	82.600	88.450	210.800	890.000

[a] Values expressed in PPB by weight.

means of a standard 5.08 cm O.D. split spoon sampler.

STATISTICAL ANALYSIS

Results of the soil gas analysis (interstitial gases) are presented in Table V. The same procedure was followed for the statistical analysis of study cases I and II by entering the data of Table V as unknowns to the data base used by Saenz and Pingitore (1989).

Figure 11 shows the mathematical classification for case III. Note that samples are arranged in discriminant space along discriminant function 1 (petroleum related gases).

To validate the statistical classification that the area is one in which there is contamination due mainly to the spillage of hydrocarbons from the underground storage tank, Figure 12 presents $C_1/\Sigma C_n$ ratios. As stated earlier in this report, ratios of 0.99 or greater are indicative of biogenic gases. Figure 12 shows no such ratios, although sample number 4 presents a relatively high value (0.97). After comparison of Figure 12 and Figure 13, the C_1–C_7 interstitial hydrocarbons plot map, these conclusions may be reached:

(1) The contaminant plume generated by the underground storage tank has been migrating mainly towards the east. Note that the highest values of interstitial hydrocarbons were found in samples 1, 2 and 9 (Table V).

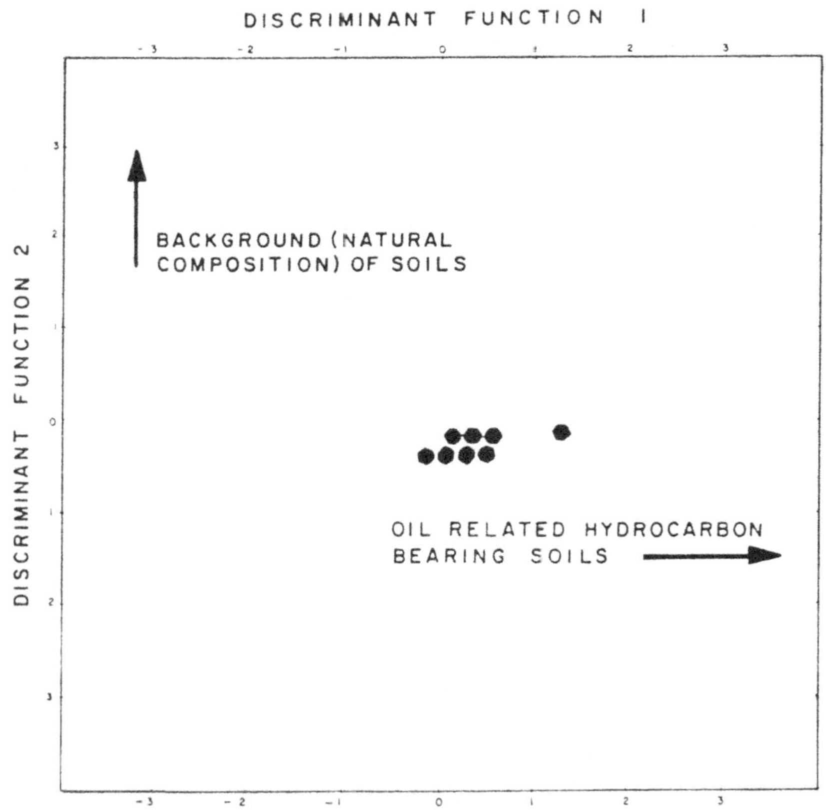

Fig. 11. Mathematical classification of Study Case III.

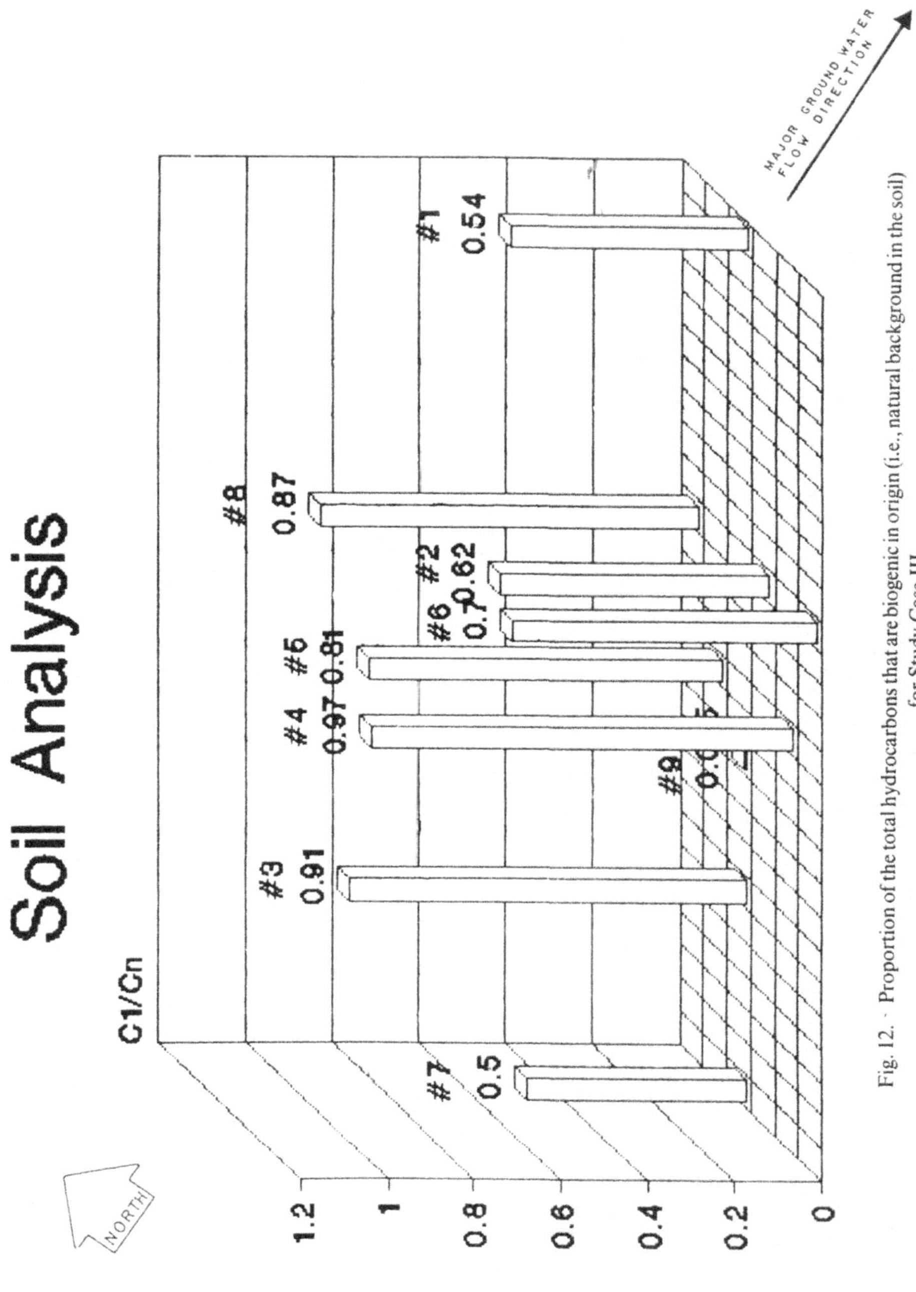

Fig. 12. · Proportion of the total hydrocarbons that are biogenic in origin (i.e., natural background in the soil) for Study Case III.

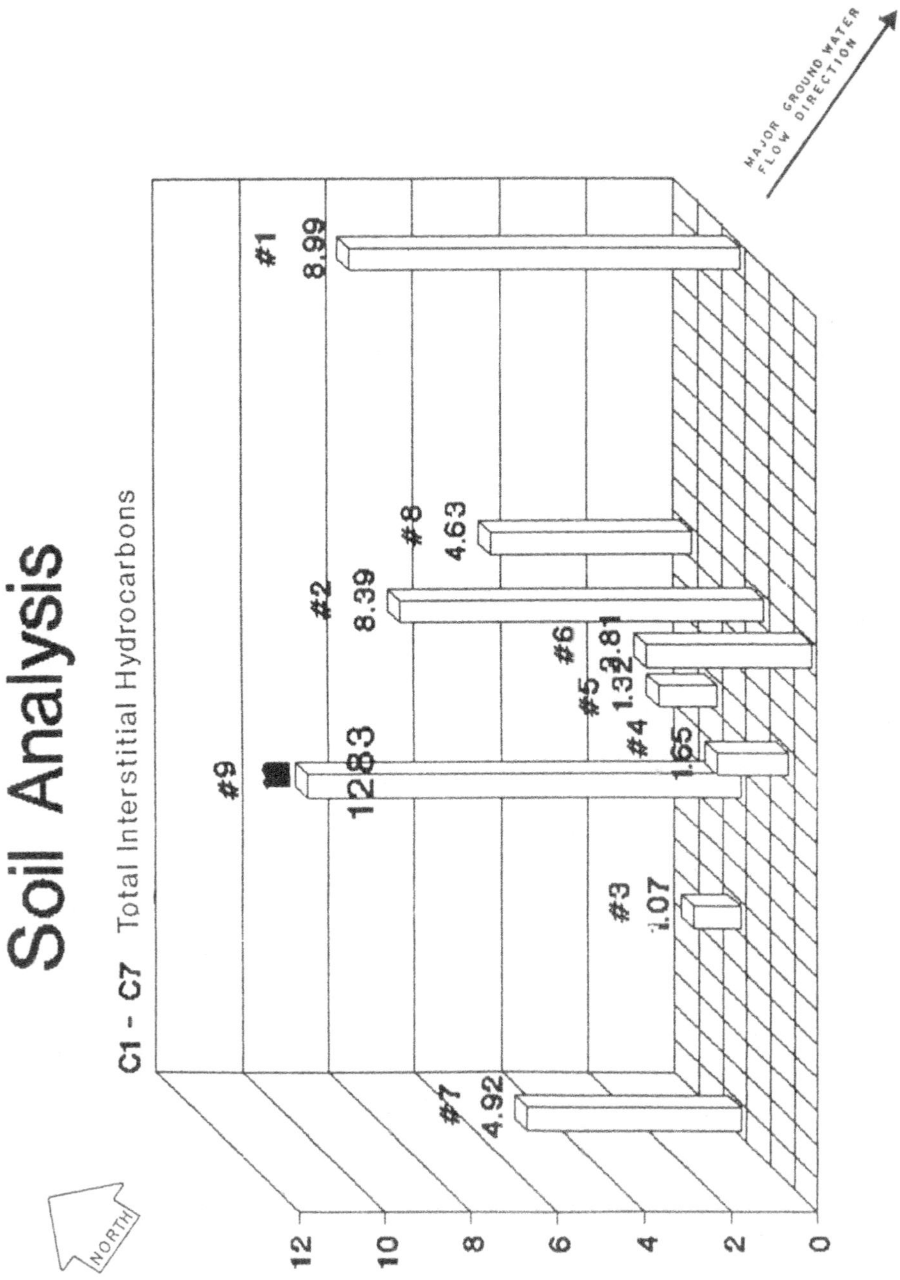

Fig. 13. Distribution of contaminant gases in soils for Study Case III.

(2) The outer part of the contaminant plume was not reached by the sampled area; sample 1 shows relatively high values. The plume presumably extends farther to the east, beyond the limits of the map.

(3) Figure 13 shows some contamination for sample 7. However, the contaminant source is different from the underground storage tank of this study. Sample 7 shows no logical upstream decreasing quantities of C_1–C_7 hydrocarbons with an increase in distance from the source of the release. Sample 3, located between sample 9 and sample 7, shows very low hydrocarbon concentration levels (background values) demonstrating that the relatively high hydrocarbon concentrations found in sample 7 have no link to the underground storage tank.

The major groundwater flow direction for the area is due east. Its influence moved hydrocarbons in that direction, displacing the contaminant plume to the east.

Conclusions

Concentration measurements of C_1–C_7 gases from near surface bulk soil samples have shown to be an alternative method to detect and delineate plumes of organic contaminants. The discriminant analysis of the C_1–C_7 data provides a way to differentiate between contaminants from man-made sources and those of background origin (i.e., biogenic gases). Discriminant analysis can also be used as a classification tool; once a data base is created, additional polluted areas may be readily predicted. Applications of this method should be useful in characterization and management of contaminated areas.

Acknowledgments

The authors express their appreciation to the city of El Paso, Texas, for allowing the use of data from two study cases. Special thanks are due to Entry-Envirosphere Geochemistry-who partially supported this investigation. Ms. B. Barnes and Mr. D. Airey from the Texas Water Commission also have the authors' sincerest thanks for the continuous help with information and comments. Thanks also to Drs. E. Springer and W. L. Polzer from the Environmental Science Group at Los Alamos National Laboratory for their general comments on case study I.

References

Albertsen, M. and Matthess, G.: 1978, 'Ground Air Measurements as a Tool for Mapping and Evaluating Organic Groundwater Pollution Zones', *International Symposium on Ground Water Pollution by Hydrocarbons:* 235–251.

Dillon, W. R. and Goldstein, M.: 1984, *Multivariate Analysis Methods and Applications*, New York, John Wiley and Sons.

Horvitz, L.: 1939, 'On Geochemical Prospecting', *Geophysics* **4**, 210–225.

Horvitz, L.: 1972, 'Vegetation and Geochemical Prospecting for Petroleum', *Am. Assoc. Pet. Geol. Bull.* **56**, 925–940.

Horvitz, L.: 1985, 'Geochemical Exploration for Petroleum', *Science* **229**, 821–827.

Hunt, J. M., Miller, R. S., and Whelan, J. K.: 1980, 'Formation of C_4–C_7 Hydrocarbons from Bacterial

Degradation of Naturally Occurring Terpenoids, *Nature* **288**, 577–588.

Lappala, E. and Thompson, G. M.: 1984, 'Detection of Groundwater Contamination by Shallow Soil Gas Sampling in the Vadose Zone: Theory and Applications', in *Proceedings of the 5th National Conference on Management of Uncontrolled Hazardous Waste Sites,* Hazardous Materials Control Research Institute. Silver Springs, MD, pp. 20–28.

Laubmeyer, G.: 1983, 'A New Geophysical Prospecting method, Specially for Deposits of Hydrocarbons', *Petroleum* **29, 1**–4.

Marrin, D. L. and Thompson, G. M.: 1984, 'Remote Detection of Volatile Organic Contaminants in Groundwater via Shallow Soil Gas Sampling', in *Proceedings of the Petroleum Hydrocarbons and Organic Chemicals in Groundwater Conference*, Houston, Texas. National Water Well Association, pp. 21-27.

Marrin, D. L.: 1985, 'Delineation of Gasoline Hydrocarbons in Groundwater by Soil Gas Analysis', in *Proceedings of the 1985 HazMat West Conference*, Long Beach, California. Tower Conference Management Company, Wheaton Ilinois, pp. 112-119.

Marrin, D. L. and Thompson, G. M.: 1987, 'Gaseous Behavior of TCE Overlying a Contaminated Aquifer', *Ground Water* **25**, 1.

Nie, N. H., Hull, C. H., Jenkins, J. G., Steinbrenner, K., and Bent, D. H.: 1975, *Statistical Package for the Social Sciences*, (2nd. ed.). New York, McGraw-Hill.

Rosaire, E. E.: 1940, 'Geochemical Prospecting for Petroleum', *Am. Assoc. Pet. Geol. Bull.* **24**, 1401–1433.

Saenz, G.: 1984, 'Geochemical Prospecting in Mexico', *Org. Geochemistry* **6,** 715–726.

Saenz, G.: 1987, 'Geochemical Exploration for Petroleum in a Marshy Area: Examination and Statistical Analysis of C_1–C_7 Hydrocarbons in near Surface Samples', Master's thesis, University of Texas at El Paso, EL Paso, Texas. 130 p.

Saenz, G. and Pingitore, N.: 1989, 'Organic Geochemical Prospecting for Hydrocarbons: Multivariate Analysis, *Jour. Geochemical Exploration* **34,** 337-349.

Saenz, G., Fuentes, H. R., and Pingitore, N. E.: 1989, 'A Discriminating Method of the Identification of Soils and Groundwater Contaminated by Hydrocarbons', *Proceeding of Petroleum Hydrocarbons and Organic Chemicals in Ground Water: Prevention, Detection and Restoration,* NWWA. **2,** 915–929.

Schoell, M.: 1983, 'Genetic Characterization of Natural Gases', *Am. Assoc. Pet. Geol. Bull.* **67,** 2225–2238.

Siegel, F. R.: 1974, 'Geochemical Prospecting for Hydrocarbons', in *Applied Geochemistry*. Wiley-Interscience, New York, pp. 228–252.

Silka, L. R.: 1988, 'Simulation of Vapor Transport Through the Unsaturated Zone. Interpretation of Soil-Gas Surveys', *GWMR Focus*, pp. 115–123.

Sokolov, V. A.: 1933, 'The Gas Survey as a Method of Prospecting for Oil and Gas Formation', *Technika.* **1.**

Spittler, T. M., Fitch, L. and Clifford, S.: 1985, 'A New Method for Detection of Organic Vapors in the Vadose Zone', in *Proceedings of the Characterization and Monitoring of the Vadose Zone Conference*, Denver, Colorado, National Water Well Association, Dublin, OH.

Stahl, W.: 1973, 'Carbon Isotope Ratios of German Natural Gases in Comparison with Isotope Data of Gaseous Hydrocarbons from Other parts of the World', in B. Tissot and F. Bienner (eds.), *Advances in Organic Geochemistry*, Paris: Editions Technip, pp. 453–461.

Walter, E. G., Pitchford, A. M. and Olhoeft, G. R.: 1987, 'A Strategy for Detecting Subsurface Organic Contaminants, in *Proceedings of the National Waterwell Assoc. Conf. on Petroleum Hydrocarbons and Organic Chemical in Ground Water*, Nov. 12–14, Houston, Texas.

FRAMEWORK FOR ENHANCING THE STATISTICAL DESIGN OF AQUATIC ENVIRONMENTAL STUDIES*

FERNANDO CAMACHO and GIAN L. VASCOTTO

Ontario Hydro Research Division, 800 Kipling Ave, Toronto, Ontario M8Z 5S4

(Received April 1990)

Abstract. Aquatic environmental studies can be categorized by the breadth of their scope and the types of desired results. The use of this categorization coupled with a clear specification of objectives and a judicious knowledge of the environmental variability should lead to more statistically efficient studies. This paper discusses the types of lacustrine studies commonly encountered in terms of their categorization. It provides examples of how the intrinsic environmental variability can influence their design and stresses the importance of properly stated objectives, the developing of testable hypotheses, the design of robust and powerful studies, and the importance of evaluating the implication of changes as critical factors for conducting effective and efficient environmental studies.

1. Introduction

All of man's activities on this planet, including this very existence, are likely to leave a mark on the environment. This mark or change is often referred to as man's environmental impact. In our era of global overcrowding and massive manipulation of natural resources, a noticeable degradation of environmental quality has taken place which has raised concerns about the future of the planet. Society is realizing that all this change has a cost, and now it must decide what type of costs (ie., environmental changes) are acceptable. To assess this damage, some times *a priori*, but most often *a posteriori*, environmental studies are carried out. These studies are referred to as impact studies or environmental assessment studies.

The intent of the environmental studies has to be applauded. However, their value may be limited since often they do not meet expectations, partly because of inadequate design, partly because of the inappropriateness of analytical techniques to deal with data which are highly variable and do not conform to main statistical assumptions, and partly because of erroneous ecological assumptions. Several authors have recognized the need for improving the quality of these studies, and have suggested several frameworks for this purpose (see Beanlands and Duinker, 1983, 1984; Rosenberg *et al.*, 1981). These frameworks tend to stress the ecological considerations rather than the statistical aspects.

The objective of this paper is to extend the concept of design stressing the importance of

* An early version of this paper was presented at the First International Conference on Environmentrics, Cairo, Egypt, April 4-7, 1989, under the title *Framework for the Design of Aquatic Environmental Studies.*

Environmental Monitoring and Assessment **17**: 303–314, 1991.

good statistical design in all aquatic environmental studies. This will be done with the help of a conceptualized framework that can be used for simplifying the problems being addressed and prioritizing their components to maximize efficiency. The environmental studies can be broken down into categories and their components so that appropriate statistical considerations can be applied to each of them. The statistical considerations include the recognition of the existence of confounding factors, the need for complex designs, the need for quantitatively defining the desired results, and the need of adequate sample sizes to achieve the desired results.

To reduce the scope of the paper to a manageable size, and because much of our group's experience has been with the Great Lakes, it was decided to discuss only the aquatic environmental studies of lacustrine systems. Therefore, many other types of important studies such as river, air, and land pollution, are not considered here. The paper is organized as follows. Section 2 classifies the studies according to the breadth of scope and the type of desired results. Section 3 discusses elements that are common among the types of studies. Section 4 discusses diverse elements found among the studies with the help of an impact assessment study example. Finally, Section 5 presents some conclusions and areas that require further consideration.

2. Classification of the Studies

Environmental aquatic investigations may be classified according to the spatial scope and the type of desired results. This classification is important since it will dictate considerations for the planning, design, and implementation of the studies. In particular, it will help in determining the amount of resources required and provide an indication of the variables that should be measured.

2.1. Classification by Breadth of Scope

Based on the spatial scope, the studies can be classified into three broad categories: large scale, partial, and local studies.

Large Scale or Whole Lake Studies

These studies are implemented over a large geographical area, usually covering a complete lake. Because of the extent of the sample area, sophisticated equipment may be required to handle the samples. These studies require intensive sampling schemes both in time and resources. To optimize the sampling effort, prior information of the spatial variability of the lake should be used to divide the lake into zones of homogeneous characteristics. It is advised to allocate the sampling effort in direct proportion to the variability of the zones.

Data collected in these studies are highly variable, both spatially and temporally. The degree of variability is likely to change with the type of lake and the zones found in it. This variability is increased during periods of unstable weather. Because of the distances covered in the larger lakes, the time spent traveling between sampling locations may have a large effect in the observed variability between samples (particularly for some nutrients, planktonic organisms, and fish) than the actual spatial variability (Esterby, 1986). Therefore, the relative merits of intensive versus extensive sampling should be carefully evaluated. In some cases it may be desirable to stratify the lake into zones, allocate the sampling effort as above, and sample each zone independently of the others over as brief a period of time as possible.

In general, sophisticated statistical techniques are required to analyze the data collected in large surveys. These techniques should be able to handle multiple variables, temporal and spatial dependency, non-normality of the distribution, and probably unequal time intervals in the sampling period.

Partial or Basin Studies

These studies are carried out over a portion of the lake that can be categorized by certain uniform characteristics. Although it may cover a large geographical area, it is usually a small proportion of the total size of the lake. Similar difficulties to those encountered on large scale studies can be found in these studies. However, the efforts required may not be as intensive.

Local Studies

These studies concentrate on a small geographical area and are usually associated with impact assessment studies. The samples are taken from near shore sites. Thus, the data are highly variable and may be affected by many other variables such as storms, winds, temperature, etc. In addition, the whole lake may be undergoing changes whose effects must be removed from the area in question. To account for part of the variability, it may be necessary to sample intensively and/or measure covariates that can be used in the analysis. The statistical methods required for these studies depend on the particular application, although tests of homogeneity, particularly ANOVA, are usually employed.

2.2. Classification by Desired Results

On the basis of the desired results, each of the above classes can be further categorized into four groups: surveys, monitoring programs, assessment studies, and general purpose studies.

Surveys

These are, in general, the first type of investigations made on a lake and are used for determining general spatial and temporal characteristics of physical, chemical, and biological variables. Later, the information may be used for making decisions about the environment or for setting baseline limits. Some of the particular applications include the following:

(i) The gathering of information on unknown environments.

(ii) The determination of the range of conditions.

(iii) The gathering of information to identify areas that meet certain conditions.

(iv) The determination of the most desirable site for a new industrial development.

The duration of the study varies depending on the extent of the survey, but it is desirable that it extends for at least a year to capture annual cycles.

Monitoring Programs

These are generally used for two main purposes:

(i) maintain historical records which may be used to identify long-term trends in a variety of parameters; and

(ii) ensure that certain environmental criteria are met.

In the first case, the monitoring program is basically a reduced survey carried out in a repetitive fashion; nevertheless, it is important to ensure good quality on the sampling programs (Kwiatkowski, 1986). In the second case, the objective of the monitoring program is to gather information to detect a (possible) change with respect to a specified base line. To meet this objective, two important requirements are necessary:

(i) a clear understanding of the base line that is considered normal or desirable (this base line should be stable, useful, and *real*); and

(ii) a clear hypothesis of the type of changes that are desirable to detect.

In order to meet these requirements, a major effort needs to be made to specify the base line condition, including an assessment of the normal (temporal and spatial) variability of the variables to be studied and an understanding of the causative factors of these changes. This has to be followed by a careful consideration of the changes that are to be detected, including a clear specification of the assumptions to be made. The design should also include considerations of the magnitude of the Type I and Type II errors that should be allowed in the study.

Assessment Studies

These are used to investigate if a particular effect is real and to establish, if possible, a direct cause and effect relationship between a given source and the observed change. Perhaps, these types of studies are the most difficult to design for several reasons:

(i) they require very specific hypotheses;

(ii) there are problems with confounding factors (in particular, the temporal and spatial variability inherent in the collection of the data) which may cause problems in the detection of certain changes; in this case, prior information and/or special designs should be used to control for such factors;

(iii) they require complex designs; and

(iv) they require a high level of replication.

The most common designs used for these studies are the controlled site studies, the pre-operational and post-operational studies and a combination of both. In the controlled site studies, data from the site where an effect is suspected is compared with data from one or more *control* sites. These designs are likely to be successful only if a complete assessment of the spatial variability is available for the study. In the pre-operational post-operational studies, data collected prior to a given intervention is compared with data collected after the intervention. Observed differences are assumed to be due to the intervention. The major difficulty in this situation is that the time period used in the study may not be sufficient to allow a complete assessment of the temporal variability of the data (see also Section 4.1). In any case, it is recommended to carry out prior deliberations to decide the scope and the type of hypotheses to be tested (or that are reasonable to test) as part of the design of the study (see for example Maher, 1984, Greig *et al.* 1984).

General Knowledge Studies

These concentrate on a specific search for basic or fundamental trends, patterns, and characteristics of the environment. These studies are usually associated with the testing of scientific hypotheses. The intensity of effort depends on the problem being investigated. The design of these studies should agree with the standards of any scientific study. The set of priorities required for these studies may be of a different nature than those used for the previous ones.

3. Common Elements of Design

3.1. Methodological considerations

The planning, implementation and analysis of each of the categories described in Section 2 have elements which are common to all, while others are unique to the problem being addressed. Among the common elements the following are basic and should be specified prior to the implementation of the study:

(i) a set of clear and concise stated objectives;

(ii) a clear idea of the nature of expected results;

(iii) a well-designed sampling plan that maximizes the effectiveness and efficiency of the study;

(iv) *a priori* strategy for analyzing the data collected; and

(v) a trade off analysis between the extent of the results and the economic issues.

It is necessary to stress the importance of these elements, particularly because some of these issues are waived in several studies.

Clear and concise objectives: Although this seems to be an obvious requirement for any study, it is surprising how seldom clear, well-defined objectives are presented. It should be pointed out that *broad objectives do not suffice*. Only concise objectives will define the focus of the study and provide an indication of the scope. Furthermore, they will be useful for developing the hypotheses that the study will address. To illustrate this point, suppose it is stated that the objective of a survey program is to determine the spatial variability of organic carbon in Lake Ontario. This statement, although seemingly clear and concise, left out some important details of the objective. For example, the statement does not specify the time frame desired for the characterization, or the precision required for the estimates.

In general, the details required in the statement of the objectives depend on the type of study being implemented. For example, in assessment studies it is necessary to specify the particular hypothesis to be tested. To clarify this point, suppose an assessment study will be implemented to determine the effects of a new power plant in Lake Ontario. Then, it should state, for example, that one of the objectives is to test the hypothesis that discharges of warm water from the station will lead to reductions *by some predetermined amounts* in round whitefish, lake whitefish, and lake trout populations. Note that the specification of the hypothesis focuses its attention to the type of changes that are considered important, and gives an indication of the scope of the expected results.

It should be recognized that any anthropogenic activity will result in environmental change. Therefore, to have socially useful studies, it is critical to identify *a priori* the type of impact that will be considered unacceptable. Furthermore, it is also necessary to identify what would be the significance of such a change to culturally desirable qualities of the system under investigation.

It could be said that the purpose in stating clear and concise objectives is to confront the researcher with questions that could be reasonably answered by the study. This is fundamental if adequate results are expected from the study.

Clear Idea of the Nature of Expected Results

This element is closely related to the specification of the objectives. In the example given above, the statement of the desired precision for the estimates clearly indicates what is expected from the study. In testing for an hypothesis, a statement of the desired confidence level and the *power level*, at a pre-established magnitude of change, are required to specify the type of expected results. Failure to do so may result in a study with a high probability of producing false positives, limiting the use of the results. Note that statements of precision will allow the researcher to foresee (and probably control) the quality of results and the level of effort that is required. Also, they will have a direct impact on the design of the sampling plan.

A Well-Designed Sampling Plan

Based on the objectives, together with the nature of the expected results, a sampling plan should be designed to maximize the effectiveness of the study. Principles of sampling design are given elsewhere (see for example Green, 1979). However, it should be noted that the plan should indicate all the variables to be measured, the frequency, the methods, and the sites. It should also assess the amount of resources required to complete the sampling program.

The selection of sites, frequency of sampling, and the variables to be measured are the tasks implicitly considered in the design of environmental studies. As can be seen from the discussion so far, this is only one of the elements of the proposed frame of design.

A Priori Strategy for Analyzing the Data Collected

This is a necessary part of a good design. Unless the data present unexpected difficulties, the prior investigation of the analysis strategy will help the researcher to foresee the amount of information that is required to achieve the desired power and to determine the variables that should be included in the sample. Also, this will reduce problems that are usually encountered during the analysis stage, making it possible to obtain more conclusive results. In the selection of the analysis technique, it should be kept in mind that usually the data have undesirable properties, such as lack of independence, non-normality of the distributions, and autocorrelation. The selected method should be able to handle these difficulties. An a priori strategy for analyzing the data should be used iteratively with the formulation of the sampling design. Such an iteration will increase the probability of meeting objectives and could result in a more economic and meaningful design.

Trade-off Analysis Between Extent of the Results and Economic Issues

On theoretical grounds it would be desirable to carry out studies that are as comprehensive as possible. However, practical and economic limitations usually determine the results that are realistically obtainable. After the proposed sampling plan has been completed, it is necessary to make an evaluation of all the resources and costs that are required to ensure that the study can be carried out. If this is not possible, it would be necessary to reduce the scope and nature of the study. In this exercise the researcher must find out by how much it is possible to reduce the precision of the results without compromising the quality of the results, or which hypotheses will have to be dropped from the study. In any case, after the economic evaluation, the researcher will have a good idea of the type and scope of results that can be reasonably obtained. Often, this type of evaluation will require a setting of priorities for the objectives to be addressed. It will also ensure that critical issues receive sufficient levels of effort at the expense of secondary or ancillary interests.

In some situations the economic considerations may reduce the amount of effort to such a degree that it may not be possible to detect the desired level of environmental change due to the associated reduction on the level of statistical power. If such studies are carried out, they will have a greater chance of having false positive results with the danger of creating false assurances. Therefore, it may be desirable not to carry out such studies at all.

3.2. General Knowledge Considerations

Prior knowledge of the environmental properties of the system being studied may reveal other common elements that could impact components of the study to varying degrees. In particular, the morphometric characteristics, the trophic state, and the spatial and temporal variability of the lakes may affect the sampling designs and the properties of the collected data. For example, take a stratified lake, if the study requires sampling of both the epilimnetic and the hypolimnetic zones, the sample effort should be larger in the epilimnetic zone because this has higher spatial and temporal variability. Likewise, sandy, homogeneous shore lines having common sediments and slopes may require less sampling effort than highly variable ones (i.e., mixtures of rock and sand). The near-shore zone, which is affected by storms, seiches, rapid temperature variations, etc., is likely to require more frequent sampling than hypolimnetic zones that are usually affected by well-defined cyclic seasonal events.

4. Diverse Elements of Design – An Impact Assesment Example

Each of the studies described in Section 2 has inherent features that need to be considered during the design. Although these influence the five elements of a good design described in Section 3, the greatest impact is in the formulation of the objectives and in the specification of the desired results. An example of a hypothetical impact assessment study will be used to illustrate how the diverse elements can be used during the design, particularly during the unfolding of the issues just mentioned. The goal of the study would be to determine the operational effects of a new generating station located on the shores of the Great Lakes (for a more general information on the ecological issues involved see Greig *et al.*, 1984). A discussion of the current practice is presented in order to contrast the differences that can be obtained with the proposed methodology.

4.1. Common approach

The most common approach is to use a control site study (see Section 2.2). These studies are based on the principle that if the two sites can be assumed to have similar biological, chemical, and morphologic characteristics, differences observed between the data sets can be attributed to the effects of the intervention. This similarity assumption is fundamental to useful conclusions from the study. However, in practice the sites are often selected only on the basis of morphometric similarities, with the implicit assumption that this is sufficient to ensure similar chemical and biological properties. This assumption is rarely tested and when it is tested, non-significant results are taken as confirmation that the two sites are similar. Unfortunately, given the large variability of the data, particularly the biological, there is a large chance of drawing false positive conclusions (i.e., of not detecting site differences when these exist). This is a real danger in many studies, unless sufficient power is ensured by adequate sampling efforts. If the similarity of the two sites is not reasonably well confirmed (i.e., by assessing the Type I and Type II errors), then the conclusions obtained by the study are of limited use.

Sometimes the option is available to initiate the study prior to the construction of the plant. In these cases, pre-operational and post-operational studies can be carried out. The implicit assumption in these studies is that the lake is not changing during the implementation of the study. Under this assumption, data from the pre-operational phase serve as control. Unfortunately, in rapidly expanding industrial areas, this assumption may not be valid making it difficult to identify the effects due to the plant.

The major drawback of traditional designs is that the assumptions related to the appropriateness of the controls may not be testable. Considerable improvements have been recently seen, particularly in the areas of placing greater emphasis on the testing of sound ecological hypothesis well bounded in time and space as recommended by Beanland and Duinker (1983, 1984). However, these are often too broad to be statistically testable, because they fail to consider and define what would be an acceptable change.

4.2. Proposed approach

Developing objectives

Before being able to obtain concise objectives, it is necessary to state a general goal. This goal is then broken down into major areas of potential effect. Each of these are analyzed and their potentially significant impacts are identified. In the current human culture, an impact is considered meaningful only if it threatens human health, a valuable resource (fish, recreational use, transportation) or an aesthetic value (the beauty of the Falls). Other ecological effects (i.e., preserving biodiversity) are now gaining in importance.

In the case of the example, the goal would be to assess the impact of the new station on the aquatic environment. Many areas of potential effect can be identified. The following three were arbitrarily chosen because they would apply to any generating facility built on the Great Lakes that uses the once-through-cooling process.

(i) The effect of removal and transfer of organisms by the intake of cooling water.

(ii) The effect of passage of organisms through the plant.

(iii) The effects on the thermal discharge of water on the use of area contained within a thermal-envelope defined, for example, by a ΔT of 1 °C.

Development of Hypotheses

The next step is to formulate testable hypotheses for each of the identified areas of impact. Methodologies for developing these must be grounded in sound ecological and limnological principles (see for example Greig *et al.*, 1984). The hypothesis must consist of a succinct statement of what is to be tested and the degree of change that is to be detected. Once the hypotheses are formulated, they can be tested by carrying out the appropriate studies. In the example this can be accomplished as follows.

For area (i)

Of all the organisms that are transferred and do not go through the plant, the only ones that may be significantly impacted are the fish. These are either completely removed, or as in modern stations, returned by the fish return systems to the vicinity of the discharge. The issue then becomes whether or not the transfer of the organisms have, in one's judgement, unacceptable consequences. What the transfer does is to move organisms from a cool zone to a hot/warm zone over a short period of time. The effects of transfer can be phrased as follows:

(i) can the organisms adapt to such a rapid temperature change?

(ii) do the stresses encountered in the passage result in significant deleterious effects to the organisms?

Once the levels of change considered acceptable are identified, (i.e. is a 10% mortality tolerable?), then a hypothesis can be formulated and addressed in a combination of laboratory experimentation and field verification. While the design of laboratory experiments do not often run into the statistical problems of field studies, the design of field verification programs requires considerable *a priori* information. To design the sampling program, sampling variability (spatial and temporal) available from either published work or from the result of a prior survey should be used to estimate the replication and frequency of sampling required to have the power necessary to detect the desired change.

As a result of these considerations, the researcher may want to test the hypothesis that the mortality (up to 24 hr following passage) of the three main commercially important species does not exceed 10%.

For Area (ii)

The issues associated with the organisms that pass through the plant (i.e., that are entrained) are similar to those associated with the organisms that are transferred, with the exception that the temperature changes are much greater (over 20 °C) and they may be also exposed to pressure changes.

The organisms of concern are planktonic or semi-planktonic. A variety of studies have investigated the effect of entrainment on phytoplankton and zooplankton in the Great Lakes, but no major effects have been detected (Dunstall, 1978, 1981). This is probably due to the rapid turn-over rates of the organisms and the high day-to-day variability. The ichthyoplankton, both the fish eggs and larvae, are vulnerable to entrainment.

The issues of concern are then:

(i) how many are entrained?

(ii) what is the viability of the entrained individuals?

(iii) what is the effect of passage?

(iv) what is the effect of relocation?

The necessity of addressing these questions ultimately rests on the ability of assessing the importance of the consequences of the changes. Questions (i) and (ii) can be addressed by a simple collection of intake and discharge samples and measurements of viability. If reduced viability is encountered, then it may be desirable to investigate questions (iii) and (iv) by carrying out laboratory experiments simulating the plant passage.

In this case a hypothesis with two components can be formulated. Is a significant proportion (for example a 10%) of the viable available larvae and eggs passing through the near shore zone in the vicinity of the plant being entrained? And, if a significant proportion is entrained, then does a significant proportion of those entrained die within a predetermined period (3 hr) after returning to ambient temperature? As above, the power necessary to test the hypothesis has to be a crucial consideration.

For Area (iii)

Several effects can be determined *a priori.* One of these is that warm water fish (gizzard shad) will move to the discharge area during the witner, while cold water species (trout, perch) will leave it during the summer.

In general, it can be assumed that the fish have preferred temperatures and distribute themselves along temperature gradients. It is then possible to formulate hypotheses to test the association between fish population and thermal gradient generated by the discharge. For example, the hypothesis that there is a dose-response relationship of the available fish with respect to the spatiotemporal temperature gradient may be formulated. To test this hypothesis it will be necessary to gather data over a grid covering the expected plume of the discharge.

Before concluding this section, it is important to note that the overall impact assessment was reduced to a series of small studies designed to address specific hypotheses, which dealt with issues that are recognizable in terms of their impact, and are bounded by acceptable and unacceptable changes. The hypotheses must be testable, both ecologically and quantitatively. What is still missing is an evaluation of the actual significance of these impacts in terms of the values that the society places on its resources. Although this is important, it is outside of the scope of the present discussion.

5. Conclusions

A way of approaching environmental studies in an effective and efficient manner has been presented. The method consists of two main components:

(i) an identification of the type of study that is to be undertaken for obtaining desired results; and

(ii) an adequate design which centers on the clear identification of the desired objectives.

With clearly identified objectives, it becomes easier to prioritize and to obtain valuable results. Also, it avoids the need for mining the data in search for possible effects.

As illustrated by the example, there are advantages in breaking down large-scale environmental studies into small parts.

- The goals of the study are better focused through the formulation of concise objectives, many of these stated as testable hypotheses.
- With small objectives, it is possible to design adequate sampling plans that will provide enough data to obtain the required precisions, and in particular, enough power to test the related hypotheses.
- The results of the study are easy to interpret.
- The researcher does not have to rely on the data to generate the hypotheses to be tested, giving more control over the type of results that can be obtained.
- The studies do not have to rely on non-testable assumptions that could limit the usefulness of the results.
- The cost of the study can often be reduced because wasted efforts are eliminated.

As a result of the ideas presented in the paper, several things have become obvious:

(i) There are many common elements among studies which may offer opportunities for better apportioning of the research efforts.

(ii) There is a critical need to improve the emphasis placed on the power of the studies carried out, on the formulation of meaningful and testable hypotheses and on the validity of the types of controls that may be used. And,

(iii) There is a need to put into perspective the ecological and economic implications of the changes detected and at the same time recognizing that all industrial activities will impose an associated environmental cost whose acceptance or rejection will depend on society's will.

References

Bealands, G. E. and Duinker, P. N., 1984: 'An Ecological Framework for Environmental Impact Assessment', *Journal of Environmental Management* **18**, 267–277.

Beanlands, G. E. and Duinker, P. N.: 1983, 'An Ecological Framework for Environmental Impact Assessment in Canada', *Institute for Resource and Environmental Studies. Dalhousie University in cooperation with the Federal Environmental Assessment Review Office.*

Dunstall, T. G.: 1981, 'Effect of Entrainment on Phytoplankton Primary Production – A Summary of Studies Conducted at Four Ontario Hydro Generating Stations, 1975–1977', Ontario Hydro Research Division Report No. 81-139-K.

Dunstall, T. G.: 1978, 'Use of a Sample Grid to Determine the Effect of Once-through Cooling on the Distribution of Zooplankton and Phytoplankton', Ontario Hydro Research Division Report No. 78-257-K.

Esterby, S. R.: 1986, 'Spatial Heterogeneity of Water Quality Parameters', in El Shaarawai, A. H. and Kwiatkowski, R. E. (eds.), *Developments in Water Science*, Elsevier.

Green, R. H.: 1979, *Sampling Design and Statistical Methods for Environmental Biologists*, John Wiley & Sons, Toronto.

Greig, A. L., Cunningham, G., Everitt, R. R., and Jones, M. L.: 1984, 'Final Report of two Workshops to Consider the Environmental Effects and Monitoring Options for the Darlington NGS', ESSA Environmental and Social Systems Analysts Ltd. Report prepared for Ontario Hydro.

Kwiatkowski, R. E.: 1986, 'The Importance of Design Quality Control to a National Monitoring Program', in El Shaarawai, A. H. and Kwiatkowski, R. E. (eds.), in *Developments in Water Science*, Elsevier.

Maher, J. F. B.: 1984, 'Outline of Environmental Pre-operational and Post-operational Studies for Darlington GS', Ontario Hydro, Environmental Studies & Assessment Department, Report No. 84252.

Rosenberg, D. M., Resh, V. H. *et al.*: 1981, 'Recent Trends in Environmental Impact Assessment', *Canadian Journal of Fisheries and Aquatic Sciences* 38, 591–624.

ANALYSIS OF TWO-WAY LAYOUT OF COUNT DATA WITH NEGATIVE BINOMIAL VARIATION

A. MAUL and A. H. EL-SHAARAWI

Département Informatique, Université de Nancy II, 2bd Charlemagne, 54000 Nancy, France.
Rivers Research Branche, National Water Research Institute, Canada Centre for Inland Waters, Burlington, Ontario, Canada L7R 4A6.

(Received xx)

Abstract. A number of methods has been proposed for dealing with single-factor or factorial experiments when the requirements for performing the normal theory analysis of variance procedure are not satisfied. This paper suggests the use of the likelihood ratio statistic for testing the main effects and the interaction between the factors in two-way layout of count data following negative binomial distributions with a common dispersion parameter. The likelihood ratio statistic for testing the equality of the dispersion parameters of several groups of count data is also derived. The method is illustrated by an example concerning the study of spatial and temporal variation of bacterial counts.

Introduction

Negative binomial models are widely used to describe count data in numerous areas of biostatistics (Anscombe, 1949; Bliss and Fisher, 1953; El-Shaarawi *et al.*, 1981; Maul *et al.*, 1985; Maul *et al.*, 1989). The normal theory analysis of variance (ANOVA) procedure is sometimes employed for analysing a single or a multiple-factor layout of count data with negative binomial variation after transforming the crude data to achieve the requirements for the application of ANOVA (Barnwall and Paul, 1988; Maul and Block, 1983). General transformations have been suggested to obtain homogeneity of the variances and approximate normality (Anscombe, 1948). However, such an approach is not always desirable since there is no evidence that stable variance and normality may be achieved simultaneously by a single transformation. It has been pointed out by Scheffé (1959) that the analysis of variance techniques are quite robust with respect to moderate deviations from normality, and that for balanced designs they are relatively robust with respect to heterogeneity of variances. These conditions are not likely to be met with data generated by the negative binomial distributions because they are likely to have very non normal distributions. It is therefore preferable to analyze the counts directly by the exact assumed probability distribution of the crude data. This prompted some recent studies on both regression analysis (Lawless, 1987) and analysis of one-way layout (Barnwal and Paul, 1988) of count data with negative binomial variation.

The present paper is concerned with testing the main effects and the interaction between the factors in a two-way layout for count data from negative binomial distributions with a common dispersion parameter k. The paper is also concerned with testing the hypothesis of a common k. Section 2 introduces the model and presents the general framework of the

Environmental Monitoring and Assessment **17**: 315–322, 1991.

situation to be examined. In Section 3 the likelihood ratio statistic is derived for testing hypotheses in a two-way layout under the assumption of a common dispersion parameter. Testing the equality of the dispersion parameters of several groups of count data is presented in Section 4. Section 5 gives an illustrative example and provides additional remarks about the method.

2. Hypotheses and models

Let R be the response variable and consider the case when the values of R, which are counts, can be classified according to the levels of two factors, A and B, and the interest is to test whether:

(i) the effect of each factor is statistically significant,
(ii) the two factors operate independently of each other.

Let $r_{ij\alpha}$ be the αth ($\alpha=1,\dots,n_{ij}$) observed value of the random variable R at level i ($i=1,\dots,\ell$) of the first factor and level j($j=1,\dots,m$) of the second factor. Let $R_{ij\alpha}$ be a random variable with a negative binomial distribution with mean m_{ij} and a dispersion parameter k. Hence

$$\Pr(R_{ij\alpha}=r_{ij\alpha}) = \frac{(k+r_{ij\alpha}-1)!\ k_k\ m_{ij}^{r}ij\alpha}{r_{ij\alpha}!\ (k-1)!(k+m_{ij})^{k+r}ij\alpha} \quad \text{for} \quad (r_{ij\alpha} = 0,1,2\dots) \tag{2.1}$$

Further, we assume that the natural logarithm of m_{ij} is expressed as an additive linear combination of the ith level of factor A, the jth level of factor B and the interaction between A and B. This can be written as:

$$\ln m_{ij} = \mu + \alpha_i + \beta_j + \gamma_{ij} \quad (i=1,\dots,\ell;\, j=1,\dots,m) \tag{2.2}$$

where μ is the general level of the process, α_i is the effect due to the ith level of the first factor, B_j is the effect associated with the jth level of the second factor, and γ_{ij} represents the interaction between them.

3. Tests of Hypotheses

3.1. Testing the significance of the overall effect of factors A and B

The competing hypotheses are

H_0: $m_{ij} = m_0$ (i.e., A and B have no effect

which is equivalent to testing that $\alpha_i=\beta_j = \gamma_{ij}=0$ for all i and j in (2.2), and

H_1: not all m_{ij}s are equal.

Maximum likelihood (ML) estimation for the parameters. The log likelihood function under the hypothesis H_1 is

$$L(m_{ij},k) = \sum_{i=1}^{l} \sum_{j=1}^{m} \sum_{\alpha=1}^{n_{ij}} \Big[\sum_{t=1}^{r_{ij\alpha}} \ln(k-1+t) + k \ln k + r_{ij\alpha} \ln m_{ij} - (k+r_{ij\alpha}) \ln(k+m_{ij}) \Big] + C(r) \tag{3.1}$$

where $C(r)$ is a function of the data only. The log likelihood function L_0 under H_0 is obtained by setting $m_{ij}=m_0$ in (3.1).

Under H_1, the ML estimate of m_{ij}, is

$$\hat{m}_{ij} = \frac{r_{ij\bullet}}{n_{ij}}, \quad \text{where} \quad r_{ij\bullet} = \sum_{\alpha=1}^{n_{ij}} r_{ij\alpha} \tag{3.2}$$

and the root $\hat{k}_{AB}$ of the equation

$$\sum_{i=1}^{l} \sum_{j=1}^{m} \sum_{\alpha=1}^{n_{ij}} \Big[\sum_{t=1}^{r_{ij\alpha}} \frac{1}{k-1+t} \Big] = \sum_{i=1}^{l} \sum_{j=}^{m} n_{ij} \ln\left(1+\frac{m_{ij}}{k}\right) \tag{3.3}$$

yields the ML estimates for k. The estimates $\hat{m}_0$ and $\hat{k}_0$ of m_0 and k under H_0 are given by

$$\hat{m}_0 = \frac{r\ldots}{n}, \tag{3.4}$$

where $r\ldots = \Sigma_{ij\alpha}\, r_{ij\alpha}$ and by solving the equation

$$\sum_{i=1}^{l} \sum_{j=1}^{m} \sum_{\alpha=1}^{n_{ij}} \Big[\sum_{k=1}^{r_{ij\alpha}} \frac{1}{k-1+t} \Big] = n \ln\left(1+\frac{\hat{m}_0}{k}\right) \tag{3.5}$$

for m_0 and k, respectively, where $n = \sum_{i=1}^{l} \sum_{j=1}^{m} n_{ij}$.

The likelihood ratio statistic for testing H_0 is

$$-2 \ln \Lambda = 2 \sum_{i=1}^{l} \sum_{j=1}^{m} \sum_{\alpha=1}^{n_{ij}} \Big[\sum_{t=1}^{r_{ij\alpha}} \ln \frac{\hat{k}_{AB}-1+t}{\hat{k}_0-1+t} + r_{ij\alpha} \ln \frac{\hat{m}_{ij}(\hat{k}_0+\hat{m}_0)}{\hat{m}_0(\hat{k}_{AB}+\hat{m}_{ij})} + \hat{k}_0 \ln\left(1+\frac{\hat{m}_0}{\hat{k}_0}\right) - \hat{k}_{AB} \ln 1+\left(\frac{\hat{m}_{ij}}{\hat{k}_{AB}}\right) \Big] \tag{3.6}$$

The asymptotic distribution of $-2 \ln \Lambda$ is chi-squared with $(lm-1)$ degrees of freedom.

3.2. Testing the significance of the effects of the factors A and B

When H_0 is rejected it becomes of interest to test the effects of A and B separately. Testing that $\alpha_i = \alpha_{ij} = 0$ for all i and j in (2.2), is the same as testing that factor A has no effect.

The likelihood ratio statistic, which is asymptotically distributed as a chi-squared with $m(l-1)$ degrees of freedom, is

$$-2\ln\Lambda = 2\sum_{i=1}^{l}\sum_{j=1}^{m}\sum_{\alpha=1}^{r_{ij}}\Big[\sum_{k=1}^{r_{ij\alpha}}\ln\frac{\hat{k}_{AB}-1+k}{\hat{k}_B-1+k}+r_{ij\alpha}\ln\frac{\hat{m}_{ij}(\hat{k}_B+\hat{m}_j)}{\hat{m}_j(\hat{k}_{AB}+\hat{m}_{ij})}+ + \hat{k}_B\ln\left(1+\frac{\hat{m}_j}{\hat{k}_B}\right)-\hat{k}_{AB}\ln\left(1+\frac{\hat{m}_{ij}}{\hat{k}_{AB}}\right)\Big], \tag{3.7}$$

where

$$\hat{m}_j = \frac{r_{.j.}}{n_{.j.}}$$

and $\hat{k}_{AB}$ is the value of k that satisfies the equation

$$\sum_{i=1}^{l}\sum_{j=1}^{m}\sum_{\alpha=1}^{n_{ij}}\Big[\sum_{k=1}^{r_{ij\alpha}}\frac{1}{k-1+t}\Big] = \sum_{j=1}^{m} n_{.j}\ln\left(1+\frac{\hat{m}_j}{k}\right).$$

The likelihood ratio test for the effect of factor B can be obtained using the above tests in a very obvious manner.

3.3. TESTING THE SIGNIFICANCE OF AN INTERACTION EFFECT BETWEEN THE FACTORS A AND B

If the above tests showed that factors A and B have significant effects, then it might be of interest in many applications to test if the two factors operate independently of each other. In this case the two null hypotheses reduces model (2.2) to

$$\ln m_{ij} = \mu + \alpha_i + \beta_j. \tag{3.8}$$

The ML estimates of the parameters of model (3.8) can be obtained by iteration using the Newton-Raphson method according to a procedure which has been described by Maul *et al.* (1989). To obtain a unique estimate for the parameters μ, $(\alpha_1,\ldots,\alpha l)$ and $(\beta_1,\ldots,\beta_m)$, it is assumed that

$$\sum_{i=1}^{l}\alpha_i = \sum_{j=1}^{m}\beta_j = 0.$$

The ML estimate, $\hat{k}_{A+B}$, for k under H_0 is the root of the equation

$$\sum_{i=1}^{l}\sum_{j=1}^{m}\sum_{\alpha=1}^{n_{ij}}\Big[\sum_{k=1}^{r_{ij\alpha}}\frac{1}{k-1+t}\Big] = \sum_{i=1}^{l}\sum_{j=1}^{m} n_{ij}\ln\left(1+\frac{\hat{m}^0_{ij}}{k}\right)$$

where $\hat{m}^0_{ij}$ is the ML estimate of m_{ij} under H_0. The likelihood ratio statistic for testing H_0 is

$$-2 \ln \Lambda = 2 \sum_{i=1}^{\ell} \sum_{j=1}^{m} \sum_{\alpha=1}^{n_{ij}} \Big[\sum_{t=1}^{r_{ij\alpha}} \ln \frac{\hat{k}_{AB}-1+t}{\hat{k}_{A+B}-1+t} + r_{ij\alpha} \ln \frac{\hat{m}_{ij}(\hat{k}_{A+B}+\hat{m}_{ij}^{0})}{\hat{m}_{ij}^{0}(\hat{k}_{AB}+\hat{m}_{ij})}$$

$$+ \hat{k}_{A+B} \ln\left(1+\frac{\hat{m}_{ij}^{0}}{\hat{k}_{A+B}}\right) - \hat{k}_{AB} \ln\left(1+\frac{\hat{m}_{ij}}{\hat{k}_{AB}}\right)\Big] \tag{3.9}$$

Under H_0, $-2 \ln \Lambda$ has a chi-squared distribution with $(\ell-1)(m-1)$ degrees of freedom.

4. Testing the Equality of the Dispersion Parameters

All the tests presented in Section 3 are based on the assumption of a common dispersion parameter k. Let $H_0 : k_{ij} = k$ for all i and j, and the alternative hypothesis is H_1 : not all the k_{ij}s are equal. Testing the null hypothesis (i.e. homogeneity of the k_{ij}s) can be performed by using the likelihood ratio test; see also Barnwal and Paul (1988).

Under H_1, the ML estimator, $\hat{k}_{ij}$, of k_{ij} is obtained as the solution to

$$\sum_{\alpha=1}^{n_{ij}} \sum_{t=1}^{r_{ij\alpha}} \frac{1}{k-1+t} = n_{ij} \ln\left(1+\frac{\hat{m}_{ij}}{k}\right) \tag{4.1}$$

where $\hat{m}_{ij}$ is given as in (3.2).

Under the null hypothesis H_0, the ML estimator, $\hat{k}_{AB}$, for k is the solution to Equation (3.3). The likelihood ratio statistic which has an asymptotic distribution as chi-squared with $\ell m-1$ degrees of freedom is

$$-2 \ln \Lambda = 2 \sum_{i=1}^{\ell} \sum_{j=1}^{m} \sum_{\alpha=1}^{n_{ij}} \Big[\sum_{t=1}^{r_{ij\alpha}} \ln \frac{\hat{k}_{ij}-1+t}{\hat{k}_{AB}-1+t} + r_{ij\alpha} \ln \frac{(\hat{k}_{AB}+\hat{m}_{ij})}{(\hat{k}_{ij}+\hat{m}_{ij})}$$

$$+ k_{AB} \ln\left(1+\frac{\hat{m}_{ij}}{\hat{k}_{AB}}\right) - \hat{k}_{ij} \ln\left(1+\frac{\hat{m}_{ij}}{\hat{k}_{ij}}\right)\Big] \tag{4.2}$$

It should be pointed out that all the ML estimates for the dispersion parameters are assumed to be positive numbers since Equations (3.3) and (4.1) have either no or only one positive solution for any data set. In particular, Equation (4.1) may have no solution for any of the combinations of i and j. This may occur more specially when n_{ij} is small and the overdispersion of the data, relative to a Poisson model for example, is not well marked. However, if Equation (4.1) has no solution for a given i and j (i.e. k_{ij} is considered infinite), then formula (4.2) is reduced to

$$\sum_{\alpha=1}^{n_{ij}} \Big[\sum_{t=1}^{r_{ij\alpha}} \ln \frac{1}{\hat{k}_{AB}-1+t} + r_{ij\alpha} \ln(\hat{k}_{AB}+\hat{m}_{ij}) + \hat{k}_{AB} \ln\left(1+\frac{\hat{m}_{ij}}{\hat{k}_{AB}}\right) - \hat{m}_{ij}\Big].$$

5. Example and Discussion

The method described above is illustrated by a numerical example of bacterial counts observed in water samples collected from three fixed locations during six surveys in a drinking water distribution system. The data presented in Table I were extracted from a large field study which was conducted to determine the spatial and temporal distribution of bacteria in a network (Maul *et al.*, 1985). Four independent observations were taken under each of the eighteen cells corresponding to the combinations of the two qualitative factors of interest: location and survey, in the 3×6 factorial experiment.

The analysis started by testing the equality of the dispersion parameters in the various cells. The ML estimates of the k_{ij}s are given in Table II. Further, the likelihood ratio statistic for testing the equality of the k_{ij}s yields the value 13.53 on 17 degrees of freedom. This indicates strong evidence that the assumption of a common dispersion parameter is reasonable. Table III shows the ML estimates and standard errors of the unknown parameters in model (3.8), that is, assuming there is no interaction effect between the main factors. The values of the likelihood ratio statistic given in Table IV indicate a significant effect for the location and a highly significant effect for the survey, thus showing a marked heterogeneity in the spatial and temporal distributions of bacteria in the network. This outcome is in agreement with the conclusion stated in Maul *et al.* (1985). Note that the ML estimates, namely $\hat{k}_0$, $\hat{k}_{AB}$, $\hat{k}_A$, $\hat{k}_B$ and $\hat{k}_{A+B}$, of the dispersion parameter which were calculated under the different hypotheses considered, are given as: 0.3286; 0.6187; 0.3410; 0.4396 and 0.4798, respectively.

The procedure presented in this paper provides a particularly convenient and useful way for analysing two-way layout of count data following the negative binomial distribution. Its interest lies in both the great versatility of the negative binomial for fitting count data which may display extra-Poisson variation and the fact that the assumption of

TABLE I

Data layout for bacterial counts

	Survey					
Location	1	2	3	4	5	6
	154	0	19	12	9	12
1	1	3	1	42	24	10
	0	0	14	3	191	17
	9	0	5	2	60	2
	1	0	1	32	2	78
	150	0	0	1	14	3
2	0	1	0	5	45	17
	2	2	3	10	82	20
	15	1	86	41	100	418
	0	2	2	18	60	29
3	0	0	25	138	14	55
	0	12	0	3	27	23

a common dispersion parameter is reasonable in many situations. The approach which is based on the use of standard maximum likelihood methods has good properties. In particular, it is appropriate to deal with nonsymmetrical or unbalanced designs and it allows testing for an interaction effect between the main factors without using the whole Fisher information matrix corresponding to all the parameters of the complete linear model.

TABLE II

Maximum likelihood estimates of the dispersion parameter for each cell in the bacterial counts example

	Survey					
Location	1	2	3	4	5	6
1	0.2299	XXXXXX	1.4918	0.8895	1.0042	3.4321
2	0.2017	XXXXXX	XXXXXX	1.0058	0.8895	1.0655
3	0.0815	0.6191	0.3264	0.8094	2.1825	0.8025

XXXXXX: Equation (4.1) has no solution; k_{ij} is considered infinite.

TABLE III

Tests of significance for the two-way layout experiment

Source	Degrees of freedom	$-2 \ln \Lambda$
Overall effect of factors *A* and *B*	17	48.05[b]
Factor *A* (location)	12	25.08[a]
Factor *B* (survey)	15	45.08[b]
*A*x*B* interaction	10	18.13

[a] Value is significant at the 5% level.
[b] Value is significant at the 0.1% level.

TABLE IV

Estimates of the parameters for the model with no interaction

Parameter	Estimate	Standard error
Intercept	2.83	0.18
Location		
α_1	–0.15	0.26
α_2	–0.51	0.26
α_3	0.70	0.27
Survey		
β_1	0.81	0.41
β_2	-2.51	0.44
β_3	–0.61	0.40
β_4	0.20	0.35
β_5	1.20	0.35
β_6	0.51	0.40

References

Anscombe, F. J.: 1948, 'The Transformation of Poisson, Binomial and Negative Binomial Data', *Biometrikla* **38**, 246-254.

Anscombe, F. J.: 1949, 'The Analysis of Insect Counts Based on the Negative Binomial Distribution', *Biometrics* **5**, 165-173.

Barnwall, R. K. and Paul, S. R.: 1988, 'Analysis of One-Way Layout of Count Data with Negative Binomial variation', *Biometrika* **75**, 215-222.

Bliss, C. I. and Fisher, R. A.: 1953, 'Fitting the Negative Binomial Distribution to Biological Data', *Biometrics* **9**, 176-200.

El-Shaarawi, A. H., Esterby, S. R., and Dutka, B. J.: 1981, 'Bacterial Density in Water Determined by Poisson or Negative Binomial Distributions', *Appl. Environ. Microbiol.* **41**, 107-116.

Lawless, J. F.: 1987, 'Negative Binomial and Mixed Poisson Regression', *Can. J. Statist.* **15**, 205-225.

Maul, A. and Block, J. C.: 1983, 'Microplate Fecal Coliform Method to Monitor Strea Water Pollution', *Appl. Environ. Microbiol.* **46**, 1032-1037.

Maul, A., El-Shaarwai, A. H., and Block, J. C.: 1985, 'Heterotrophic Bacteria in Water Distribution Systems, I. Spatial and Temporal Variation, II. Sampling Design for Monitoring', *Sci. Total Environ.* **44**, 201-224.

Maul, A., El-Shaarawi, A. H., and Ferard, J. F.: 1989, 'Application of Negative Binomial Regression Models to the Analysis of Quantal Bioassays Data', *Environmetrics* (to appear).

Scheffé, H.: 1959, *The Analysis of Variance*, John Wiley, New York.

AN OVERVIEW OF THE ACIDIFICATION OF LAKES IN ATLANTIC CANADA

GEOFF HOWELL and A. H. EL-SHAARAWI

Water Quality Branch, Atlantic Region, Moncton, New Brunswick

and

National Water Research Institute, Canada Centre for Inland Waters, Burlington, Ontario, Canada

(Received August 1990)

Abstract. Analysis of water chemistry from a sample of lakes ($\simeq$ 1300) in Atlantic Canada has indicated that lakes in geologically sensitive portions of Nova Scotia and Newfoundland have been acidified due to the combined effects of natural organic acids and anthropogenically derived mineral acids. Principal component analysis of six measured variables (pH, Ca, Conductance, SO_4*, Alkalinity, Colour) and one computed variable (Alk/Ca*+Mg*) for each province result in four components which retain at least 89% of the original variability. Cluster analysis of the four principal components resulted in 6 lake groups for New Brunswick and 8 groups for Nova Scotia and 7 groups for Newfoundland. Geograghic ordination of these clusters indicates that there is good correspondence between cluster group and the underlying bedrock geology of the region.

Introduction

Atlantic Canada lies downwind of major industrial emission sources in central Canada and the midwestern United States and presently receives wet sulphate deposition at or below the prescribed 20 kg $ha^{-1}yr^{-1}$ target loading (MOI, 1983). However, due to a preponderance of volcanic bedrock and thin soil overburdens, much of the region is extremely sensitive to acidification. This is particularly evident for parts of Nova Scotia and Newfoundland where chemical responses to acid rain has been documented (Howell and Brooksbank, 1987) and (Thompson, 1986). In addition to long range transport of anthropogenic mineral acids, surface waters in Atlantic Canada are characteristically highly coloured and thus are influenced by naturally produced organic acids. Recent studies in highly organic systems in southwestern Nova Scotia (Kerekes *et al.*, 1986, Gorham *et al.*, 1984, Howell, 1989) indicate that strong mineral acids serve to further acidify these naturally acidic systems. Principal component analysis of selected lakes from Nova Scotia and Newfoundland (Esterby *et al.*, 1989) identified a component with high eigenvalues for pH, sulphate, alkalinity and colour which further illustrates the importance of both mineral and organic acids in the overall acidification process.

Thompson (1986) has shown that the rivers in southern Newfoundland have been impacted by acid rain. Howell (1989) in a study of 456 Nova Scotian lakes noted that 35% of the lakes have been acidified by acid rain while a further 14% have been impacted by either natural organic acids or by local sources of strong mineral acids. In addition to the chemical responses to the acid rain, some biological effects have also been documented. Watt *et al.* (1983) have indicated that acid rain has resulted in a 9% loss of Atlantic

Environmental Monitoring and Assessment **17**: 323–338, 1991.

salmon (*Salmo salar*) biological reproductive capacity for the Maritimes as a whole and up to a 50 percent loss for the sensitive portions of Nova Scotia.

In this paper a water chemistry data set of 1378 lakes sampled from 1979 to 1988 is analyzed in order to provide a spatial overview of the acidification status in Atlantic Canada. Principal component and subsequent cluster analysis of the chemistry data is employed to identify the importance of underlying acidification processes and to group the lakes into homogeneous clusters based on water chemistry. The derived clusters will then be related to bedrock geology in order to assess the efficacy of the clustering from an acid rain perspective.

Materials and Methods

The lake water chemistry data utilized in this paper were extracted from the NAQUADAT data base (WQB, 1979) and subsequently validated using a standard ion balance calculation. Following validation, median values for each Lake in the data set were then calculated. The final data set includes 1378 lakes of which 596, 340, 185 and 257 were lacated in Nova Scotia, Newfoundland, Labrador and New Brunswick respectively. The dominant bedrock geology for each lake was determined from the provincial geological maps and classified into three sensitivity classes (highly sensitve, moderate and unsenitive).

Variables selected for inclusion into the statistical analysis include pH, water colour, specific conductance, calcium, sulphate, alkalinity and the $Alk/\Sigma Ca^* + Mg^*$ ratio. For each province, median values and ranks were computed at the tertiary watershed level and used to construct median rank needle plots for six variables.

For each province, a correlation matrix was calculated and subsequently used in the principal component analysis. For Nova Scotia, a small number of lakes significantly influenced the analysis due to elevated calcium and sulphate levels from terrestrial gypsum deposits. To overcome this difficulty the Nova Scotia lakes were clustered by calcium and sulphate prior to the principal component analysis. The first four principal components were selected for performing cluster analysis as they retained more than eighty-five percent of the variabilities.

A non-hierarchial clustering (El-Shaarawi *et al.*, 1989) procedure was used to group the lakes using the principal components. The cluster means were calcualted for each variable and for each cluster a graph of the ordinal values for each variable was constructed. Each station was geographically referenced and plotted by cluster number on both provincial base maps and surficial geology maps. Histograms of the relative frequency of sites underlain by geological structures that are highly sensitive, moderately sensitive, and insensitive to acid rain were constructed for each cluster.

Results and Discussion

Boxplots (Figure 1) give a good overview of the general chemical characteristics of lakes in Nova Scotia, New Brunswick and Newfoundland. The calcium and Alk values are

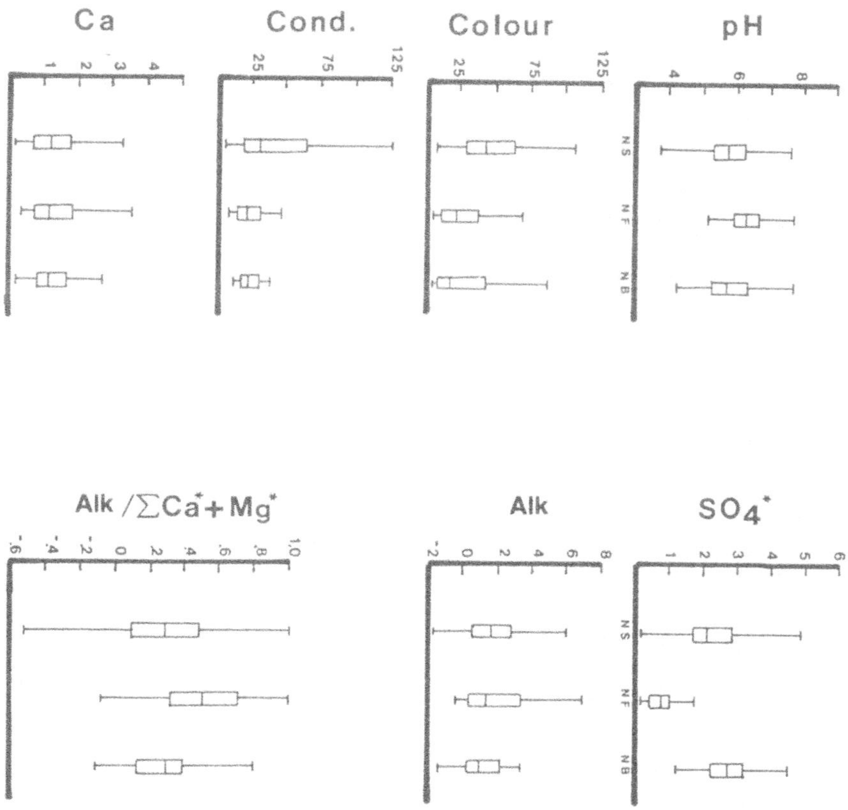

Fig. 1. Boxpots for the chemical characteristics of lakes in Nova Scotia (NS), Newfoundland (NFLD) and New Brunswick (N.B.).

rather similar for all three data sets suggesting similar terrain sensitivity to acid rain. From the sulphate boxes, it appears that the Nova Scotia and New Brunswick lakes have the greatest mineral acid influence whereas the water colour data indicates that natural organic acids are also highly implicated in the acidification of lakes in Nova Scotia. It should be noted that the Newfoundland lakes have much lower sulphate concentration which is consistent with a west to east sulphate deposition gradient. The Alk/ΣCa* + Mg* ratio's indicate that the greatest acidification responses have been observed in the Nova Scotia and New Brunswick lakes and that some lakes in each province have lost all of their original Alk.

Median rank plots for Nova Scotia, New Brunswick, insular Newfoundland and Labrador (Figures 2 to 5) provide an indication of which tertiary watersheds in the various provinces have been the most severely impacted by acid rain. The greatest acidification response in Nova Scotia is observed for the extremely sensitive, highly coloured basins in southwestern Nova Scotia. These basins all exhibit low pH, Alk and Alk/ΣCa*

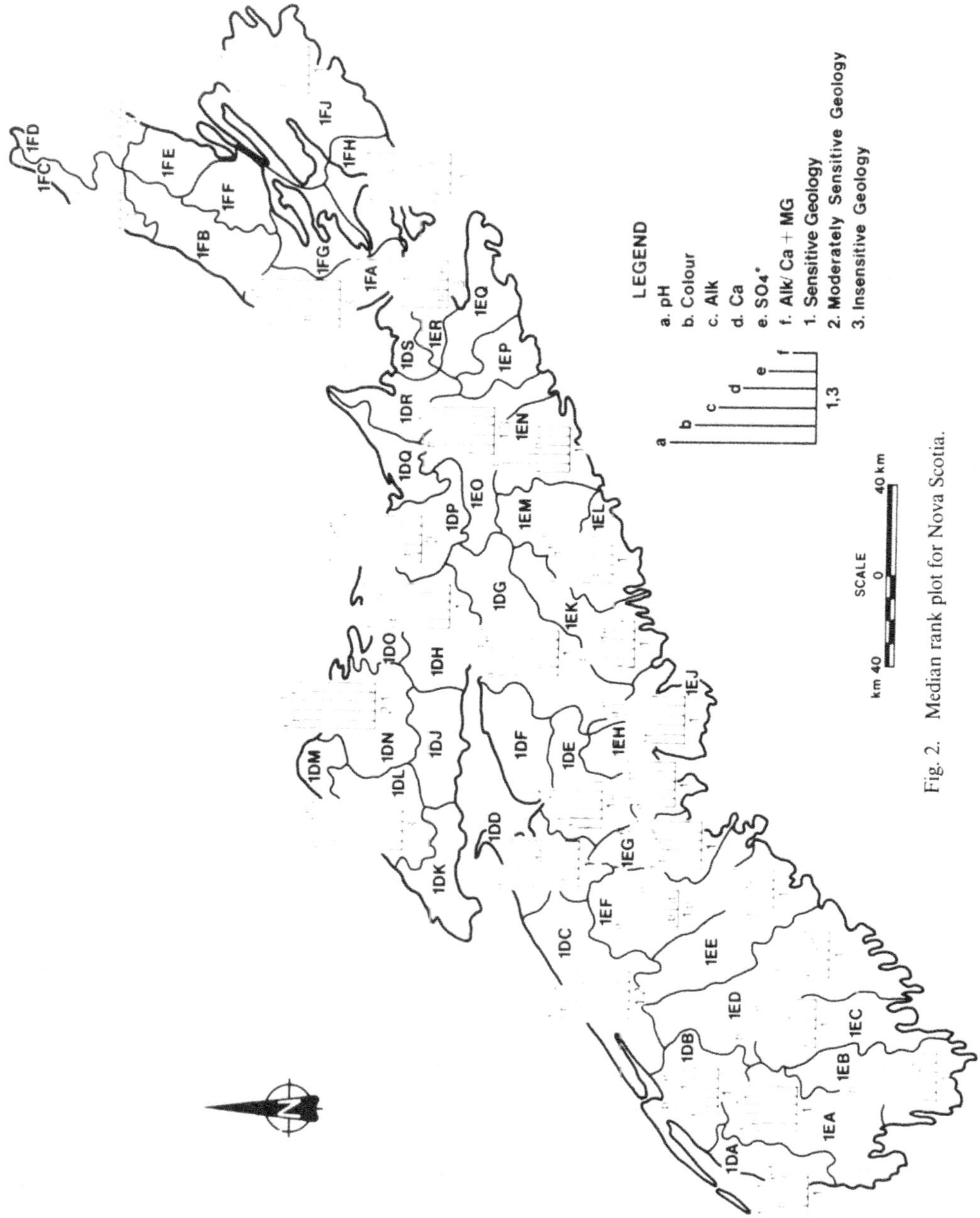

Fig. 2. Median rank plot for Nova Scotia.

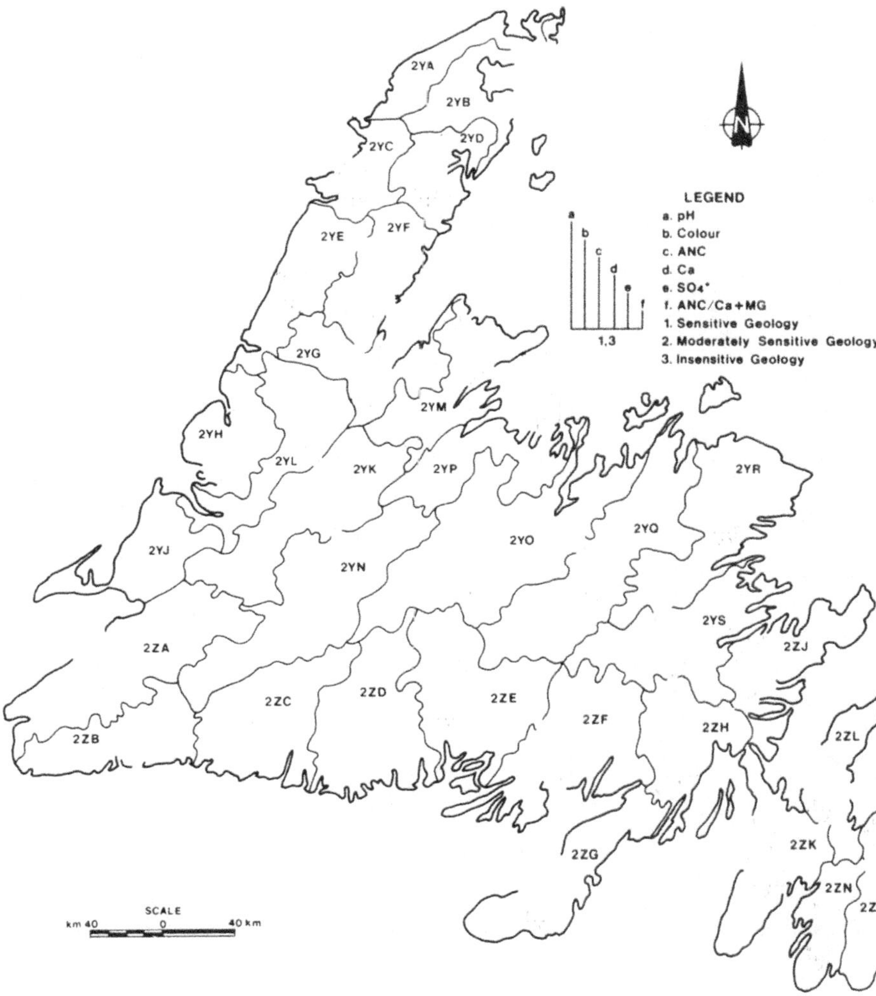

Fig. 3. Median rank plot for insular Newfoundland.

+ Mg* levels due to the combined effects of organic and mineral acid loading. The southwestern Nova Scotia coastal watersheds (DA, Ea, DB) also exhibit significant acidification responses despite somewhat less terrain sensivity. The bassins have lower water colours but higher excess sulphates than the inland watersheds, suggesting that Long Range Transport of Air Pollution (LRTAP) plays a more dominant role in the acidification process. The local emission and mineralized slate influence in the vicinity of Halifax-Dartmouth is also evident with several basins (EK, EJ, DE) exhibiting lower than

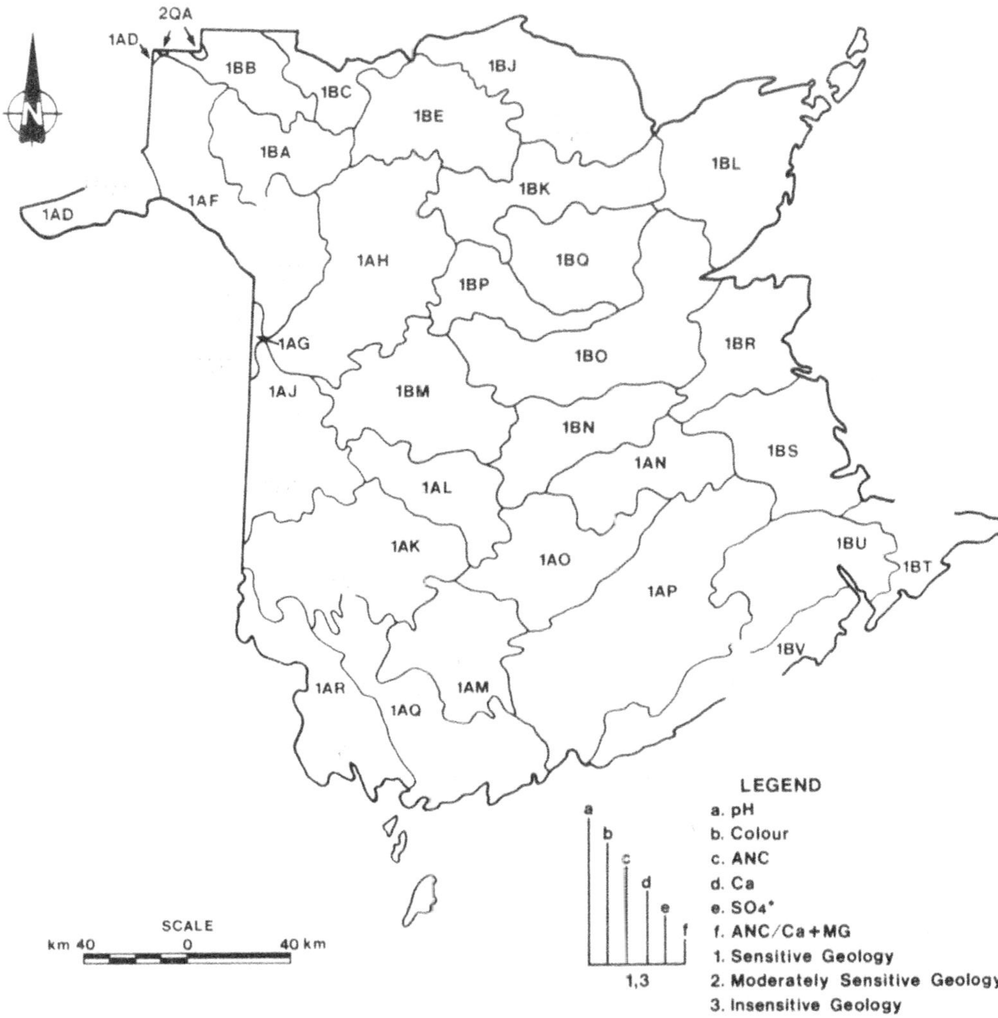

Fig. 4. Median rank plot for New Brunswick.

expected pH, Alk and Alk/ΣCa* + Mg* levels due primarily to high mineral acid loads as indicated by elevated excess sulphates.

The insular Newfoundland tertiary watersheds which exhibit the greatest response to acidification are located on the east coast of the Great Northern Peninsula (YD and YF), the south coast (ZB, ZC, ZD) and on the northeastern portion of the island (YR). These watersheds have small pH and Alk bars and Alk/Σca* + Mg* rations that are indicative of major losses in theoretical pre-acidification alkalinity. The sulphate and the water colour bars are both dominant in basins YD, ZB, YF, ZC, ZD and thus both mineral and organic acids, play a major role in the acidification of these systems. However, the

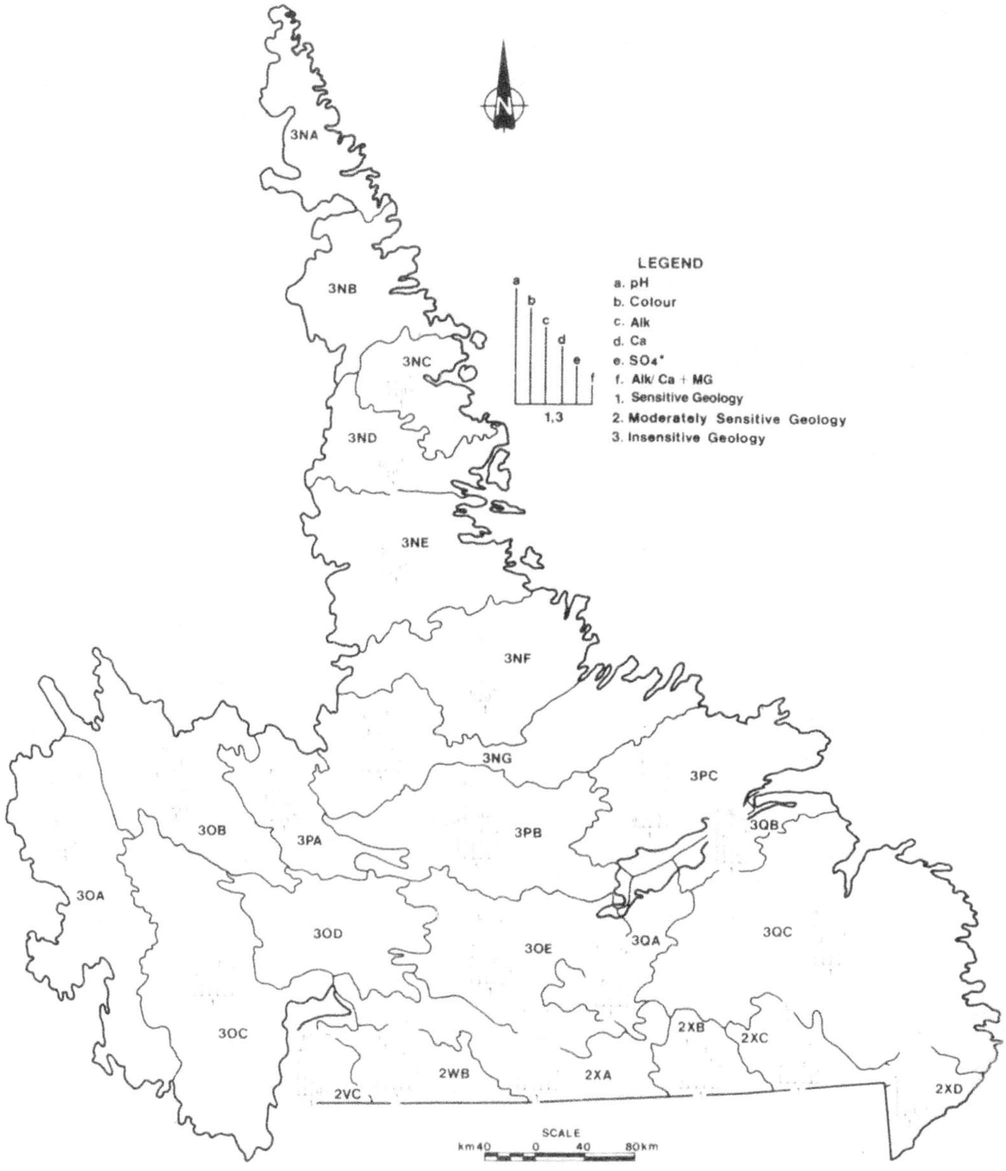

Fig. 5. Median rank plot for Labrador.

northeastern basin (YR) has a low sulphate bar and a high water colour bar which emphasizes the importance of natural organic acidity.

The most severely acidified watersheds in New Brunswick are located in the southwestern portion of the pronvince (AR, AQ, AM, AP). These basins have low pH,

Alk and Alk/ΣCa* + Mg* bars and high sulphate bars which indicates a response to mineral acidification.

In Labrador the greater acidification response is observed in basin 'PC' in the vicinity of Goose Bay and in the tertiary watersheds (WB, XA, XO, XC) located near the Labrador-Quebec border. These basins show low ordinal values of pH, Alk and Alk/ΣCa* + Mg* and elevated values for water colour and sulphate which indicates that both organic and mineral acids are important.

Results of the principal component analysis for Nova Scotia, Newfoundland and New Brunswick are presented in Tables I to III respectively. In all cases the first components retain at least 87% of the original variability while reducing the total number of variables considered. The principal components coefficients are standardized so that their values fall in the inveterval (-1,1). This will permit the evaluation of the contribution of the original variables to th principal components.

The first component for the Nova Scotia data set has high standardized coefficients for pH, calcium, Alk, Alk/ΣCa* + Mg*, and explains 41% of the total variation. These variables are dependent on supply from the terrestrial watershed and thus the first component indicates the importance of the terrestrial weathering process. Previous principal component analysis of small sets of lake data from Nova Scotia (Esterby *et al.*, 1989 has also resulted in a firs component which was highly weighted by variables associated with terrestrial weathering. The second component explains 17% of the variation and has high coefficients for sulphate, colour and the Alk/ΣCa* + Mg* ratio. These variables are consistent with a mineral acidification process. The third and fourth principal components explain 16 and 13% of the varaition respectively and in both cases conductivity and water colour have the highest coefficients.

The first component for the Newfoundland data set explains 51% of the variation and is dominated by variables associated with terrestrial weathering. As was the case for Nova Scotia, water colour had a low coefficient in the first component, presumably as a reslut of the biological and chemical control in the production and subsequent release of organic carbon compounds into surface waters. The second principal component for Newfoundland explains 21% of the variation and is highly weighted for variables associated with organic acidification. Water colour and Alk/ΣCa* + Mg* have high coefficients with

TABLE I

Principal components coefficients for Nova Scotia

	PC1	PC2	PC3	PC4
pH	-1.0	0.48	-0.27	-0.11
Colour	0.48	-0.98	-0.60	-0.88
Sp. Cond.	0.18	0.08	1.0	-1.0
Calcium	-0.94	-0.61	-0.16	-0.48
Suphate	-0.69	-1.0	0.35	0.60
Alk	-0.97	-0.24	0.36	0.04
Alk/CA*+Mg*	-0.76	0.78	-0.27	-0.52
% Variation	41	17	15	13

TABLE II

Principal components coefficients for Newfoundland

	PC1	PC2	PC3	PC4
pH	-0.88	0.55	-0.15	0.47
Colour	0.31	-1.0	-0.10	-1.0
Sp. Cond.	-0.93	-0.59	0.06	-0.18
Calcium	-1.0	-0.42	0.14	-0.08
Suphate	-0.09	0.50	1.0	0.41
Alk	-0.99	-0.43	0.14	-0.18
Alk/CA*+Mg*	-0.69	0.79	-0.47	0.41
% Variation	51	17	15	9

TABLE III

Principal components coefficients for New Brunswick

	PC1	PC2	PC3	PC4
pH	0.79	0.66	0.08	-1.0
Colour	-0.15	-0.38	1.0	-0.05
Sp. Cond.	0.97	-0.21	0.18	0.05
Calcium	0.98	-0.30	-0.06	0.11
Suphate	0.91	-0.39	-0.24	0.35
Alk	1.0	-0.15	0.08	-0.05
Alk/CA*+Mg*	0.43	1.0	0.26	0.82
% Variation	51	19	15	2

opposite signs which is consistent with this process. Specific conductance and pH also have relatively high coefficients. The third principal component is highly weighted for excess sulphate and to a lesser extent for the $Alk/\Sigma Ca^* + Mg^*$ ratio and explains 15% of the variation. As was observed for the second explains 15% of the coefficients for these two variables suggests that this component is indicative of a mineral acidification process. The coefficient for pH in this component is extremely low suggesting that although the present level of mineral acidification in Newfoundland is resulting in significant losses of alkalinity, it is not sufficient to result in major reductions in pH. The fourth principal component is highly weighted for water colour and to a lesser extent pH, sulphate and $Alk/\Sigma Ca^* + Mg^*$ ratio and accounts for 10% of the variation.

The first principal component for the New Brunswick lakes explains 63% of the variation and as was observed for both Newfoundland and Nova Scotia is dominated by variables associated with terrestrial weathering. Excess sulphate is highly weighted in this component suggesting that terrestrially derived sulphates are important in the New Brunswick lakes. The second component explains 19% of the variation and has high coefficients for both pH and the $Alk/\Sigma Ca^* + Mg^*$ ratio. The coefficients for sulphate and water colour have signs which are consistent with acidification but given the low values for these two variables it appears that acidification is not a dominant process. The third principal component explains 15% of the variation and is completely dominated by water colour. The fourth component only explains 7 percent of the variation and has high coefficients of opposite sign for pH and the Alk/ΣCa* + Mg* ratio.

The results of the cluster analysis are summarized in the following. Figure 6 persents the centroid mean values of pH, water colour, specific conductance, calcium, excess sulphate, alkalinity and the Alk/ΣCa* + Mg* ratio for the eight Nova Scotia clusters along with a relative frequency histogram of the terrain sensitivity for the cluster membership. The most acidic group of lakes comprise cluster # 2 and are characterized by low pH, Alk, sulphates, Alk/ΣCa* + Mg* ratios, and high water colours. Ninety-six percent of these fifty-two lakes are underlain by sensitive granitic or slate bedrock while the final four percent are on unsensitive geological formation. These lakes can be considered to be acidified primarily by natural organic acids and they are concentrated in southwestern Nova Scotia and the Northern tip of Cape Breton Island. The lakes which comprise cluster groups 3 and 4 are also acidic but given the relatively low mean water colours of these groups, it appears that strong mineral acids play a major role in the acidification process. The mean excess sulphate concentration of cluster group 3 is higher than would be expected from atmospheric deposition which suggests that local acid sources may be present. A large number of the cluster 3 lakes are concentrated in the vicinity of Halifax, Dartmouth and thus are subject to local emission sources. In addition, many of the lakes in this cluster are underlain by mineralized slate bedrock which when exposed to the atmosphere by anthropogenic activity can result in acidification due to oxidation of pyrite bearing minerals. The cluster 4 mean sulphate concentration is at a level which is consistent with atmospheric deposition and thus it appears that LRTAP is strongly implicated in the acidification of these thirty-two lakes. Some of these lakes are located near to the coast which explains the high mean specific conductance for this group and also suggests some potential for sea-salt acidification. Cluster group 5 has the largest membership of all the clusters and is comprised of sensitive lakes which have been moderately influenced by LRTAP. The lakes are widely distributed throughout the province but are predominantly situated in areas with highly sensitive geologies. The remaining cluster groups (1,6,7,8) include forty-five lakes which have shown little acidification response. Although many of these lakes are underlain by moderate or unsensitive geology, others are in highly sensitive areas and thus probably reflect localized deposits of calcareous minerals.

Figure 7 presents the centroid ranks and mean values of the seven variables for the Newfoundland water quality clusters as well as the relative frequency histogram of the terrain sensitivity for the cluster membership. The most acidic group of lakes (cluster group # 4) is comprised of 15 sites which have low pH, conductance, calcium Alk and Alk/ΣCa* + Mg*. The lakes have a low water colour ordinal and a high excess sulphate ordinal indicating that mineral acids play a major role in the acidification process. Fourteen (93%) of these lakes are located in an area of low terrain sensitivity. These lakes are grouped in the Long Range Mountains of the Great Northern Peninsula and on the extreme southwest coast of the island. Cluster group # 3 has a membership of 12 acidified

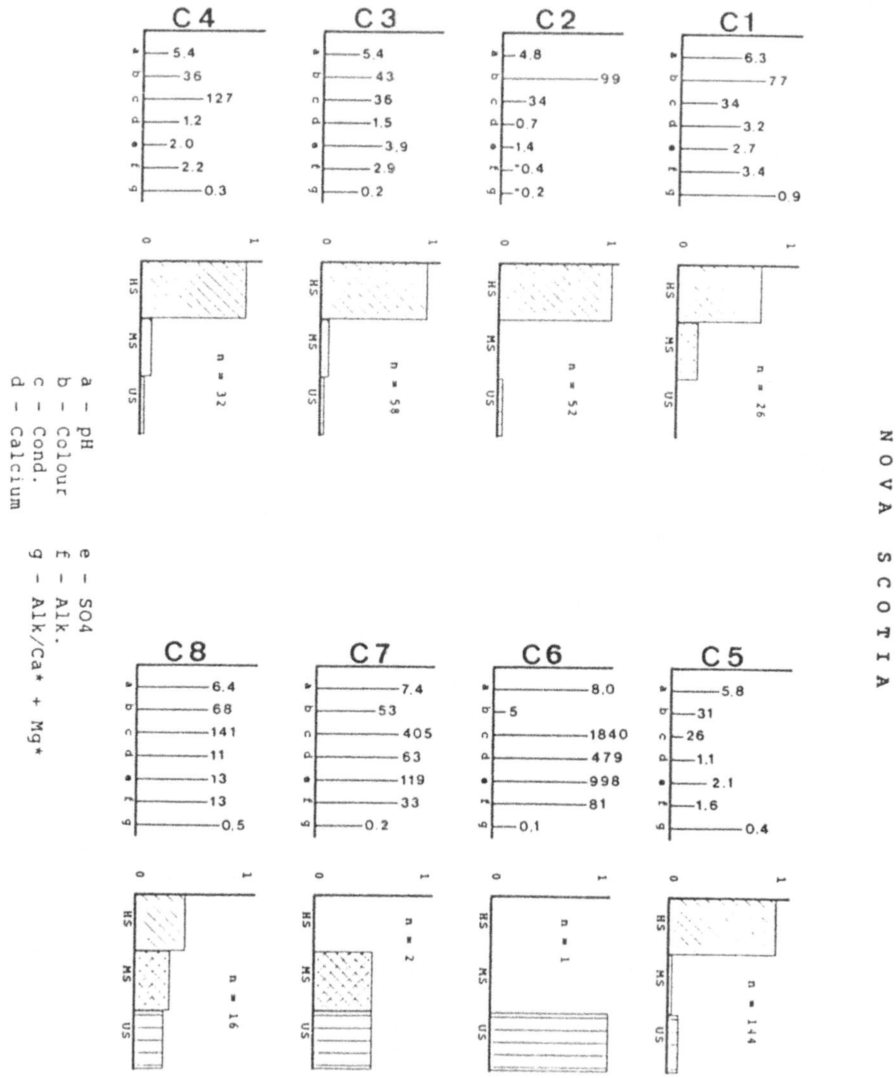

Fig. 6. Summary statistics for the Nova Scotia clusters.

lakes characterized by low pH, conductance, calcium, Alk, excess sulphate and Alk/ΣCa* + Mg* ratio and high water colour. The low excess sulphate ordinal and the high water colour ordinal indicates that natural organic acids are primarily responsible for the acidification. Of the 12 lakes, 7 (58%) are situated on moderately sensitive bedrock. Geographically, this cluster group is much more widely spread than the cluster 4 lakes except for a concentraion of six sites (50%) in the northeastern part of the province. This part of insular Newfoundland is known to have lakes with high water colours and

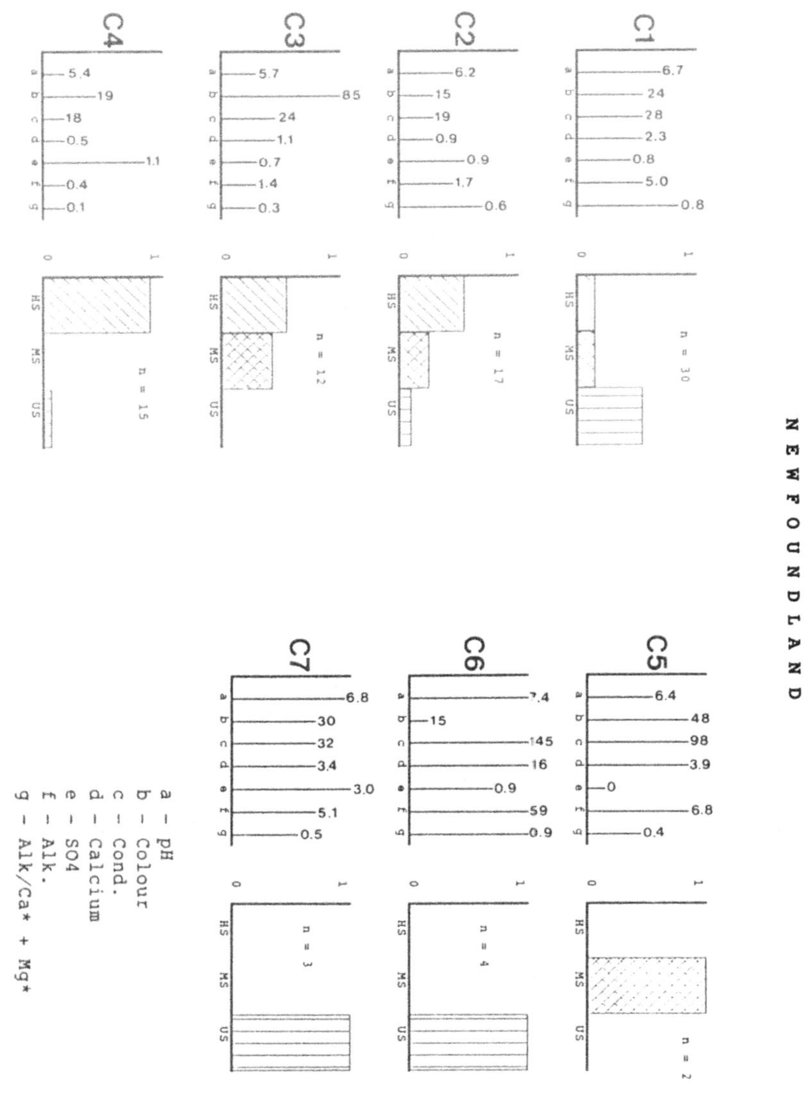

Fig. 7. Summary statistics for the Newfoundland clusters.

dissolved organic carbon concentrations. The cluster group #2 contains lakes which have rather similar chemistry to those in cluster group #4. However, as a consequence of higher calcium and lower excess sulphate, acidification effects are at present rather minimal. Given a pH centroid of 6.2, an Alk centroid of 1.7 and an approximate 40% loss in theoretical pre-acidification alkalinity, this cluster group represents a series of extremely sensitive lakes. Of the seventeen lakes in this cluster, ten (59%) are situated on highly sensitive terrain, five (29%) are located on moderately sensitivity. Lakes in this cluster are also geographically well distributed but many are concentrated along the south coast and in Terra Nova National park on the east coast. Cluster # 1 includes twenty lakes which exhibit limited acidification response having a centroid pH of 6.4 and a theoretical pre-acidification alkalinity loss of less than 20%. These lakes are located from west to east but are concentrated into a band in the central portion of the province. Of these twenty lakes, twelve (70%) are underlain by unsensitive sedimentary bedrock while the remaining eight sites are evenly spread between highly and moderately sensitive terrains. The final three clusters groups (5, 6 and 7) all have small memberships and include lakes that have not experienced any significant acidification. These cluster lakes are underlain by either moderately sensitive or unsensitive geological structures and tend to be concentrated in well defined areas. The cluster # 5 lakes are both located in Terra Nova National park and given their slightly elevated conductance and extreme low excess sulphate concentrations, may be influenced by road salt applications. The cluster group 6 lakes are situated on the extreme tip of the Great Northern Peninsula while the three cluster 7 are located on the south coast.

The Labrador lakes fall into four of the seven Newfoundland cluster groups (C1, C2, C3, C4) with the majority of the lakes falling in the unacidified cluster group 1 or the highly sensitive cluster group 2. The unacidified lakes tend to be concentrated in the northern in the central and southern portions of Labrador. Only two of the lakes are classed into cluster group 4, which is indicative of mineral acidification.

Figure 9 presents the centroid ranks and mean values of the six New Brunswick water quality clusters as well as the relative frequency histogram of the terrain sensitivity for the cluster membership. The most acidic lakes comprise cluster groups 2 and 5 and given the high water colour ordinals are indicative of natural organic acidification. Cluster group 3 also includes acidified lakes but given the low water colour ordinal it appears that mineral acids have a major role in the acidification of these lakes. Without exception, the lakes in cluster groups 2, 3 and 5 underlain by geologies with either high or moderate sensitivity to acid rain. The other three cluster groups include lakes which are more highly buffered and thus exhibit little acidification response.

The majority of the lakes considered in this analysis are concentrated in the most sensitive portion of the province and as such tend to overestimate the extent of acidification in New Brunswick. It should be noted that acidified and unacidified lakes are situated in close proximity which emphasizes the moderating influence of localized soil deposits in areas of highly sensitive geology.

Conclusions

Principal component analysis indicates that both organic and mineral acidification processes are important for the Nova Scotia and Newfoundland lake data sets. Subsequent cluster analysis of the Nova Scotia lakes has resulted in three groups of acidified lakes and one group of moderately impacted lakes. The lakes which comprise these clusters are generally situated on highly sensitive geological structures and represent acidification by various sources including organic acids, LRTAP, local mineral acids and

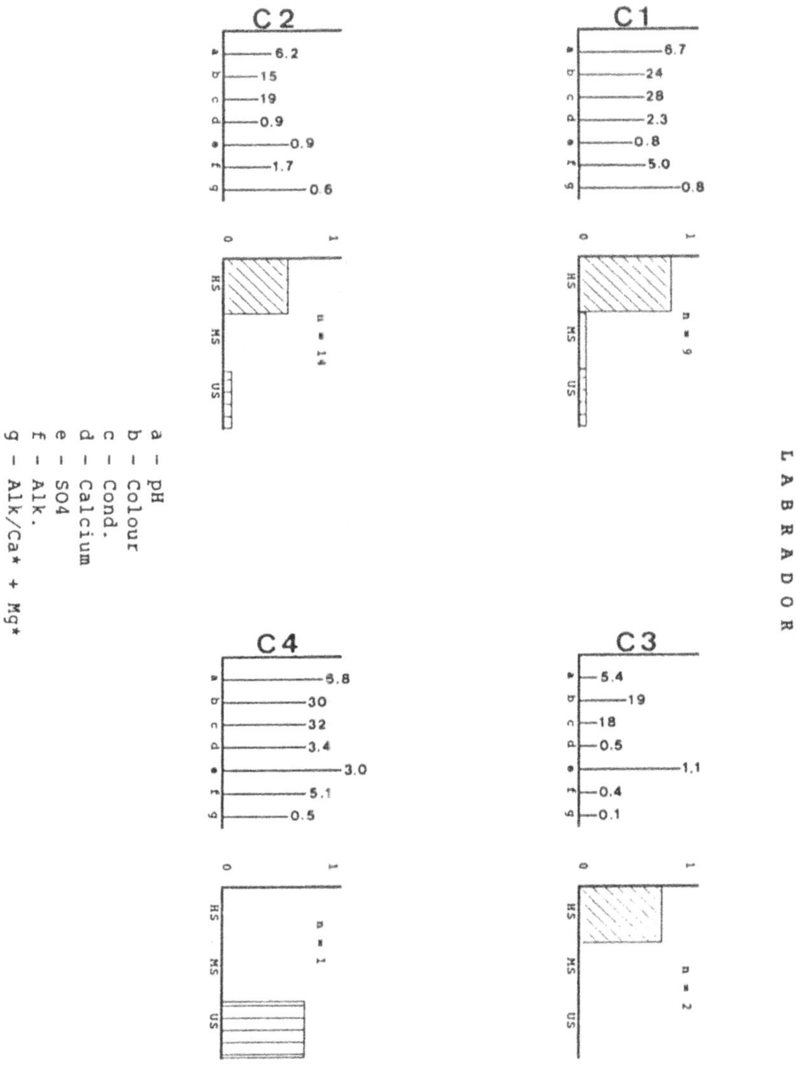

Fig. 8. Summary statistics for the labrador clusters.

possibly sea-salt induced acidification. Acid rain has been shown to have a major role in the acidification of many lakes particularly those situated in the southwestern portion of the province. Both insular Newfoundland and New Brunswick have cluster groups which are indicative of lakes acidified by organic and mineral acids. In insular Newfoundland, acid rain effects are concentrated to the sensitive portions of the south coast and the Great Northern Peninsula whereas the New Brunswick acid rain influence is centered in the southwestern portions of the province. The Labrador clusters indicate that although the

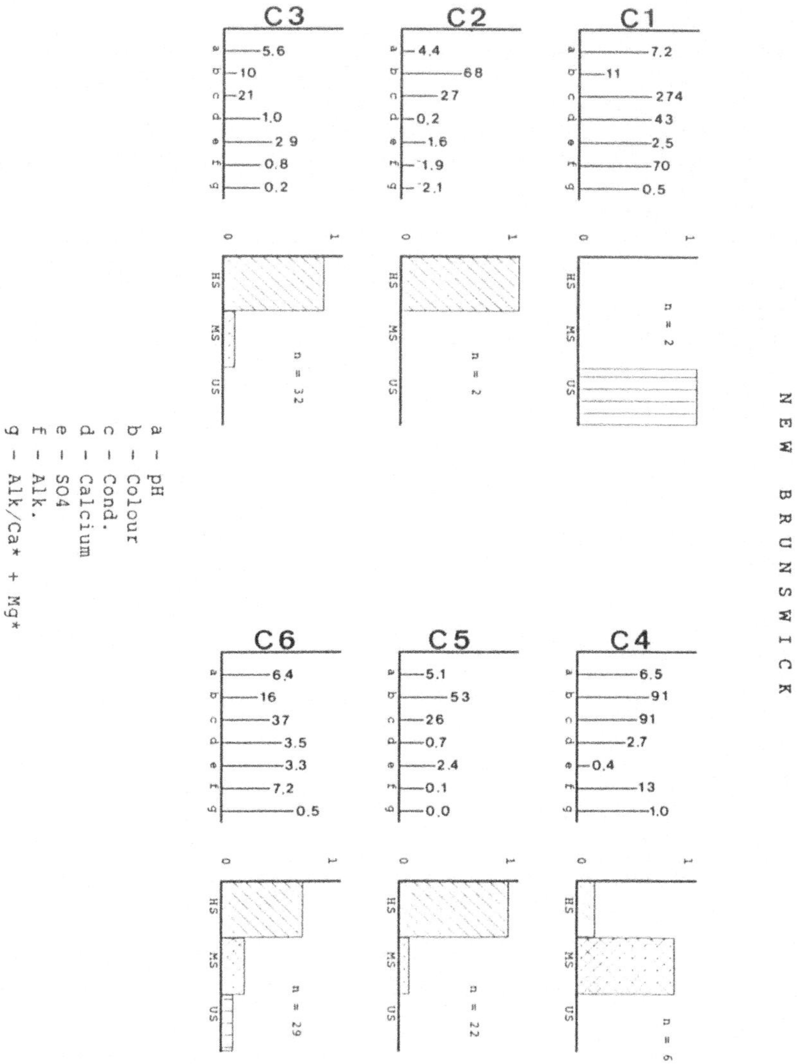

Fig. 9. Summary statistics for the New Brunswick clusters.

lakes are extremely sensitive, atmospheric loading is presently insufficient to result in appreciable acidification.

References

El-Shaarawi, A. H., Esterby S. R., and Howell, G. D.: 1989, *Water, Air, and soil Pollut.* **46,** 305.

Esterby, S. R., El-Shaarawi, A. H., Howell, G. D., and Clair, T. A.: 1989, *Water, Air, and Soil Pollut.* **46,** 289.

Gorham, E., Bayley, S. E., and Schindler, D. W.: 1984, *Can J. Fish. Aquat. Sic.* **41,** 1256.

Howell, G. D. and Brooksbank, P.: 1987, 'An Assessment of LRTAP Acidification of Surface Waters in Atlantic Canada', Inland Waters Directorate. Water Quality Branch Report IW/L-AR-WQB-87-121. 292 pp.

Howell, G. D.: 1989, *Water, Air, and Soil Pollut.* **46,** 165.

Kerekes, J. J., Beauchamp, S., Tordon, R., Tremblay, C., and Pollock, T.: *1986, Water, Air, Soil Pollu.* **31,** 165.

M.O.I.: 1983, United Sttes-Canada Memorandum of Intent Impact Assessment, Work Group 1, Final Report, January, 1983.

Thompson, M.: 1986, Water, *Air, Soil Pollut.* **31,** 17.

Watt, W. D., Scott, D., and White, W. J.: 1983, *Can J. Fish. Aquat. Sci.* **401** 462.

STATISTICAL INFERENCE FROM MULTIPLY CENSORED ENVIRONMENTAL DATA

A. H. EL-SHAARAWI and A. NADERI

National Water Research Institute, Canada Centre for Inland Waters, Burlington, Ontario, L7R 4A6

(Received August 1990)

Abstract. Maximum likelihood estimation for multiply censored samples are discussed. Approximate confidence intervals for the lognormal mean are obtained using both Taylor expansion method and direct method. It is shown that the direct method performs noticeably better than the Taylor expansion method. Simulation results and applications are provided.

1. Introduction

In routine water and air quality monitoring of toxic contaminants and trace metals, it frequently happens that a certain portion of the observations examined, have concentrations that cannot be measured. It is only possible to determine that the concentrations for those observations fall within certain intervals. The endpoints of these intervals are detection limits determined by analytical methods. If $D_1 < \ldots < D_{k-1} < D_k$ are such detection limites, then a censored observation occurs when its value falls below D_k. Approaches adopted by environmental scientists for estimating the mean and standard deviation in the presence of a single censoring limit D_1, ranges from assigning a value to an observation reported as less than D_1, to the use of the log regression method (Gilliom and Helsel, 1986). Assuming the normal or lognormal distribution for the observations, El-Shaarawi (1989) and El-Shaarawi and Dolan (1989) discussed the use of the method of maximum likelihood for estimating the mean and standard deviation when $k = 1$. In addition, Shumway *et al.* (1989) considered the possibility of using the Box and Cox (1964) transformation to normalize the data. The general problem of maximum likelihood estimation of the parameters of a censored normal sample has been considered by many authors. Cohen (1950) used the maximum likelihood method to estimate the parameters of type I singly and doubly censored normal samples. Gupta (1952) found maximum likelihood equations to estimate the parameters of type II censored normal samples. Cohen (1950) and Gupta (1952) also formulated the asymptotic variances and covariances. Harter and Moore (1966) and Harter (1970) considered the maximum likelihood estimators for type II censoring and performed a simulation study which showed that maximum likelihood estimators had mean square errors smaller than the variances of the best linear unbiased estimators for $n \geq 10$. Tiku (1967) modified the maximum likelihood equations from a type II censored normal sample so that an explicit formula for the estimators could be obtained. The general results concerning censored normal samples have been summarized and extensively studied by Schneider (1986). Progressively censored samples from normal, exponential, Weibull and lognormal

Environmental Monitoring and Assessment **17**: 339–347, 1991.

distributions have also received previous attention from Herd, Robert, Cohen and Ringer and Sprinkle, (Cohen, 1976).

The present paper first discusses the maximum likelihood estimation for multiply censored samples and then studies large sample confidence intervals for the lognormal mean. Two large sample confidence intervals for the lognormal mean are obtained using both Taylor expansion method and direct method. A simulation study indicates that the direct method performs noticeably better than the Taylor expansion method. The simulation results are provided and an application using the concentrations (nanograms per litre) of Fluoranthene in water samples from the Niagara River is presented.

2. Estimation of the Mean and Standard Deviation from Type I Multiply Censored Normal Samples

Let $D_0 = -\infty$, and consider k detection limits $D_1, ..., D_k$. Let the random variable N_i ($i=0$, 1, ..., $k-1$) denote the number of observations that fall in the interval (D_i, D_{i+1}). Furthermore, let the random variables $X_1, ..., X_n$ represent the n uncensored observations ($X_i > D_k$, $i = 1, ..., n$). The observed values of X_i and N_i are denoted respectively, by x_i and n_i. Under the assumption that X_i's are independent and normally distributed with mean μ and variance σ^2, the likelihood function is:

$$L = C_0\, \sigma^{-n} \prod_{i=0}^{k-1} (\Phi(\eta_{i+1}) - \Phi(\eta_i))^{n_i} \prod_{i=1}^{n} \phi\left(\frac{x_i - \mu}{\sigma}\right),$$

where C_0 is a constant, $\eta_i = (D_i - \mu)/\sigma$ ($i=1, ..., k$),

$$\phi(x) = \frac{1}{\sqrt{2\pi}} \exp\left(\frac{-x^2}{2\sigma^2}\right), \quad \text{and} \quad \Phi(x) = \int_{-\infty}^{x} \phi(t)\, dt.$$

The maximum likelihood estimators for the mean and variance of the normal distribution from interval censored data can be obtained using one of the many available packages (i.e. SAS, CENSOR, SYSTAT, etc.). The following closed form approximate maximum likelihood estimators for the mean and variance of normal distribution are developed for the case when values of $\Delta_i = D_{i+1} - D_i$ ($i=1,..., k-1$) are small.
By the mean value theorem, L can be written as:

$$L = C_0\, \sigma^{-\left(n + \sum_{i=1}^{k-1} n_i\right)} (\Phi(\eta_1))^{n_0} \prod_{i=1}^{k-1} \left\{ \phi\left(\frac{\xi_i - \mu}{\sigma}\right)(D_{i+1} - D_i) \right\}^{n_i} \prod_{i=1}^{n} \phi\left(\frac{x_i - \mu}{\sigma}\right), \tag{1}$$

where ζ_i ($i=1, ..., k-1$) is between D_i and D_{i+1}. Let

$$N = n + \sum_{i=0}^{k-1} n_i, \; \bar{x} = \sum_{i=1}^{n} x_i/n,$$

and define

$$M = \left(\sum_{i=1}^{k-1} n_i \zeta_i + n\bar{x} \right) / (N-n_0),$$

$$S^2 = \left\{ \sum_{i=1}^{n} x_i^2 + \sum_{i=1}^{k-1} n_i \zeta_i^2 - (N-n_0)M^2 \right\} / (N-n_0),$$

and $g(x) = \phi(x)/\Phi(x)$, where x is a real number. From (1) it follows that the maximum likelihood estimates of μ and σ satisfy the following equations:

$$\mu = M - \sigma \frac{n_0}{N-n_0} g(\eta_1), \tag{2}$$

$$\sigma^2 = S^2 + (M-\mu)(M-D_1). \tag{3}$$

Replacing $g(\eta_1)$ by Tiku's (1967) linear approximation $\alpha+\beta\eta_1$, where

$$\beta = \{g(t_2) - g(t_1)\}/(t_2-t_1), \quad \alpha = g(t_1)-t_1\beta,$$

$$t_1 = \Phi^{-1}\{q-\sqrt{q(1-q)/N}\}, \quad t_2 = \Phi^{-1}\{q + \sqrt{q(1-q)/N}\}, \quad q = n_0/N,$$

(2) becomes

$$\mu = \{M(N-n_0) - n_0\alpha\sigma - n_0\beta D_1\}/\{N-n_0(1+\beta)\}. \tag{4}$$

For small values of Δ_i, ζ_i can be approximated by the midpoint $x_{m_i} = (D_i + D_{i+1})/2$. Setting $\zeta_i = x_{m_i}$, Equations (3) and (4) provide approximate explicit solutions $\hat{\mu}$ and $\hat{\sigma}$ for μ and σ. As the total number of observations tends to infinity, and as Δ_i (i=1, ..., k-1) approaches zero, these estimates approach the maximum likelihood estimates for μ and σ.

The asymptotic variance - covariance matrix of $\hat{\mu}$ and $\hat{\sigma}$ is denoted by the matrix $\begin{bmatrix} V_{11} & V_{12} \\ V_{21} & V_{22} \end{bmatrix}$ and is obtained by nothing that:

$$E\left(\frac{\partial^2 \ln L}{\partial \mu^2}\right) \simeq -\frac{N}{\sigma^2}\{1 - (1+\beta)\Phi(\eta_1)\},$$

$$E\left(\frac{\partial^2 \ln L}{\partial \mu \partial \sigma}\right) = E\left(\frac{\partial^2 \ln L}{\partial \sigma \partial \mu}\right) \simeq \frac{2N}{\sigma^3}[(\mu + \sigma g(-\eta_k))\Phi(\eta_k) +$$

$$\{(\beta\eta_1 + \frac{\alpha}{2})\sigma - \mu\}\Phi(\eta_1) - (\sigma g(-\eta_k) + A)],$$

where $$A = \sum_{i=1}^{k-1} x_{m_i} (\Phi(\eta_{i+1}) - \Phi(\eta_i)),$$

and $$E\left(\frac{\partial^2 \ln L}{\partial \sigma^2}\right) \simeq \frac{N}{\sigma^4}[3\{\sigma^2(1 + \eta_k g(-\eta_k)) - \mu^2\}\Phi(\eta_k) +$$

$$\{3\mu^2 + \sigma^2(2\alpha\eta_1 + 3\beta\eta_1^2 - 1)\}\Phi(\eta_1) - \sigma^2(2+3\eta_k g(-\eta_k)) - 3(B-2\mu A)],$$

where $$B = \sum_{i=1}^{k-1} x_{m_i}^2 \left(\Phi(\eta_{i+1}) - \Phi(\eta_i)\right).$$

Let $\hat{L}$ be obtained by substituting $\hat{\mu}$ and $\hat{\sigma}$ directly into L and let $R(\mu, \sigma) = \ln L - \ln \hat{L}$. Then, by noting that $-2R(\mu, \sigma)$ is approximately $\chi^2(2)$, approximate confidence regions for (μ, σ) may be obtained.

We now proceed to obtain the conditional bias in the estimators. Consider the Taylor expansion:

$$\left[\frac{\partial \ln L}{\partial \mu}\right]_{\hat{\mu}} = \left[\frac{\partial \ln L}{\partial \mu}\right]_{\mu} + (\hat{\mu} - \mu)\left[\frac{\partial^2 \ln L}{\partial \mu^2}\right]_{\overset{*}{\mu}},$$

where $\overset{*}{\mu}$ is between $\hat{\mu}$ and μ. We have

$$E(\hat{\mu}|\sigma) - \mu = -\mathrm{E}\left(\frac{\partial \ln L}{\partial \mu}|\sigma\right)_{\mu} \Big/ E\left(\frac{\partial^2 \ln L}{\partial \mu^2}|\sigma\right)_{\overset{*}{\mu}}.$$

Hence, the conditional bias for $\hat{\mu}$ is

$$E(\hat{\mu}|\sigma) - \mu \simeq \frac{-(\sigma g(-\eta_k)+\mu)\Phi(\eta_k)+\{\mu-\sigma(\alpha+\beta\eta_1)\}\Phi(\eta_1) + A + \sigma g(-\eta_k)}{1-(1+\beta)\Phi(\eta_1)}. \tag{5}$$

Similarly, the conditional bias for σ is obtained by

$$E(\hat{\sigma}|\mu) - \sigma \simeq \frac{\sigma[\{\mu^2-\sigma^2(1+\eta_k g(-\eta_k))\}\Phi(\eta_k)-\{\mu^2+\sigma^2(\alpha\eta_1+\beta\eta_1^2-1)\}\Phi(\eta_1) +\sigma^2\eta_k g(-\eta_k)+(B-2\mu A)]}{3\{\mu^2-\sigma^2(1+\eta_k g(-\eta_k))\}\Phi(\eta_k)-\{3\mu^2+\sigma^2(2\alpha\eta_1+3\beta\eta_1^2-1)\}\Phi(\eta_1) +3(B-2\mu A)+\sigma^2(2+3\eta_k g(-\eta_k))}. \tag{6}$$

3. Estimation Under Transformations

When the normality assumption of the X_i's cannot be justified, it is appropriate to find a suitable transformation so that the transformed data satisfy the normality assumption. Box and Cox (1964) suggested the use of the transformation,

$$\begin{aligned} g_\lambda(x) &= (x^\lambda - 1)/\lambda \quad & \lambda \neq 0 \\ &= \ln x & \lambda = 0 \end{aligned} \tag{7}$$

with λ chosen so that the distribution of $g_\lambda(X_1)$ is normal with mean μ_λ and variance σ_λ^2.

Given detection limits $D_1, \ldots, D_k$ and observations $x_1, \ldots, x_n$, $(x_i > D_k, i=1, \ldots, n)$, we may obtain the estimates $\hat{\mu}_\lambda$ and $\hat{\sigma}_\lambda$ by the methods of the previous section. The transformation parameter λ is then chosen as the value $\hat{\lambda}$ that maximizes:

$$h(\lambda, \hat{\mu}_\lambda, \hat{\sigma}_\lambda) = -n \ln \hat{\sigma}_\lambda + \sum_{i=0}^{k-1} n_i \ln(F(D_{i+1}) - F(D_i))$$

$$+ \lambda \sum_{i=1}^{n} \ln(x_i) - \frac{1}{2} \sum_{i=1}^{n} \left(\frac{g_\lambda(x_i) - \hat{\mu}_\lambda}{\hat{\sigma}_\lambda} \right)^2 ,$$

where

$$F(y) = \frac{1}{\sqrt{2\pi}\hat{\sigma}_\lambda} \int_0^y x^{\lambda-1} \exp \left\{ \frac{-(g_\lambda(x) - \hat{\mu}_\lambda)^2}{2 \quad \hat{\sigma}_\lambda{}^2} \right\} dx.$$

Let

$$A_i = \frac{1}{\sqrt{2\pi}\sigma_{\hat{\lambda}}} \int_{-1/\hat{\lambda}}^{\infty} x^i (1 + \hat{\lambda} x)^{1/\hat{\lambda}} \exp \left\{ \frac{-(x - \mu_{\hat{\lambda}})^2}{2\sigma_{\hat{\lambda}}^2} \right\} dx, \quad i = 0, 1, 2. \tag{8}$$

Then $\gamma = E(X_1) = A_0$. The estimate $\hat{\gamma}$ of γ may be obtained by substituting $\hat{\mu}_{\hat{\lambda}}$ and $\hat{\sigma}_{\hat{\lambda}}$ directly into λ above. Note that using the Taylor expansion, $\hat{\gamma}$ may be approximated by,

$$\hat{\gamma} \simeq \gamma + (\hat{\mu}_{\hat{\lambda}} - \mu_{\hat{\lambda}})\, \gamma_\mu + (\hat{\sigma}_{\hat{\lambda}} - \sigma_{\hat{\lambda}}) \gamma_\sigma,$$

where
$$\gamma_\mu = \frac{1}{\sigma_{\hat{\lambda}}{}^2} (A_1 - \mu_{\hat{\lambda}} A_0),$$

and
$$\gamma_\sigma = \frac{1}{\sigma_{\hat{\lambda}}{}^3} \{A_2 + (\mu_{\hat{\lambda}}{}^2 - \sigma_{\hat{\lambda}}{}^2)\, A_0 - 2\mu_{\hat{\lambda}} A_1\}.$$

Let $\tilde{V}_{11}$, $\tilde{V}_{22}$, and $\tilde{V}_{12}$ be the asymptotic variances of $\hat{\mu}_{\hat{\lambda}}$ and $\hat{\sigma}_{\hat{\lambda}}$, and the covariance of $\hat{\mu}_{\hat{\lambda}}$ and $\hat{\sigma}_{\lambda\beta+C}$/ ρεσπεφτιψελθ! Τηεν/ $\hat{\gamma}$ is approximately normal with mean γ and variance $V_\gamma^2 = \gamma_\mu^2 \tilde{V}_{11} + 2\gamma_\mu \gamma_\sigma \tilde{V}_{12} + \gamma_\sigma^2 \tilde{V}_{22}$ where V_γ^2 may be approximated by $\hat{V}_\gamma^2$ obtained by substituting $\hat{\mu}_{\hat{\lambda}}$ and $\hat{\sigma}_{\hat{\lambda}}$ directly into V_γ^2.

An approximate 100 (1–a)% confidence interval for γ may then be obtained by ($\hat{\gamma} - c_a \hat{V}_\gamma$, $\hat{\gamma} + c_a \hat{V}_\gamma$), where c_a is determined from $\Phi(-c_a) = a/2$.

4. The Lognormal Case

The special case of $\lambda = 0$ in transformation (7) leads to the assumption that $X_1, ..., X_n$ have lognormal distributions. Since the lognormal distribution is frequently used as a model for the environmental data, and since the numerical integrations (8) can be avoided for this case, this section is devoted to the lognormal case. Note that the estimation results for $\hat{\lambda}$ close to zero can be approximated by the methods of this section.

Let the mean and variance of $g_0(X_1)$ be $\mu_\circ$ and $\sigma_\circ^2$, respectively. Then the mean $\gamma_\circ$ of X_1 is given by,

$$\gamma_\circ = \exp(\mu_\circ + \sigma_\circ{}^2/2).$$

In order to estimate the mean of a multiply censored lognormal sample, one may obtain the estimation $\hat{\mu}_\circ$ and $\hat{\sigma}_\circ$ of the mean and the standard deviation of the corresponding normal sample and then substitute these estimates directly into the expression for $\gamma_\circ$ to

obtain the estimator $\hat{\gamma}_\circ$. This estimator, however, is biased as the following argument shows:
Let $\mathring{V}_{11}$, $\mathring{V}_{22}$, and $\mathring{V}_{12}$ be the asymptotic variances of $\hat{\mu}_\circ$ and $\hat{\sigma}_\circ$ and the covariance of $\hat{\mu}_\circ$ and $\hat{\sigma}_\circ$, respectively. Then, since $a\hat{\mu}_\circ + b\hat{\sigma}_\circ^2$ is approximately normal with mean $a\mu_\circ + b\sigma_\circ^2$ and variance $a^2\mathring{V}_{11} + 4ab\sigma_\circ\mathring{V}_{12} + 4b^2\sigma_\circ^2\mathring{V}_{22}$,

$$E(\exp(a\hat{\mu}_\circ + b\hat{\sigma}_\circ^2)) = \exp\{(a\mu_\circ + b\sigma_\circ^2) + \frac{1}{2}(a^2\mathring{V}_{11} + 4ab\sigma_\circ\mathring{V}_{12} + 4b^2\sigma_\circ^2\mathring{V}_{22})\}$$
$$= \exp(a\mu_\circ + b\sigma_\circ^2)\, h(a, b).$$

In particular,

$$E(\hat{\gamma}_\circ) = E(\exp(\hat{\mu}_\circ + \frac{\hat{\sigma}_0^2}{2})) = \gamma_\circ h(1, \frac{1}{2}) = \gamma_\circ \tau.$$

As a result the estimator $\hat{\gamma}_\circ$ can be modified to yield approximately an unbiased estimator for $\gamma_\circ$ as follows:

$$\tilde{\gamma}_\circ = \hat{\gamma}_\circ / \tilde{\tau},$$

where $\tilde{\tau}$ is obtained by replacing the parameters $\mu_\circ$ and $\sigma_\circ$ in τ by $\hat{\mu}_\circ$ and $\hat{\sigma}_\circ$ respectively.

An approximate confidence interval for $\gamma_\circ$ may be obtained based on the fact that $\hat{\mu}_\circ + 1/2\,\hat{\sigma}_\circ^2$ is approximately normal with mean $\mu_\circ + 1/2\,\sigma_\circ^2$ and variance $\hat{V}\gamma_\circ^2 = \mathring{V}_{11} + 2\sigma_\circ\mathring{V}_{12} + \sigma_\circ^2\mathring{V}_{22}$. By a similar argument used in Land (1972) for complete samples, an approximate 100 $(1-a)\%$ confidence interval for $\gamma_\circ$ is directly obtained by

$$(\hat{\gamma}_\circ \exp(-c_a \hat{V}_{\gamma\circ}),\ \hat{\gamma}_\circ \exp(c_a \hat{V}_{\gamma\circ})),$$

where $\hat{V}_{\gamma\circ}$ is the value of $V_{\gamma\circ}$ evaluated at $\hat{\mu}_\circ$ and $\hat{\sigma}_\circ$, and c_a is as defined earlier.

5. Simulation Results and Applications

Simulation experiments were conducted to evaluate the performance of the methods of this paper and their sensitivity to small-sample effects. For a given sample size N and two detection limits D_1 and D_2, samples from the standard normal distribution and from the lognormal distribution with mean 7.389 and standard deviation 54.096 were generated using the International Mathematical and Statistical Libraries (IMSL, 1987). The values of the detection limits reflect both low and high levels of censoring. The results summarized in Tables Ia and Ib are the averages over 1000 repetitions. The estimates of the mean and the standard deviation for the normal samples along with their asymptotic variance-covariance are listed in Table Ia. These estimates, as is expected, appear to be uncorrelated for low levels of censoring. Table Ia shows that the elements of the asymptotic variance-covariance matrix decline with the increase in sample sizes. Table Ib gives the estimates of the lognormal means along with their 95% confidence interval as well as the probabilities of coverage for both low and high levels of censoring. It can be seen that the width of the confidence intervals decreases as the number of observations increases. Figure 1 also presents the probabilities of coverage for both low and high levels of censoring using both direct and Taylor expansion methods. This figure also reflects the dependency of both methods on the standard deviations of the corresponding normal

TABLE Ia

Simulation results for the standard normal distribution

N	n	$\hat{\mu}$	$\hat{\sigma}$	V_{11}	V_{12}	V_{22}	$E(\hat{\mu}\|\sigma)-\mu$	$E(\hat{\sigma}\|\mu)-\sigma$	(D_1, D_2)
30	27	-0.004	0.98	0.033	-0.001	0.018	0.0006	-0.007	(-1.6,-1.5)
60	55	-0.01	0.994	0.017	-0.0004	0.009	-0.0004	-0.003	(-1.6,-1.5)
120	111	0.001	0.997	0.008	-0.0002	0.004	0.0009	-0.002	(-1.6,-1.5)
240	223	0.0003	0.999	0.004	-0.00008	0.002	0.0004	-0.001	(-1.6,-1.5)
30	13	-0.022	0.989	0.057	-0.025	0.046	-0.005	0.00001	(0., 0.1)
60	27	-0.016	0.996	0.027	-0.011	0.022	-0.003	-0.0003	(0., 0.1)
120	55	-0.003	0.999	0.013	-0.005	0.011	-0.002	-0.0001	(0., 0.1)
240	110	-0.002	1.0003	0.006	-0.002	0.005	-0.001	-0.0001	(0., 0.1)

TABLE Ib

Simulation results for the lognormal distribution with mean 7.389 and standard deviation 54.096. The mean and the standard deviation for the corresponding normal distribution are 0 and 2 respectively

N	n	$\hat{\mu}_o$	$\hat{\sigma}_o$	$\hat{\gamma}_0$	$\tilde{\gamma}_o$	Approximate 95% confidence interval for γ_0	Probability of Coverage	(D_1, D_2)	$(\ln D_1, \ln D_2)$
30	28	-0.008	1.960	8.746	6.631	(2.211, 39.939)	91.7	(0.04076, 0.04505)	(-3.2, -3.1)
60	56	-0.020	1.989	8.016	7.109	(3.135, 21.012)	93.6	(0.04076, 0.04505)	(-3.2, -3.1)
120	112	0.002	1.994	7.788	7.364	(4.070, 15.000)	95.2	(0.04076, 0.04505)	(-3.2, -3.1)
240	225	0.0007	1.998	7.612	7.409	(4.837, 11.999)	93.6	(0.04076, 0.04505)	(-3.2, -3.1)
30	14	-0.044	1.978	11.523	6.021	(1.804, 444.237)	89.2	(1., 1.10517)	(0., 0.1)
60	28	-0.033	1.992	8.549	7.021	(2.737, 30.722)	92.5	(1., 1.10517)	(0., 0.1)
120	57	-0.006	1.998	7.996	7.369	(3.723, 17.661)	94.9	(1., 1.10517)	(0., 0.1)
240	115	-0.004	2.0007	7.719	7.436	(4.554, 13.161)	93.8	(1., 1.10517)	(0., 0.1)

distributions. The results indicate that both methods provide good probabilities of coverage for small values of the standard deviations. For moderate and large values of the standard deviations, however, the Taylor expansion method is less satisfactory and the direct method performs noticeably better. The results for both methods become more satisfactory as the number of observations increases, and confirm the earlier results obtained by Land (1972) for complete samples. The methods of this paper were also applied to the concentrations (nanograms per litre) of Fluoranthene in water samples from the Niagara River collected by Environment Canada at the Niagara-on-the-Lake station (Data Interpretation Group, 1989). The values for the number of observations and detection limits as well as the estimation results are presented in Table II.

TABLE II

The results for the Fluoranthene data

Data	N	n	$\hat{\lambda}$	$\hat{\mu}_{\hat{\lambda}}$	$\hat{\sigma}_{\hat{\lambda}}$	$\hat{\gamma}$	Confidence interval for γ	(D_1, D_2)
Fluoranthene	44	27	0.16	-0.660	0.662	0.618	(0.480, 0.755)	(0.35, 0.4)

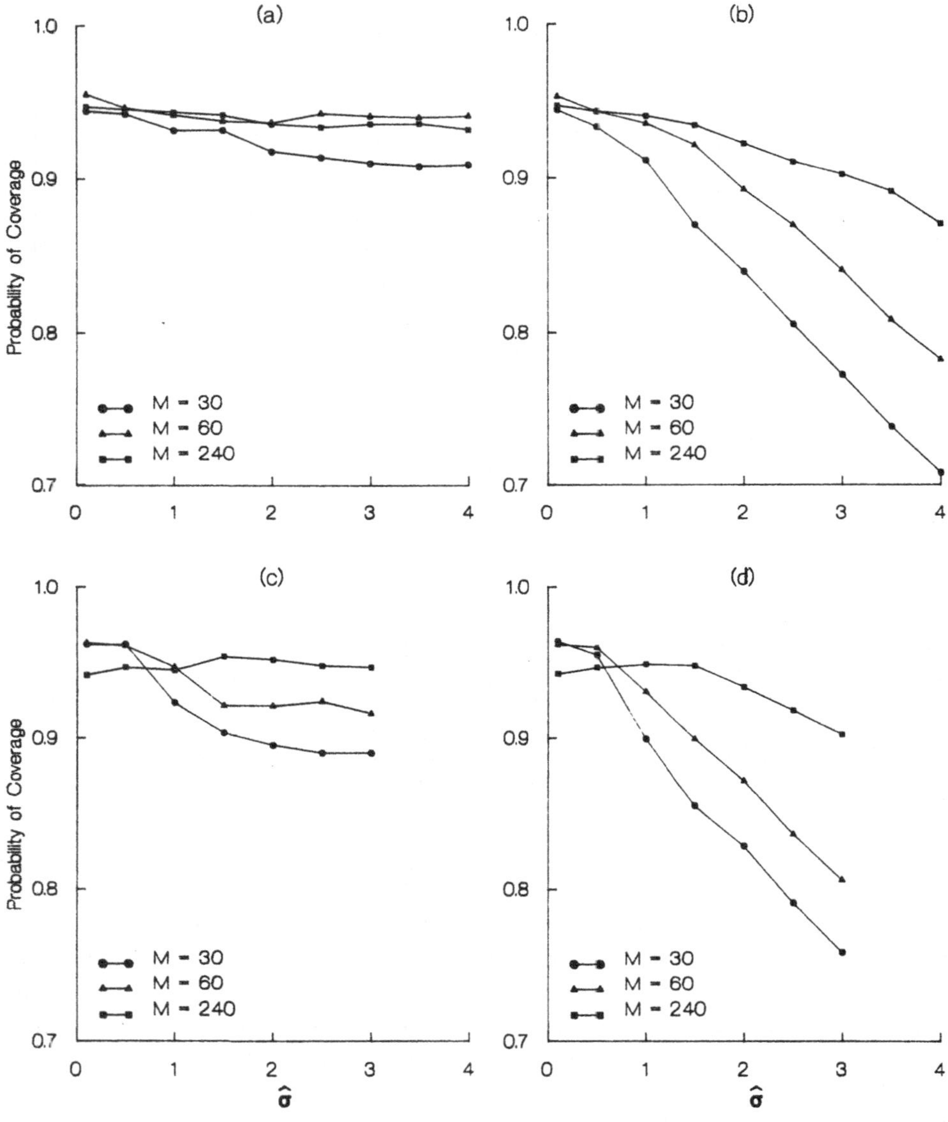

Fig. 1. Probability of coverage of the 95% confidence interval for the lognormal mean: (a) low-level censoring – direct method (b) low-level censoring – Taylor expansion method (c) high-level censoring – direct method (d) high-level censoring – Taylor expansion method.

References

Box, G. E. P. and Cox, D. R.: 1964, 'An Analysis of Transformations (with discussion)', *Journal of the Royal Statistical Society*, Ser. B, **39**, 211-252.

Cohen, A. C.: 1950, 'Estimating the Mean and Variance of Normal Populations from Singly Truncated and Doubly Truncated Samples', *Annuals of Mathematical Statistics* **21**, 557-569.

Cohen, A. C.: 1976, 'Progressively Censored Sampling in the Three Parameter Lognormal Distribution', *Technometrics* **18**, 99-103.

Data Interpretation Group: 1989, *Joint Evaluation of Upstream/Downstream Niagara River Monitoring Data, 1987-1988,* A Joint Publication of EC, USEPA, MOE and NYSDEC.

El-Shaarawi, A. H.: 1989, 'Inferences about the Mean from Censored Water Quality Data, *Water Resour. Res.* **25**, 685-690.

El-Shaarawi, A. H. and Dolan, D. M.: 1989, 'Maximum Likelihood Estimation of Water Quality Concentrations from Censored Data', *Canadian Journal of Fisheries and Aquatic Sciences* **46**, 1033-1039.

Gilliom, R. J. and Helsel, D. R.: 1986, 'Estimation of Distributional Parameters for Censored Trace Level Water Quality Data, 1, Estimation Techniques', *Water Resour. Res.* **22**, 135-146.

Gupta, A. K.: 1952, 'Estimation of the Mean and Standard Deviation of a Normal Population from a Censored Sample', *Biometrika* **39**, 260-273.

Harter, H. L.: 1970, *Order Statistics and Their Use in Testing and Estimation*, Vol. 2, Washington, U.S. Government printing office.

Harter, H. L. and Moore, A. H.: 1966, 'Iterative Maximum Likelihood Estimation of the Parameters of Normal Populations from Singly and Doubly Censored Samples', *Biometrika* **53**, 205-213.

IMSL: 1987, Math/Library and Stat/Library, IMSL, Inc., Houston: Texas.

Land, C. E.: 1972, 'An Evaluation of Approximate Confidence Interval Estimation Methods for Lognormal Means', *Technometrics* **14**, 145-158.

Schneider, H.: 1986, *Truncated and Censored Samples from Normal Populations,* Marcel Dekker, Inc.: New York.

Shumway, R. H., Azari, A. S., and Johnson, P.: 1989, 'Estimating Mean Concentrations Under Transformation for Environmental Data with Detection Limits', *Technometrics* **31**, 347-356.

Tiku, M. L.: 1967, 'Estimating the Mean and Standard Deviation from a Censored Normal Sample', *Biometrika* **54**, 155-165.